P9-CQL-300

WITHDRAWN
UNIVERSITY LIBRARY
THE UNIVERSITY OF TEXAS RIO GRANDE VALLEY

LIBRARY
THE UNIVERSITY OF TEXAS
PAN AMERICAN AT BROWNSVILLE
Brownsville, TX 78520-4991

Progress in Systems and Control Theory
Volume 2

Series Editor

Christopher I. Byrnes, Washington University

Associate Editors

S.-I. Amari, University of Tokyo, Japan
B.D.O. Anderson, Australian National University, Canberra, Australia
Karl J. Åström, Lund Institute of Technology, Lund, Sweden
Jean-Pierre Aubin, CEREMADE, Paris, France
H.T. Banks, University of Southern California, Los Angeles, California
John S. Baras, University of Maryland, College Park, Maryland
A. Bensoussan, INRIA, Paris, France
John Burns, Virginia Polytechnic Institute, Blacksburg, Virginia
Han-Fu Chen, Beijing University, People's Republic of China
M.H.A. Davis, Imperial College of Science and Technology, London, England
Wendell Fleming, Brown University, Providence, Rhode Island
Michel Fliess, CNRS-ESE, Gif-sur-Yvette, France
Keith Glover, University of Cambridge, Cambridge, England
D. Hinrichsen, University of Bremen, Federal Republic of Germany
Alberto Isidori, University of Rome, Italy
B. Jakubczyk, Academy of Sciences, Warsaw, Poland
Hidenori Kimura, Osaka University, Japan
Arthur Krener, University of California, Davis
H. Kunita, Kyushu University, Japan
Alexandre Kurzhansky, IIASA, Laxenburg, Austria
Harold M. Kushner, Brown University, Providence, Rhode Island
Anders Lindquist, Royal Institute of Technology, Stockholm, Sweden
Andrzej Manitius, George Mason University, Fairfax, Virginia
Clyde F. Martin, Texas Tech University, Lubbock, Texas
Sanjoy Mitter, Massachusetts Institute of Technology, Cambridge, Massachusetts
Giorgio Picci, LADSEB-CNR, Padova, Italy
Hector Sussman, Rutgers University, New Brunswick, New Jersey
T.J. Tarn, Washington University, St. Louis, Missouri
Pravin P. Varaiya, University of California, Berkeley
Jan C. Willems, University of Gröningen, Sweden
W.M. Wonham, University of Toronto, Canada

B. Jakubczyk K. Malanowski W. Respondek

Perspectives in Control Theory

Proceedings of the
Sielpia Conference,
Sielpia, Poland,
September 19–24, 1988

With 10 Illustrations

1990

Birkhäuser
Boston · Basel · Berlin

B. Jakubczyk
Institute of Mathematics
Polish Academy of Sciences
00-950 Warsaw
Poland

K. Malanowski
Systems Research Institute
Polish Academy of Sciences
01-447 Warsaw
Poland

W. Respondek
Institute of Mathematics
Polish Academy of Sciences
00-950 Warsaw
Poland

Library of Congress Cataloging-in-Publication Data
Perspectives in control theory: proceedings of the Sielpia Conference, Sielpia, Poland, 1988 / B. Jakubczyk, K. Malanowski, W. Respondek.
p. cm.—(Progress in systems and control theory: v. 2)
Includes bibliographical references.
ISBN 0-8176-3456-8 (alk. paper)
1. Control theory—Congresses. I. Jakubczyk, Bronislaw. II. Malanowski, Kazimierz. III. Respondek, W. IV. Series.
QA402.3.P39 1990
629.8'312—dc20 89-18337

Printed on acid-free paper.

© Birkhäuser Boston, 1990
All rights reserved. No part of this publication may be reproduced, stored in a retrieval system, or transmitted, in any form or by any means, electronic, mechanical, photocopying, recording or otherwise, without prior permission of the copyright owner. Permission to photocopy for internal or personal use, or the internal or personal use of specific clients, is granted by Birkhäuser Boston, Inc., for libraries and other users registered with the Copyright Clearance Center (CCC), provided that the base fee of $0.00 per copy, plus $0.20 per page is paid directly to CCC, 21 Congress Street, Salem, MA 01970, U.S.A. Special requests should be addressed directly to Birkhäuser Boston, Inc., 675 Massachusetts Avenue, Cambridge, MA 02139, U.S.A.

ISBN 0-8176-3456-8
ISBN 3-7643-3456-8

Camera-ready text provided by the individual authors.
Printed and bound by Edwards Brothers, Inc., Ann Arbor, Michigan.
Printed in the U.S.A.

9 8 7 6 5 4 3 2 1

Preface

The volume contains papers based on lectures delivered during the school "Perspectives in Control Theory" held in Sielpia, Poland on September 19–24, 1988.

The aim of the school was to give the state-of-the-art presentation of recent achievements as well as perspectives in such fields of control theory as optimal control and optimization, linear systems, and nonlinear systems. Accordingly, the volume includes survey papers together with presentations of some recent results. The special emphasis is put on:

— nonlinear systems (algebraic and geometric methods),

— optimal control and optimization (general problems, distributed parameter systems),

— linear systems (linear-quadratic problem, robust stabilization).

An important feature of the school (and consequently of the volume) was its really "international" character since it brought together leading control theoriests from West and East. All together the school was attended by 108 participants from 18 countries. During the school 21 one-hour invited lectures were delivered. Moreover, five half-an-hour talks were given and 30 contributions were presented in frames of poster sessions.

The school was organized and supported by:

— Institute of Mathematics of the Polish Academy of Sciences,

— Committee of Automatic Control and Robotics of the Polish Academy of Sciences,

— Institute of Automatic Control, Warsaw University of Technology (as Coordinator of the Basic Research Program R.P.I.02 "Theory of Control of Continuous Dynamic Systems and Discrete Processes").

The organizing committe consisted of: B. Frelek, B. Jakubczyk, T. Kaczorek, M. Kocięcki, K. Malanowski (vice-chairman), M. Niezgódka, A. Olbrot, C. Olech (chairman), W. Respondek (secretary), A. Sosnowski, A. Wierzbicki.

We would like to thank Ms. M. Wolińska for her excelent typing of some of the manuscripts.

B. JAKUBCZYK

K. MALANOWSKI

W. RESPONDEK

Warsaw, August 1989.

Contents

List of Invited Speakers

Dirk Aeyels, University of Gent

Jozsef Bokor, Hungarian Academy of Sciences

Hector O. Fattorini, University of California

Michel Fliess, Ecole Superieure d'Electricité

Matheus L.J. Hautus, Eindhoven University of Technology

D. Hinrichsen, University of Bremen

Alexander D. Ioffe, Technion*

Velimir Jurdjevic, University of Toronto

Karl Kunisch, Technische Universität Graz

I.A.K. Kupka, University of Toronto

A.B. Kurzhanski, I.I.A.S.A.

F. Lamnabhi-Lagarrique, Ecole Superieure d'Electricité

Irena Lasiecka, University of Virginia

W.M. Marchenko, Kirov Byelorussian Institute of Technology

Riccardo Marino, Seconda Universita di Roma

Henk Nijmeijer, University of Twente

B.T. Polyak, Institute of Control Problems, Moscow

Boris N. Pshenichnyj, Academy of Sciences of the Ukrainian S.S.R.

A.J. van der Schaft, University of Twente

Hector J. Sussmann, Rutgers University

V.M. Tikhomirov, Moscow State University

---

* Contribution not received

Addresses of Contributors

Dirk Aeyels	Department of Systems Dynamics University of Gent Grotesteenweg Nord 2, 9710 Gent, Belgium
Jozsef Bokor	Computer and Automation Institute Hungarian Academy of Sciences Kende U. 13-17, P.O. Box 63 H-1502 Budapest, Hungary
Peter E. Crouch	Department of Electrical and Computer Engineering Arizona State University Tempe, Arizona 85287, U.S.A.
Hector O. Fattorini	Department of Mathematics University of California Los Angeles, California 90024, U.S.A.
Michel Fliess	Laboratoire des Signaux et Systèmes Ecole Superieure d'Electricité Plateau du Moulon 91190 Gif-sur-Yvette, France
A.H.W. Geerts	Department of Mathematics and Computing Science Eindhoven University of Technology 5600 MB Eindhoven, The Netherlands
Matheus L.J. Hautus	Department of Mathematics and Computing Science Eindhoven University of Technology 5600 MB Eindhoven, the Netherlands
D. Hinrichsen	Institute for Dynamical Systems University of Bremen D-2800 Bremen 33, FRG

I. Ighneiwa — Department of Electrical
and Computer Engineering
Arizona State University
Tempe, Arizona 85287, U.S.A.

Velimir Jurdjevic — Department of Mathematics
University of Toronto
Toronto, M5S 1A1 Canada

L. Keviczky — Computer and Automation Institute
Hungarian Academy of Sciences
Kende U. 13-17, P.O. Box 63
H-1502 Budapest, Hungary

Karl Kunisch — Institut für Mathematik
Technische Universität Graz
Kopernikusgasse 24, A-8010 Graz, Austria

I.A.K. Kupka — Department of Mathematics
University of Toronto
Toronto, M5S 1A1 Canada

A.B. Kurzhanski — International Institute
for Applied Systems Analysis
A-2361 Laxenburg, Austria and
Institute of Mathematics and Mechanics
Sverdlovsk, U.S.S.R.

F. Lamnabhi-Lagarrique — Laboratoire des Signaux et Systèmes
Ecole Superieure d'Electricité
Plateau du Moulon
91190 Gif-sur-Yvette, France

Irena Lasiecka — Department of Applied Mathematics
University of Virginia
Charlottesville, Virginia 22903, U.S.A.

W.M. Marchenko — Department of Mathematics
Kirov Byelorussian Institute of Technology
Swierdlova 13a, 200630 Minsk, U.S.S.R.

Riccardo Marino — Dipartimento di Ingegneria Elettronica
Seconda Universita di Roma
Via Orazio Raimondo, 00173 Roma, Italia

Henk Nijmeijer — Department of Applied Mathematics
University of Twente, P.O. Box 217
7500 AE Enschede, the Netherlands

O.I. Nikonov	International Institute for Applied Systems Analysis A-2361 Laxenburg, Austria and Institute of Mathematics and Mechanics Sverdlovsk, U.S.S.R.
B.T. Polyak	Institute of Control Problems Profsojuznaja 65 117-342 Moscow, B279, U.S.S.R.
A.J. Pritchard	Control Theory Centre, University of Warwick Coventry, England
Boris N. Pshenichnyj	Institute of Cybernetics Academy of Sciences of the Ukrainian S.S.R. Prospiekt 40-letija Oktiabria 142/144 25-2207 Kiev 207, U.S.S.R.
A.J. van der Schaft	Department of Applied Mathematics University of Twente, P.O. Box 217 7500 AE Enschede, the Netherlands
Hector J. Sussmann	Department of Mathematics Rutgers University New Brunswick, New Jersey 08903, U.S.A.
V.M. Tikhomirov	Department of Mathematics and Mechanics Moscow State University Leninskije Gory 117-234 Moscow, U.S.S.R.
Roberto Triggiani	Department of Applied Mathematics University of Virginia Charlottesville, Virginia 22903, U.S.A.

REMARKS ON THE STABILIZABILITY OF NONLINEAR SYSTEMS BY SMOOTH FEEDBACK

Dirk Aeyels

Abstract

In this paper we discuss the known result that if a system is smoothly stabilizable then adding an integrator does not change this property. For this extended system stabilizing feedbacks depending on data explicitly available from the original system are proposed.

1. Introduction

Consider the following system

$$\dot{x} = f(x,u)$$

with $f: R^n x R \to R^n$ smooth, $f(0,0) = 0$.

Suppose that this system is **smoothly stabilizable**, i.e. there exists a smooth function $k: R^n \to R$ such that $k(0)=0$ and the origin is asymptotically stable for

$$\dot{x} = f(x,k(x)) \tag{1}$$

It is well known that the "extended" system

$$\dot{x} = f(x,y)$$
$$\dot{y} = u \tag{2}$$

is also smoothly stabilizable.

The proof of this result is given in [4], where additional references also containing the proof are given. It will be repeated in the next section.It will be seen that the feedback which stabilizes the extended system depends on a Lyapunov fuction for the stabilized system (1). This Lyapunov function exists by the inverse Lyapunov theorem [2], but an explicit expression is in general not available for implementation.

In this paper we discuss the stabilization of (2) by means of a procedure independent of the Lyapunov function corresponding to (1). The proposed feedback --although ensuring local stability--does not guarantee global stability. This will be shown by means of a counterexample.

However for some classes of nonlinear systems an indication will be given showing that the proposed feedback is globally stabilizing. The results concerning the global stabilization issue are incomplete. They will be the subject of a forthcoming paper.

2. Stabilization of the extended system

In this section we recall the proof (taken from [1]) that if the original system (1) is smoothly stabilizable, then the extended system is also smoothly stabilizable .

Assume that $\dot{x} = f(x,u)$ is smoothly stabilizable. Let

$$f_0(x) := f(x,k(x))$$

be the closed-loop system.

By the inverse Lyapunov theorem [2], there exists a positive definite function V such that $L_{f_0} V(x) < 0$ for all $x \neq 0$

Since f and k are smooth, there exists a smooth function g defined on R^{n+1} such that for all x,z

$$f(x,k(x)+z) = f_0(x)+zg(x,z)$$

Introduce the positive definite function on R^{n+1}:

$$W(x,y) := V(x)+ \tfrac{1}{2}(y-k(x))^2$$

Take the feedback for (2)

$$u(x,y) = -y+k(x)+ \quad k(x).f(x,y)- \quad V(x).g(x,y-k(x))$$

Take the derivative of W along trajectories of (2) with the feedback just defined

$$\begin{aligned} \text{Then } \dot{W} &= \nabla V(x).f(x,y)+(y-k(x))(\dot{y}-\nabla k(x).\dot{x}) \\ &= \nabla V(x).f(x,y)+(y-k(x)(-y+k(x)-\nabla V(x).g(x,y-k(x)) \\ &= \nabla V(x).f(x,y)-(y-k(x))^2-(y-k(x)).\nabla V(x)g(x,y-k(x)) \end{aligned}$$

Since $f(x,y) = f(x,k(x))+(y-k(x))g(x,y-k(x))$

$$\dot{W} = L_{f0}\ V(x)-(y-k(x))^2 < 0$$

for all nonzero (x,y). This assures stability.

It is remarked that the proposed feedback assures local stability of the extended system if the original system is locally stabilizable. It also assures **global** stability of the extended system if the original closed loop system is **globally** stable.

Notice that the expression for the feedback contains a term $\nabla V(x).g(x,y-k(x))$ which depends on a Lyapunov function V for the asymptotically stable system $\dot{x} = f(x,k(x))$. This renders the feedback u(x,y) hard to implement .

Therefore we consider (3) with the ∇V term left out ,i.e.

$$u(x,y) = -y+k(x)+\nabla k(x).f(x,y)$$

and investigate its stabilizing potential for the extended system.
In fact, under some extra conditions on f(x,u) (e.g. f(x,u) contains no linear terms in u) it follows rather immediately by means of the center manifold approach [1] that the new feedback stabilizes the extended system. The case f(x,u) containing no linear terms at all was communicated to me by Sontag & Sussmann.

We will show in what follows that -- without extra assumption on f -- the feedback locally stabilizes the extended system if the original closed loop system is asymptotically stable.

The proof which will be given in section 3 is an application of the center manifold approach and is rather straightforward.

3. Local stabilizability

Consider again the system

$$\dot{x} = f(x,u)$$

with f: $R^n xR \text{-->} R^n$.

Let u=k(x) be a smooth stabilizing feedback, i.e.

$$\dot{x} = f(x,k(x))$$

is asymptotically stable.
Consider the extended system

$$\dot{x} = f(x,y)$$
$$\dot{y} = u \qquad (2)$$

We will show by means of the center manifold approach [1] that (2) is locally stabilized by

$$u = -y+k(x)+\nabla k(x).f(x,y)$$

i.e.the system

$$\dot{x} = f(x,y)$$
$$\dot{y} = -y+k(x)+\nabla k(x).f(x,y) \qquad (3)$$

is asymptotically stable.
We perform a coordinate change:

x unchanged
$$z = y-k(x)$$

then (3) becomes

$$\dot{x} = f(x,k(x)+z)$$
$$\dot{z} = -z \qquad (4)$$

We want to show that (4) is asymptotically stable in $(x,z)=(0,0)$ knowing that

$$\dot{x} = f(x,k(x))$$

is asymptotically stable.
Rewrite (4) as

$$\dot{x} = Ax+bz+h(x,z)$$
$$\dot{z} = -z \qquad (5)$$

with A and b the appropriate Jacobians and h the higher order terms.

The system

$$\dot{x} = Ax+h(x,0) \tag{6}$$

is stable by assumption.

First, we blockdiagonalize (5) by means of a linear transformation

$$x_1 = Tx+r.z$$
$$z = z$$

In these coordinates, the system is

$$\dot{x}_1 = TAT^{-1}x_1+Th(T^{-1}x_1-T^{-1}rz,z)$$
$$\dot{z} = -z$$

It is remarked that

$$\dot{x}_1 = TAT^{-1}x_1+Th(T^{-1}x_1-T^{-1}rz,z)$$

with $z = 0$ is asymptotically stable in $x_1 = 0$ since (6) is asymptotically stable.

The matrix T can be taken such that TAT^{-1} consists of two diagonal blocks A_1 (with eigenvalues on the imaginary axis) and A_2 (with eigenvalues in the left half plane).
The system (4) is then represented by

$$\dot{x}_1 = A_1x_1+h_1(x_1,x_2,z)$$
$$\dot{x}_2 = A_2x_2+h_2(x_1,x_2,z)$$
$$\dot{z} = -z \tag{7}$$

with

$$\dot{x}_1 = A_1x_1+h_1(x_1,x_2,0)$$
$$\dot{x}_2 = A_2x_2+h_2(x_1,x_2,0) \tag{8}$$

asymptotically stable.

Therefore there exists a center manifold

$x_2 = C(x_1)$ such that

$$\dot{x}_1 = A_1 x_1 + h_1(x_1, C(x_1), 0) \tag{9}$$

is asymptotically stable with the center manifold defined by

$$A_2 C(x_1) + h_2(x_1, C(x_1), 0) = \nabla C(A_1 x_1 + h1(x_1, C(x_1), 0) \tag{10}$$

Returning to the system (7) there is a center manifold

$x_2 = C_2(x_1)$
$z = C_1(x_1)$

such that the stability behavior of (7) is determined by

$$\dot{x}_1 = A_1 x_1 + h_1(x_1, C_2(x_1), C_1(x_1)) \tag{11}$$

with C_1 and C_2 determined by

$$\nabla C_2(A_1 x_1 + h_1(x_1, C_2(x_1), C_1(x_1)) = A_2 C_2(x_1) + h_2(x_1, C_2(x_1), C_1(x_1)))$$

$$\nabla C_1(A_1 x_1 + h_1(x_1, C_2(x_1), C_1(x_1)) = -C_1(x_1)$$

This set of equations is satisfied by

$C_1 \equiv 0$
$C_2 \equiv C$ with C defined by (10)

Therefore (11) becomes

$$\dot{x}_1 = A_1 x_1 + h_1(x_1, C(x_1), 0)$$

which is asymptotically stable, cfr.(9). This ends the proof.

Remarks

1.The following class of feedbacks all locally stabilize the system(2):

$u = -y+k(x)+\nabla k(x).f(x,k(x)+l(y-k(x)))$

with $l:R \to R$ smooth and $l(0) = 0$.

By taking l the identity function one recovers the feedback featuring above.

By taking l identical to zero, one obtains

$$u = -y+k(x)+\nabla k(x).f(x,k(x))$$

as a stabilizing feedback.

2. Recall that after introducing an appropriate feedback and coordinate transformations one ended up with the system

$$\dot{x} = Ax+bz+h(x,z) \qquad (5)$$
$$\dot{z} = -z$$

with

$$\dot{x} = Ax+h(x,0) \qquad (6)$$

asymptotically stable.

That (5) is asymptotically stable under the condition (6) also appears in [3]. It can be seen as a specialization ($k(x)\equiv 0$) of the result proven above.

3.We have proposed a stabilizing feedback:

$u = k_2(x,y)$ for the system

$$\dot{x} = f(x,y)$$
$$\dot{y} = u$$

derived from a known stabilizing feedback

$u = k_1(x)$

for the system

$$\dot{x} = f(x,u)$$

This procedure can of course be extended to higher order systems, e.g. to the construction of a stabilizing feedback

$u = k_3(x,y,z)$ for the system

$\dot{x} = f(x,y)$

$\dot{y} = z$

$\dot{z} = u$

3.Global stabilizability

In this section we will give some indications on the stabilizing potential of the feedback

$u = -y+k(x)+\nabla k(x).f(x,y)$

for the system

$\dot{x} = f(x,y)$

$\dot{y} = u$

where $k(x)$ is a stabilizing feedback for

$\dot{x} = f(x,u)$.

Again introducing new coordinates as above the extended closed loop system is written as

$\dot{x} = f(x,k(x)+z)$

$\dot{z} = -z.$

This system is in general not globally stable. Indeed consider the system defined on R^2

$\dot{x} = -x^3+ux^3$

which is asymptotically stable for $u = k(x) \equiv 0$

Consider the extended closed loop system

$$\dot{x} = -x^3-zx^3 \tag{12}$$

$\dot{z} = -z$

Then with initial condition $(x,z)(0) = (x_0, z_0)$ the solution to (12) is

$z = e^{-t} z_0$

and $x(t)$ defined by

$\dot{x} = -x^3(1-e^{-t}z_0)$

or $x(t)$ given by

$$\tfrac{1}{2}x^{-2} = \tfrac{1}{2}x_0^{-2} + (t+z_0(e^{-t}-1)) \qquad (13)$$

Define the function

$F(t) \equiv t+z_0(e^{-t}-1)$

which is monotonically increasing if the initial condition $z_0 < 1$. This implies that the extended closed loop system is asymptotically stable for x_0 arbitrary and $z_0 < 1$.
However $F(t)$ has for $z_0 > 1$, at $t - \ln z_0$ a minimum

$\ln z_0 + 1 - z_0$

which is negative for $z_0 > 1$.
This implies that for an initial condition (x_0, z_0) such that

$\tfrac{1}{2}x_0^2 < |1+\ln z_0 - z_0|$

there is a time $t > 0$ which makes the right hand side of (13) negative, i.e. the system (12) has a finite escape time.
We have therefore shown that the proposed feedback is not globally stabilizing for the extended system.

Consider again the original system

$\dot{x} = f(x,u)$

which is assumed to be linear in u, i.e.

$\dot{x} = f_1(x)+u.g(x)$

Let $u = k(x)$ be a stabilizing feedback.

Consider the extended system

$$\dot{x} = f_1(x) + y.g(x)$$

$$\dot{y} = u$$

and apply the feedback proposed above, then after a change of coordinates one obtains

$$\dot{x} = f_1(x) + k(x)g(x) + zg(x)$$

$$\dot{z} = -z$$

with $f_1(x)+k(x)g(x)$ asymptotically stable.

Assume that $g(x)$ is a constant vector b. Then it can be shown that the system is globally stable. Extensions of this result will be the subject of a forthcoming paper.

References

[1] Aeyels, D., Stabilization of a class of nonlinear systems by a smooth feedback control, Systems and Control Letters, 5(1985), pp.281-294.

[2] Hahn, W., Stability of Motion, Springer-Verlag, Berlin-Heidelberg 1967.

[3] Isidori, A., and C.I. Byrnes, Local stabilization of minimum-phase nonlinear systems, Systems and Control Letters, 11(1988):9-17.

[4] Sontag, E.D. and H.J. Sussmann, Further comments on the stabilizability of the angular velocity of a rigid body, to appear in Systems and Control Letters, 12(1989):No.3.

SOME REMARKS ON PONTRYAGIN'S MAXIMUM PRINCIPLE FOR INFINITE DIMENSIONAL CONTROL PROBLEMS

H. O. Fattorini

Abstract

During the last three decades, numerous infinite dimensional versions of Pontryagin's maximum principle have been obtained, mostly for particular equations or restricted classes of equations. Recently there have been some efforts to obtain a unified version of the maximum principle, which have (among other things) clarified the interaction between necessary conditions and controllability and opened the way to a simplified treatment of problems with state constraints.

1. The finite dimensional maximum principle.

For reference below, we state Pontryagin's maximum principle for a system described by an ordinary differential system

$$y'(t) = f(t,y(t),u(t)) \quad (0 \leq t \leq T) \tag{1.1}$$

$$y(0) = y^o, \tag{1.2}$$

where $y(\cdot)$ is a n-dimensional vector function, $u(\cdot)$ is a m-dimensional vector function and y^o is a point in $\mathbb{R}^n$. The control constraint is

$$u(t) \in U \subseteq \mathbb{R}^m, \tag{1.3}$$

the cost functional

$$y_o(t,u) = \int_0^t f_o(s,y(s),u(s))ds , \tag{1.4}$$

and the target condition

$$y(\bar{t}) \in Y, \tag{1.5}$$

where the time $\bar{t}$ may or may not be fixed in advance. The corresponding optimal control problem is

$$\text{minimize} \quad y_o(t,u) \quad (u \in W(0,t,U)) \tag{1.6}$$

$$\text{subject to} \quad y(t,u) \in Y . \tag{1.7}$$

Here $W(0,\bar{t},U)$ denotes the space of all admissible controls in the interval $0 \leq t \leq \bar{t}$ (measurable vector functions satisfying the control constraint (1.3)) and $y(t,u)$ denotes the trajectory of (1.1)-(1.2) corresponding to an admissible control $u = u(\cdot)$.

Under adequate assumptions on U, f, f_o, Pontryagin's maximum principle holds ([3], [36]). To formulate it we use the Hamiltonian

$$(1.8) \qquad H(t,\tilde{y},\tilde{z},u) = (\tilde{z},\tilde{f}(t,y,u)) = z_o f_o(t,y,u) + (z,f(t,y,u))$$

where

$$(1.9) \qquad \tilde{y} = \{y_o,y\},\ \tilde{z} = \{z_o,z\},\ \tilde{f}(t,y,u) = \{f_o(t,y,u),f(t,y,u)\}$$

and where we denote by $(\ ,\ )$ the ordinary scalar product in $\mathbb{R}^n$ and by $\{x_o,x\}$ the vector with "coordinates" x_o, x. Let $0 \leq t \leq \bar{t}$ be the control interval, $\bar{u}(t)$ the optimal control, $\bar{y}(t) = y(t,\bar{u})$ the optimal trajectory, $\tilde{y}(t) = \{y_o(t,\bar{u}),y(t,\bar{u}))\}$. Consider the Hamiltonian equations

$$(1.10) \qquad \tilde{y}'(s) = \mathrm{grad}_z H(s,\tilde{y}(s),\tilde{z}(s),\bar{u}(s)),$$

$$(1.11) \qquad \tilde{z}'(s) = -\ \mathrm{grad}_y H(s,\tilde{y}(s),\tilde{z}(s),\bar{u}(s)) .$$

The first is the original equation (1.1) plus the cost functional equation (1.4) in differentiated form. The second splits into a scalar equation implying that the first coordinate $\bar{z}_o(s)$ of $\tilde{z}(s)$ is constant and into the vector equation

$$(1.12) \qquad \bar{z}'(s) = -\ \partial_y f(s,y(s,\bar{u}),\bar{u}(s))^* \bar{z}(s) - \bar{z}_o \partial_y f_o(s,y(s,\bar{u}),\bar{u}(s))$$

for the second coordinate $\bar{z}(s)$ of $\tilde{z}(s)$.

Let $\tilde{y}(t,\bar{u}) = \{y_o(t,\bar{u}),y(t,\bar{u})\}$. Pontryagin's maximum principle asserts the existence of a solution $\tilde{z}(s) = \{\bar{z}_o(s),\bar{z}(s)\}$ of (1.11) in the control interval $0 \leq s \leq \bar{t}$ satisfying

$$(1.13) \qquad \bar{z}_o \leq 0,\ (\bar{z}_o,\bar{z}(\bar{t})) \neq 0$$

such that the two following properties hold.

(A) Dependence of the Hamiltonian on the control parameter:

$$(1.14) \qquad H(s,\tilde{y}(s),\tilde{z}(s),\bar{u}(s)) = \max_{v \in U} H(s,\tilde{y}(s),\tilde{z}(s),v) .$$

(B) Dependence of the Hamiltonian on time: For autonomous systems $(f = f(y,u),\ f_o = f_o(y,u))$ and fixed terminal time $\bar{t}$ we have

$$(1.15)\qquad H(\tilde{y}(s),\tilde{z}(s),\bar{u}(s)) = \text{constant} \qquad (0 \le s \le \bar{t}).$$

For free terminal time $\bar{t}$ we have

$$(1.16)\qquad H(\tilde{y}(s),\tilde{z}(s),\bar{u}(s)) = 0 \qquad (0 \le s \le \bar{t}).$$

Suitable versions of (1.15) and (1.16) can also be obtained for non-autonomous systems: see [36].

2. *Early versions of the infinite dimensional maximum principle.*

The study of control systems in infinite dimensional spaces has been motivated mainly by:

(1) Delay differential equations (more generally, functional differential equations). Here, infinite dimensionality stems from the fact that solutions are viewed through sections in certain intervals, thus in function spaces. Certain problems, however (such as those involving Euclidean target conditions) are essentially finite dimensional. The theory was initiated in [35]; for more recent references, see [9], [10].

(2) Partial differential equations. Here, trajectories live in function spaces for time dependent systems, the same is true of solutions of stationary systems. Some of the first works on the subject were [4], [5], [11], [12].

(3) Differential equations in Banach spaces. The study of control problems for these systems was motivated mainly by the treatment of nonstationary partial differential equations. Some of the early works are [13], [14], [17]. In the first two, a version of Pontryagin's maximum principle is obtained following the approach in [36] (separation of variation-generated cones from target points).

During the sixties and seventies, a number of results appeared dealing with the linear system

$$(2.1)\qquad y'(t) = Ay(t) + Bu(t),$$

$$(2.2)\qquad y(0) = y^0,$$

where A is the infinitesimal generator of a strongly continuous semigroup $S(t)$ in a Banach space E and $B : F \to E$ is a bounded operator from a second Banach space F into E. In [1] and [34] a simple generalization of the finite dimensional approach to the maximum principle (developed in [2]) is used in the time optimal case ($f_0 = 1$). Since the difficulties encountered here reappear in much more general nonlinear situations, we outline this method below. The target condition (1.5) is a point target condition: $Y = \{y\}$, so that

$$(2.3) \qquad y(\bar{t},u) = \bar{y}.$$

Denote by $\bar{t}$ the optimal time, and let $K(\bar{t})$ be the reachable set in time $\bar{t}$, consisting of all elements of the form

$$(2.4) \qquad y = \int_0^{\bar{t}} S(\bar{t} - s)Bu(s)ds$$

with $u(\cdot)$ in $W(0,\bar{t},U)$. If the control set U is convex so is $K(\bar{t})$; moreover, it is possible to prove that the target point $\bar{y}$ (which obviously belongs to $K(\bar{t})$) is a boundary point of $K(\bar{t})$. Assuming that $\bar{y}$ *can be separated from* $K(\bar{t})$ *by a hyperplane*, we have a non-zero functional z in the dual space E^* such that

$$(2.5) \qquad (z,y) \leq (z,\bar{y})$$

for all y in $K(\bar{t})$. This inequality can be rewritten as

$$(2.6) \qquad \int_0^{\bar{t}} (B^*S(\bar{t} - s)^*z,u(s))ds \leq \int_0^{\bar{t}} (B^*S(\bar{t} - s)^*z,\bar{u}(s))ds$$

where $\bar{u}$ is the optimal control, and implies the v-dependence part (A) of the maximum principle in Section 1:

$$(2.7) \qquad (B^*S(\bar{t} - s)^*z,\bar{u}(s)) = \max_{v \in U} (B^*S(\bar{t} - s)^*z, v) .$$

The key point is whether or not $\bar{y}$ can be separated from $K(\bar{t})$. This is always true in finite dimensional spaces [2]. In infinite dimensional spaces separation is possible if $F = E$, $B = I$ and $S(t)$ is a group, for in this case, $K(\bar{t})$ will contain interior points if U does (see [34]).In [1], separation is insured by requiring $\bar{y}$ to be a support point of $\bar{y}$; however, this property (which is equivalent to the separation property) does not seem to be verifiable except in the case where $K(\bar{t})$ has interior points.

3. *The linear maximum principle.*

If $S(t)$ is not a group, $K(\bar{t})$ cannot contain interior points (see [18]) and separation (thus the maximum principle) is in doubt. A way out of this difficulty was found in [18] and consists in working not in E but in a subspace $E_0 \subseteq E$ such that $K(\bar{t}) \cap E_0$ contains interior points. This is done in [18] for the case $B = I$, where E_o is $D(A)$, the domain of A, endowed with its graph norm. The maximum principle takes the following form: there exists a nonzero functional z such that $S(\bar{t} - s)^* z$ belongs to $D(A^*)$ for all $s < \bar{t}$ and

$$(3.1) \qquad (A^* S(\bar{t} - s)^* z, \bar{u}(s)) = \max_{v \in U} (A^* S(\bar{t} - s)^* z, v) .$$

A minor drawback is that it is not known whether the maximum principle holds in some form for target points $\bar{y}$ that do not belong to $D(A)$.

Essentially the same method (to search for a subspace E_o where separation can be carried out) works in other situations. This has been done in [21] for a distributed parameter system described by a wave equation and in [20] for a boundary control system described by the heat equation: here, an important role in the construction of the space E_o is played by the controllability results in [37]. All the cases where a suitable E_o can be found can be chacterized roughly speaking as follows: "E_o has nice linear functionals" and "E_o is closely related to A" (for instance, it is of the form $E_o = D(f(A))$ for a suitable function f.

It is natural to ask whether a subspace E_o having the required properties can be found in every case. The answer to this question must be in the negative, since it is not difficult to construct linear control systems of the form (2.1)-(2.2) where the maximum principle fails for a condition of the form (2.3). We show this below.

Example 3.1 Let E be a separable Hilbert space, $\{\phi_j\}$ a complete orthonormal system in E. We consider the system (2.1)-(2.2) with $F = \mathbb{R}$, $Bu = ub$, where the operator A and the element b of E are given by

$$A(\sum c_j \phi_j) = -\sum j c_j \phi_j , \quad b = \sum j^{-1} \phi_j .$$

With this choice of A, b it is not difficult to check that if $u(\cdot)$, $v(\cdot)$ are two arbitrary controls in $W(0,\bar{t},U)$ such that we have $y(\bar{t},u) = y(\bar{t},v)$ for the corresponding trajectories then we must have

$$(3.2) \qquad \int_0^{\bar{t}} e^{-js} u(\bar{t} - s) ds = \int_0^{\bar{t}} e^{-js} v(\bar{t} - s) ds$$

for $j = 1,2,\ldots$ which, since the exponentials e^{-js} are complete in any finite interval, implies that

(3.3) $u(t) = v(t)$ a.e. in $0 \leq t \leq \bar{t}$.

Assume now the maximum principle (2.7) holds for the time optimal problem for the system (2.1)-(2.2) with $y^o = 0$ and control set U the interval [-1,1]. Obviously, (2.7) implies that

(3.4) $\bar{u}(s) = \text{sign}(S(\bar{t} - s)z, b)$

so that $\bar{u}(s)$ must satisfy the bang-bang principle $|u(s)| = 1$ a.e. However, it is easy to construct time optimal controls that do no satisfy the bang-bang principle. To do this, choose an arbitrary function $\bar{u}(t)$ with $\bar{u}(t) \neq 0$, $|\bar{u}(t)| < 1$ a.e. in $0 \leq t \leq \bar{t}$. If the point $\bar{y} = y(t,\bar{u})$ can be reached by means of an admissible control $u(t)$ in a time $\bar{t} - h < \bar{t}$, define a control $v(t)$ by the formula $v(t) = 0$ $(0 \leq t < h)$, $v(t) = u(t - h)$ $(h < t \leq \bar{t})$. Then, by translation invariance $y(\bar{t},\bar{u}) = y(\bar{t},v)$ so that (3.2) holds and $\bar{u}(t) = v(t)$, a contradiction since $v(t)$ vanishes in an interval.

Although this example only shows that the maximum principle in its original form (2.7) cannot be satisfied, the same holds for any form of the maximum principle that implies the bang-bang principle.

4. *The linear maximum principle: set targets.*

The difficulties involved in separating the target point $\bar{y}$ from the reachable set $K(\bar{t})$ disappear if we consider a set target condition

(4.1) $y(t,u) \in Y$

with Y open and convex (or, more generally, of finite codimension in a suitable sense): in fact, here we only have to separate Y from $K(\bar{t})$, which is always possible.

5. *Nonlinear infinite dimensional versions of the maximum principle.*

At the beginning of the eighties, infinite dimensional generalizations of the maximum principle existed in a multitude of particular cases, mostly linear, and where the proofs were obtained usually by separation theorems. Many abstract treatments of control problems (for instance, as nonlinear programming problems in linear spaces) were also available, notably those due to Neustadt, Halkin and Neustadt, Dubovitskii and Milyutin, Ioffe and Tikhomirov, Zowe and Kurcyusz and others. Most of these theories (but not all) were restricted to finite dimensional valued functions and were heavily geometric in nature.

LIBRARY
THE UNIVERSITY OF TEXAS
PAN AMERICAN AT BROWNSVILLE
Brownsville, TX 78520-4991

Interestly enough, what made possible a general theory of the maximum principle (admittedly, not the only one!) was an earlier major breakthrough in the finite dimensional theory due to Clarke [6], [7] and Ekeland [15], [16]. This completely new approach to optimal control problems (which, in a very vague description, consists in viewing their solutions as approximate solutions of "neighboring" much simpler minimization problems) opened the way not only to improved versions of the classical theorems but to new directions of research, such as nonsmooth finite dimensional systems (see [8]). It proved of key importance in treating infinite dimensional problems. In [23] (and in [22] in a more restricted version) control problems for general *input-output maps* control or input $u(\cdot) \to$ output or trajectory $y(t,u(.))$ in infinite dimensional spaces are considered. The setup is the following. E and F are Hilbert spaces; $U \subseteq F$ is the control set. The control or input space $W(0,T,U)$ is the set of all F-valued strongly measurable functions defined in $0 \le t \le T$ with the distance

$$(5.1) \qquad d(u(\cdot),v(\cdot)) = \text{meas}\{t;\ u(t) \neq v(t)\}$$

We call E the output space. The trajectory space is the space $C(0,T,E)$ of continuous E-valued functions defined in $0 \le t \le T$. A *system* is a map

$$(5.2) \qquad u(\cdot) \to y(\cdot,u)$$

from $W(0,T,U)$ into $C(0,T,E)$ that satisfies three postulates: *Causality:* $y(t,u)$ in $0 \le t \le \bar{t}$ depends only on $u(t)$ in the same interval. *Continuity:* The map $u(\cdot) \to y(\bar{t},u)$ is continuous from $W(0,T,U)$ (endowed with the distance (5.1)) into E. The third postulate involves interaction of the map (5.2) with the traditional *spike variations* of control theory: given a control u in $W(0,\bar{t},U)$ and an element v of the control set U we define

$$u_{s,h,v}(t) = v \quad (s - h < t < s)$$

$$u_{s,h,v}(t) = u(t) \quad \text{elsewhere}$$

Differentiability: There exists a set e of full measure in the interval $0 < s < \bar{t}$ and a function $\xi(\bar{t},s,u,v)$ (the *derivative* of the system) such that

$$\lim_{h \to 0+} h^{-1}(y(\bar{t},u_{s,h,v}) - y(\bar{t},u)) = \xi(\bar{t},s,u,v) \ .$$

Example 5.1 Consider the quasilinear differential system

$$(5.3) \qquad y'(t) = Ay(t) + f(t,y(t),u(t)) \quad (0 \leq t \leq T)$$

$$(5.4) \qquad y(0) = y^o$$

where A is the infinitesimal generator of a strongly continuous semi-group in the Hilbert space E and $f(t,y,u)$ is a smooth function. Then, if $y(t,u)$ denotes the solution of (5.3)-(5.4) corresponding to the control $u(\cdot)$, the map (5.2) is a system. The derivative is given by

$$(5.5) \qquad \xi(\bar{t},s,u,v) = S(\bar{t},s,u)\{f(s,y(s,u),v) - f(s,y(s,u),u(s))\}$$

where $S(t,s,u)$ is the solution operator of the linearized equation

$$(5.6) \qquad z'(t) = (A + \partial_y f(t,y(t,u),u(t))z(t).$$

Control problems for systems are formulated in a standard way. Given a system $u(\cdot) \to y(\cdot,u)$ with arbitrary output space E and a second system $u(\cdot) \to y_o(\cdot,u)$ with output space $\mathbb{R}$ (the system y_o is the cost functional). The optimal control problem is that of finding (or, rather, derive properties of) optimal controls $\bar{u}(\cdot)$ in the space $W(0,\bar{t},U)$, optimality meaning that

$$(5.7) \qquad y_o(\bar{t},\bar{u}) = \min y_o(t,u)$$

among all controls $u(\cdot)$ in $W(0,t,U)$ satisfying the target condition

$$(5.8) \qquad y(t,u) \in Y \ .$$

Using Ekeland's variational principle [15], [16] we can prove a "weak" or "sequential" version of the maximum principle. In the particular case where the target set Y is closed and convex, this sequential maximum principle takes the following form: given a sequence $\{\delta_n\}$ of positive numbers tending to zero, there exists a sequence of controls $\{u^n\}$ in $W(0,\bar{t},U)$ such that $u^n \to \bar{u}$ in $W(0,\bar{t},U)$ ($\bar{u}$ the optimal control) and a sequence of Lagrange multipliers $\{\mu_n,z_n\}$ in $\mathbb{R} \times E$ with

$$(5.9) \qquad \mu_n\xi_o(\bar{t},s,u^n,v) + (z_n,\xi(\bar{t},s,u^n,v)) \geq - \delta_n$$

for $n = 1,2,..,$ where $\xi_o(t,s,u,v)$ denotes the derivative of y_o.

The multipliers $\{\mu_n, z_n\}$ satisfy

(5.10) $\quad \|\{\mu_n, z_n\}\| = 1 \quad (n = 1,2,\ldots)$.

6. *The point target case.*

When the target set Y reduces to a point, (5.10) is the only information we have on the $\{\mu_n, z_n\}$. If the space E is finite dimensional, we can apply the Bolzano-Weierstrass theorem and deduce that, for a subsequence,

(6.1) $\quad \{\mu_n, z_n\} \to \{\mu, z\}$.

Hence, assuming continuity of the derivatives ξ, ξ_o with respect to u, we obtain

(6.2) $\quad \mu\xi_o(\bar{t},s,\bar{u},v) + (z,\xi(\bar{t},s,\bar{u},v)) \geq 0 \quad (\{\mu,z\} \neq 0)$

for almost all s in the control interval $0 \leq s \leq \bar{t}$ and every v in the control set U; here $\bar{u}$ is the optimal control. This is exactly what is done in [16], and, modulo a few manipulations, it is easily seen to imply the v-dependence part (A) of the maximum principle (Section 1, (1.14)) for finite dimensional systems of the form (1.1)-(1.2) with a cost functional of the form (1.4).

In infinite dimensional spaces, the argument breaks down, since (6.1) will only hold in the weak topology and the limit multiplier $\{\mu,z\}$ may be zero, thus rendering (6.2) empty. However, it is possible to show that, if the derivatives $\xi(t,s,u,v)$ of y are a "large enough" set as s and v move, the limit multiplier will be nonzero and (6.2) is obtained again. The precise condition is: for n large enough, the reachable sets $K_n(\bar{t})$ of the linearized systems

(6.3) $$z'(s) = (A + \partial_y f(s,y(s,u^n),u^n(s))z(s) + \\ + \{f(s,y(s,u^n),v(s)) - f(s,y(s,u^n),u^n(s))\}$$

(where $v(\cdot)$ belongs to $W(0,\bar{t},U)$) contain a common open set. This condition is of course a descendant of the controllability assumptions in the linear case in Section 2, where $K(\bar{t})$ was required to possess interior points in order to apply separation theorems. For details, see [23] or [29].

Results of this type can be applied to quasilinear hyperbolic systems, where the controllability assumptions can be verified. In the case of quasilinear parabolic systems, the required controllability property never takes place, but, in certain cases, it is verified in a subspace E_o, which makes possible to obtain generalized versions of the maximum principle (6.2). See [22] for such an instance.

7. *The set target case.*

When the target set Y is "sufficiently large" (for instance, when it has nonempty interior), additional information is available on the multipliers $\{\mu_n, z_n\}$ besides (5.10). This information suffices to place the multipliers in a cone with sufficiently small aperture, and thus prevent their weak convergence to zero. This is treated in a rudimentary way in [23] and in a much improved fashion (using nonsmooth set analysis) in [28], [29]. The results can be considered descendants of the linear results referred to in Section 4.

8. *Dependence of the Hamiltonian on time.*

The nonlinear results outlined in Sections 5, 6 and 7 are obtained by consideration of $y(\bar{t},u)$ for $\bar{t}$ fixed, thus no additional information follows when $\bar{t}$ is not fixed in advance. In other words, it is only part (A) of the maximum principle that is generalized. To fix ideas, we make a few comments on systems in Hilbert space of the form (5.3)-(5.4). We note first that the Hamiltonian must be defined as

$$(8.1) \qquad H(t,\tilde{y},\tilde{z},u) = z_o f_o(t,y,u) + (z,Ay) + (z,f(t,y,u)).$$

However, the generalized solutions of (5.3)-(5.4) necessary in the treatment of control problems may not belong to $D(A)$ for any t, thus the second term on the right hand side of (8.1) may not make sense. This does not affect the v-dependence part (A) of the maximum principle since the term does not depend on v and can be dropped from both sides of (1.14). Unfortunately, this is not the case for part (B) (equalities (1.15) or (1.16)). In some cases (such as for abstract parabolic equations) the Hamiltonian (8.1) exists, but it may not be directly related to the maximum principle. The only results on the generalization of (1.15) and (1.16) to infinite dimensional systems that we know are in [25] for a very particular case.

9. *Infinite dimensional nonlinear programming problems.*

An apparently far reaching generalization of the ideas above has been carried out in [26], [27]. Here, the functions $y(\bar{t},u)$, $y_o(\bar{t},u)$

are replaced by arbitrary functions f, f_o and the following nonlinear programming problem is considered:

(9.1) minimize $f_o(u)$ $(u \in V)$

(9.2) subject to $f(u) \in Y$

where V is a complete metric space, E is a Hilbert space, $f : V \to E$ and $f_o : V \to \mathbb{R}$ are continuous and the target set Y is a subset of E. The role of the derivative of a system is played here by the *variations* of a function g from a metric space V into a linear topological space E. We say that $\xi \in E$ is a variation of g at u (in symbols, $\xi \in \partial g(u)$) if there exists $\delta > 0$ and a function $u(h)$ defined in $0 < h < \delta$ such that

(9.3) $d(u,u(h)) \leq h,\ h^{-1}(g(u(h)) - g(u)) \to \xi$ as $h \to 0+$.

The sequence maximum principle (5.9) generalizes here to a "sequence Kuhn-Tucker theorem". Let $\bar{u}$ be a solution of the nonlinear programming problem (9.1)-(9.2), and let $\{\delta_n\}$ be a sequence of positive numbers tending to zero. Then there exists a sequence $\{u^n\}$ in V such that $u^n \to \bar{u}$ in V and a sequence of Lagrange multipliers $\{\mu_n, z_n\}$ in $\mathbb{R} \times E$ such that (5.10) holds and

(9.4) $\mu_n \eta^n + (z_n, \xi^n) \geq - \delta_n$

for $n = 1,2,\ldots$ for all $\{\eta^n, \xi^n\}$ in the variation set $\partial\{f_o, f\}(u^n)$.

To take limits in (9.4), the remarks in Sections 6 and 7, suitably generalized, apply in full: if f has "sufficiently many variations" (a condition of controllability type) or if the target set Y is "sufficiently large", we can take limits in (9.4) under adequate assumptions on f_o, f and obtain a theorem of Kuhn-Tucker type,

(9.5) $\mu\eta + (z,\xi) \geq 0$ $(\{\mu, z\} \neq 0)$

for all variations $\{\eta, \xi\}$ in the variation set $\partial\{f_o, f\}(\bar{u})$.

10. Conclusions.

The approach of Clarke and Ekeland to optimal control problems has proved very suitable for the treatment of infinite dimensional problems. It has produced very general versions of the maximum principle, as well as a number of related new results, such as theorems on the convergence of the Lagrange multipliers in the maximum principle and explicit estimates for close-to-optimal controls [28], [29] that generalize finite dimensional results [32] [33]. Although the method is somewhat "nongeometric", this may in fact be an advantage, since geometric intuition in infinite dimensional spaces is sometimes unreliable.

In its last incarnation as a theory of nonlinear programming problems in infinite dimensional spaces, this approach can handle stationary optimal control problems (such as those stemming from elliptic equations) and variations much more general than those included in the theory of systems.

Finally, some recent results (such as the extension of the theory from the Hilbert space to the Banach space setting by Frankowska in [30] and [31]) make it possible to treat, among other things, optimal control problems with state constraints.

This work was supported in part by the National Science Foundation under grant DMS-00645

References

[1] A. V. BALAKRISHNAN, Optimal control problems in Banach spaces, SIAM J. Control 3 (1965) 152-180

[2] R. BELLMAN, I. GLICKSBERG and O. GROSS, On the bang-bang control problem, Quart. Appl. Math. 14 (1956) 11-18

[3] V. G. BOLTYANSKII, R. V. GAMKRELIDZE and L. S. PONTRYAGIN, On the theory of optimal processes (Russian), Dokl. Akad. Nauk SSSR 110 (1956) 7-10

[4] A. G. BUTKOVSKII, Optimal processes in systems with distributed parameters (Russian), Avtomatika i Telemekhanika 22 (1961) 23-55

[5] A. G. BUTKOVSKII, *Theory of optimal control of distributed parameter systems* (Russian), Izdatelstvo "Nauka", Moscow, 1965

[6] F. CLARKE, Necessary conditions for a general control problem, *Calculus of Variations and Control Theory*, Academic Press, New York (1976) 257-278

[7] F. CLARKE, The maximum principle with minimum hypotheses, SIAM J. Control 14 (1976) 1078-1091

[8] F. CLARKE, *Optimization and Nonsmooth Analysis*, Wiley - Interscience, New York, 1983

[9] F. COLONIUS, The maximum principle for relaxed hereditary differential systems with function space end condition, SIAM J. Control 20 (1982) 695-712

[10] F. COLONIUS and D. HINRICHSEN, Optimal control of functional differential systems, SIAM J. Control 16 (1978) 861-879

[11] Yu. V. EGOROV, Certain problems in the theory of optimal control (Russian), Dokl. Akad. Nauk SSSR 145 (1962) 1080-1084

[12] Yu. V. EGOROV, Some problems in the theory of optimal control (Russian), Z. Vycisl. Mat. Fiz 5 (1962) 887-904

[13] Yu. V. EGOROV, Optimal control in Banach spaces (Russian), Dokl. Akad. Nauk SSSR 150 (1963) 241-244

[14] Yu. V. EGOROV, Necessary conditions for optimal control in Banach spaces (Russian), Mat. Sbornik 64 (1964)

[15] I. EKELAND, On the variational principle, J. Math. Anal. Appl. 47 (1974) 324-353

[16] I. EKELAND, Nonconvex minimization problems, Bull. Amer. Math. Soc. 1 (1979) 443-474

[17] H. O. FATTORINI, Time optimal control of solutions of operational differential equations, SIAM J. Control 2 (1964) 54-59

[18] H. O. FATTORINI, The time optimal control problem in Banach spaces, Appl. Math. Optimization 1 (1974) 163-188

[19] H. O. FATTORINI, Exact controllability of linear systems in infinite dimensional spaces, Springer Lecture Notes in Mathematics 466 (1975) 166-183

[20] H. O. FATTORINI, The time optimal control problem for boundary control of the heat equation, *Calculus of Variations and Control Theory*, Academic Press, New York (1976) 305-319

[21] H. O. FATTORINI, The time optimal problem for distributed control of systems described by the wave equation, *Control Theory of Systems Governed by Partial Differential Equations*, Academic Press, New York (1977) 151-175

[22] H. O. FATTORINI, The maximum principle for nonlinear nonconvex systems in infinite dimensional spaces, Springer Lecture Notes in Control and Information Sciences 75 (1986) 162-178.

[23] H. O. FATTORINI, A unified theory of necessary conditions for nonlinear nonconvex control systems, Applied Math. Optimization 15 (1987) 141-185

[24] H. O. FATTORINI, Convergence of suboptimal elements in infinite dimensional nonlinear programming problems, Springer Lecture Notes in Control and Information Sciences 114 (1989) 23-34

[25] H. O. FATTORINI, Constancy of the Hamiltonian in infinite dimensional control problems, to appear in Proceeding of 4th. International Conference on Distributed Parameter Systems, Vorau, 1988.

[26] H. O. FATTORINI and H. FRANKOWSKA, Necessary conditions for infinite dimensional control problems, to appear in Proceedings of 8th. International Conference on Analysis and Optimization of Systems, Antibes - Juan Les Pins 1988

[27] H. O. FATTORINI and H. FRANKOWSKA, Necessary conditions for infinite dimensional nonlinear nonconvex control problems, to appear

[28] H. O. FATTORINI and H. FRANKOWSKA, Explicit estimates for suboptimal controls, to appear in Proceedings of Conference on 30 Years of Modern Optimal Control, Kingston, 1988

[29] H. O. FATTORINI and H. FRANKOWSKA, Explicit convergence estimates for suboptimal controls I, II, to appear

[30] H. FRANKOWSKA, A general multiplier rule for infinite dimensional problems with constraints, to appear

[31] H. FRANKOWSKA, Some inverse mapping theorems, to appear

[32] H. FRANKOWSKA and Cz. OLECH, R-convexity of the integral of set-valued functions, Amer. J. Math. (1982) 156-165

[33] H. FRANKOWSKA and Cz. OLECH, Boundary solutions of differential inclusions, J. Diff. Equations 44 (1982) 156-165

[34] A. FRIEDMAN, Optimal control in Banach spaces, J. Math. Anal. Appl. 18 (1967) 35-55

[35] G. L. KARATISHVILI, The maximum principle in the theory of optimal processes with a delay, Dokl. Akad. Nauk SSSR 136 (1961) 39-42

[36] L. S. PONTRYAGIN, V. G. BOLTYANSKII, R. V. GAMKRELIDZE, E. F. MISCHENKO, *The Mathematical Theory of Optimal Processes*. (Russian) Gostekhizdat, Moscow, 1961

[37] D. L. RUSSELL, A unified boundary controllability theory for hyperbolic and parabolic partial differential equations, Studies Appl. Math. 52 (1973) 189-211

[38] LI XUNJING and YAO YUNLONG, Optimal control of distributed parameter systems, Proc. 3rd. IFAC Triennial Word Congress, Kyoto (1981) 207-221

STATE-VARIABLE REPRESENTATION REVISITED, APPLICATION TO SOME CONTROL PROBLEMS

Michel Fliess

Abstract

A new philosophy on state-variable realization is presented via methods from differential algebra. Some applications to specific control problems are briefly discussed.

Table of contents

0. Introduction

The state-space representation of linear systems, following the work of Kalman [20, 21], and the book by Zadeh and Desoer [33], is certainly the main ingredient of what should perhaps be termed *modern* control theory. Its nonlinear generalization, which really started at the end of the sixties via the language of differential geometry, has produced a wealth of interesting and remarkable results (see Isidori [16] for a splendid survey). Twenty years ago, however, Rosenbrock [27] pointed out that Kalman's representation might be of little value when dealing with complex interconnections of linear systems. He therefore introduced the notion of *pseudo-state* (see also Blomberg and Ylinen [1] for some interesting and provocative comments).

We have recently proposed a completely new approach to linear and nonlinear realization [7,8,9] by employing differential algebra. Remember that in 1985 we introduced this part of mathematics into control theory in order to give a clear-cut answer to the long-standing problem of understanding input-output inversion of nonlinear multivariable systems [6]. A minimal state is now a non-differential transcendence basis of a finitely generated differentially algebraic extension of ordinary differential fields. Two major consequences can be drawn:

- We arrive at implicit equations. This means that explicit equations are in general only "locally" valid, which is confirmed by many realistic case studies.

- Two minimal states are related by formulae involving the control variables and a finite number of their derivatives. This viewpoint, which is quite opposed to what theoreticians are now doing, is supported by some instances of more practically oriented works.

Our renewed approach to realization has several implications when considering some classic control problems:

- Since Kalman [20, 21], controllability and observability are major concepts in system theory. Thanks to works by Hermann [14], Lobry [25], Sussmann and Jurdjevic [31], and many others, the nonlinear generalization of controllability via Lie brackets of vector fields was the real starting point of the differential geometric love story. A slight change in the way of looking at the controllability of constant linear systems [11] makes it clear that its field theoretic counterpart boils down to the fact that the ground field is differentially algebraically closed in the differential field defining the dynamics (cf. Pommaret [26]). This implies that there are no elements which are not influenced by the input.

- Thanks to the differential analogue of the theorem of the primitive element, a generalized controller canonical form is proposed [7,10,12] which is valid for any system except linear constant multivariable systems and autonomous differential equations (see [12] for more details). With the help of our new viewpoint on realization, it is then possible to show that any nonlinear dynamics can exactly be linearized via a dynamic state feedback. This is a rather unexpected result when the abundant literature on this lively topic is taken into account (see Charlet, Levine and Marino [2]).

Acknowledgements. The author would like to express his deepest thanks to the organizers of this most valuable and pleasant Conference. It gave him another opportunity of trying to persuade some of his colleagues that many basic control concepts should be re-examined in depth and that serious mathematical difficulties arising today in system theory are perhaps caused by an inadequate philosophy.

I. Some elementary facts about differential algebra(1)

I.1. We assume the reader to be familiar with basic properties of commutative fields as they are now taught in many textbooks. For the sake of simplicity, all the fields will be of characteristics zero.

I.2. A *differential field* is a commutative field K which is equipped with a single derivation $\frac{d}{dt}$ = "$\cdot$" which satisfies the usual properties:

---

(1) See Kolchin's book [23] for more details.

$$\forall\ a, b \in K,\quad \frac{d}{dt}(a+b) = \dot{a} + \dot{b}$$
$$\frac{d}{dt}(ab) = \dot{a}b + a\dot{b}.$$

A *constant* in K is an element $c \in K$ such that $\dot{c} = 0$. The set of constants of K is a subfield of K, which is called the *field of constants*.

I.3. **Remark.** We restrict ourselves to systems with lumped parameters, i.e., to systems described by ordinary differential equations and therefore to *ordinary* differential fields with a single derivative. *Partial* differential fields are equipped with several pairwise commuting derivatives.

I.4. **Examples**

(i) The classical fields $\mathbb{Q}$, $\mathbb{R}$, $\mathbb{C}$ of rational, real and complex numbers are trivial fields of constants.

(ii) The field of not necessarily proper rational transfer functions in one variable s is a differential field with respect to the derivation $\frac{d}{ds}$.

(iii) The set of complex meromorphic functions in one variable z, defined in a connected domain of the complex plane, is a differential field with respect to the derivation $\frac{d}{dz}$.

I.5. A differential field extension L/K is given by two differential fields K, L such that $K \subset L$. As in the classic non-differential case, only two situations are possible:

(i) An element of L is said to be *differentially algebraic* over K if, and only if, it satisfies an algebraic differential equation with coefficients in K. The extension L/K is said to be differentially algebraic if, and only if, any element of L is differentially algebraic over K.

(ii) An element of L is said to be *differentially transcendental* over K if, and only if, it is not differentially algebraic over K, i.e., if, and only if, it does not satisfy any algebraic differential equations with coefficients in K. The extension L/K is said to be *differentially transcendental* if, and only if, at least one element of L is differentially transcendental over K.

I.6. A set $\{\xi_i \mid i \in I\}$ of elements in L is said to be *differentially K-algebraically independent* if, and only if, the set of derivatives of any order $\{\xi_i^{(\nu_i)} \mid i \in I, \nu_i = 0,1,2,\ldots\}$ is K-algebraically independent. Such an independent set, which is maximal with respect to inclusion, is called a *differential transcendence basis* of L/K. Two such bases have the same cardinality which is called the *differential transcendence degree* of L/K, and denoted by diff tr d° L/K.

I.7. **Examples**

(i) The extension L/K is differentially algebraic if, and only if, diff tr d° L/K = 0.

(ii) Assume that L/K is differentially transcendental. Take a differential transcendence basis $u = \{u_i \mid i \in I\}$. Denote by K<u> the differential field generated by K and by the components of u. The extension L/K<u> is then differentially algebraic.

I.8. The next result will be most useful.

Theorem. A finitely generated differential extension L/K is differentially algebraic if, and only if, its (non-differential) transcendence degree is finite.

In more down-to earth, but less precise, language, this non-differential transcendence degree is nothing less than the minimum number of initial conditions which must be known for computing the solution of the differential system.

I.9. The classic theorem of the *primitive element* can be generalized to differential fields.

Theorem. Assume that

- L/K is a finitely generated differentially algebraic extension,
- K is not a field of constants.

There then exists a *differential primitive element* $\pi \in$ L such that $L = K\langle\pi\rangle$, i.e., L is generated by K and π.

I.10. Let k be a given differential ground field. A *differential* k-*vector space* V is a k-vector space on which $\frac{d}{dt}$ = "$\cdot$" operates subject to the condition:

$$\forall\ a, b \in k, \forall\ v, w \in V, \ \frac{d}{dt}\,(av + bw) = \dot{a}v + a\dot{v} + \dot{b}w + b\dot{v}.$$

I.11. A set $\{v_i \mid i \in I\}$ of elements in V is said to be *differentially* k-*linearly dependent* if, and only if, the set of derivatives $\{v_i^{(\nu_i)} \mid i \in I, \nu_i = 0,1,2, \dots\}$ is k-linearly dependent. Such an independent set, which is maximal with respect to inclusion, is called a *differential basis* of V. Two such bases have the same cardinality, which is the *differential dimension* of V.

I.12. **Example**. The set $\mathbb{R}[t]$ of real polynomials in the variable t is a differential $\mathbb{R}$-linear vector space with respect to the derivation $\frac{d}{dt}$. Since, for any $p \in \mathbb{R}[t]$, $\frac{d^\nu p}{dt^\nu} = 0$ for ν large enough, the differential dimension of $\mathbb{R}[t]$ is zero, although it is well known that its (non-differential) dimension is infinite.

I.13. Here is a linear analogue of Section I.9.

Proposition. The differential dimension of a finitely generated differential k-vector space is zero if, and only if, its (non-differential) dimension is finite.

I.14. Take two differential k-vector spaces V, W such that $V \subseteq W$. Assume that W/V is finitely generated and of differential dimension zero, and that k is not a field of constants. As in I.9, there exists a *cyclic* element $\gamma \in W$ such that $W = V + [\gamma]$, where $[\gamma]$ is the differential k-vector space spanned by γ.

I.15. **Remark.** For studying linear differential equations, a related notion of *module with connection* has been introduced in the literature (see Deligne [4]). From there we have borrowed the word cyclic.

II. Input-output systems and dynamics [7,8,9,12]

II.1. Let k be a given differential ground field. Take two finite sets of differential quantities $u = (u_1, ..., u_m)$ and $y = (y_1, ..., y_p)$. A *system* with input u and output y is defined as a differentially algebraic extension k<u,y>/k<u>, where k<u> (resp. k<u,y>) denotes the differential field generated by k and the components of u (resp. u and y).

II.2. When compared to an input-output system, a *dynamics* dos not specify the output. We therefore define a dynamics as a finitely generated differentially algebraic extension K/k<u>.

II.3. The input is said to be *independent* if, and only if, it is a differential transcendence basis of k<u,y>/k<u> or of K/k<u>.

II.4. **Remark.** The preceding differential fileds k<u,y> and k are finitely generated differentially transcendental extensions of the ground field k. They are therefore the differential analogue of the well-known fields of algebraic functions (cf. Zariski and Samuel [34]). The role of independent variables is now played by an independent input.

II.5. **Remark.** We are fortunately not limited to algebraic differential equations. We easily check, for example, that any solution of the pendulum equation

$$\ddot{y} + \omega^2 \sin y = 0 \qquad (m = 0, p = 1, \omega \in \mathbb{R})$$

also satisfies the following algebraic differential equation

$$(y^{(3)})^2 + (\dot{y}\ddot{y})^2 = \omega^2 \ddot{y}^2$$

It can now be argued that realistic case studies are always described by differential equations, which might not be algebraic, but the coefficients of which are solutions of algebraic differential equations. It is rather routine work to extend the previous elimination procedure to the latter case (see Rubel [28]).

II.6. Denote by $[u]$ (resp. $[u,y]$) the differential k-vector space spanned by the components of u (resp. u and y). A *linear system* with input u and output y is a differential k-vector space $[u,y]$ such that the quotient $[u,y]/[u]$ has differential dimension zero. A *linear dynamics* is a finitely generated differential k-vector space V which contains $[u]$ and such that $V/[u]$ has differential dimension zero. The input u is said to be independent if, and only if, it is a differential basis of $[u,y]$ or of V.

II.7. **Remark.** When k is a field of constants, our definition of linear systems can be shown to be equivalent to the generalized linear systems considered by Blomberg and Ylinen [1].

III. State-variable realization [7,8,9,12]

III.1. Consider an input-output system $k\langle u,y\rangle/k\langle u\rangle$. According to I.8, the (non-differential) transcendence degree of $k\langle u,y\rangle/k\langle u\rangle$ is finite, say n. Take a (non-differential) transcendence basis $x = (x_1, \ldots, x_n)$ of $k\langle u,y\rangle/k\langle u\rangle$. The derivatives $\dot{x}_1, \ldots, \dot{x}_n$ and the components of y are $k\langle u\rangle$-algebraically dependent on x. This reads:

$$
\text{(I)}\quad \begin{cases}
A_1(\dot{x}_1, x, u, \dot{u}, \ldots, u^{(\alpha)}) = 0 \\
\text{- - - - - - - - - - - - - - - - - -} \\
A_n(\dot{x}_n, x, u, \dot{u}, \ldots, u^{(\alpha)}) = 0 \\
B_1(y_1, x, u, \dot{u}, \ldots, u^{(\alpha)}) = 0 \\
\text{- - - - - - - - - - - - - - - - - -} \\
B_p(y_p, x, u, \dot{u}, \ldots, u^{(\alpha)}) = 0
\end{cases}
$$

where $A_1, \ldots, A_n, B_1, \ldots, B_p$ are polynomials with coefficients in k. Note that equations (I) are *implicit* with respect to the derivatives of the *state* x and with respect to the components of the output y.

III.2. In order to apply the implicit functions theorem(2), we shall assume that the quantities appearing in (I) take real or complex values. When the two Jacobian matrices

(2) The use of the implicit functions theorem in our algebraic context has only a heuristic value. One should preferably employ some sophisticated tools from algebraic geometry.

$$\begin{pmatrix} \partial A_1/\partial x_1 & & (0) \\ & \ddots & \\ (0) & & \partial A_n/\partial y_n \end{pmatrix}$$

$$\begin{pmatrix} \partial B_1/\partial y_1 & & (0) \\ & \ddots & \\ (0) & & \partial B_p/\partial y_p \end{pmatrix}$$

are non-singular, we obtain the "local" *explicit* equations:

$$(E) \begin{cases} \dot{x}_1 = a_1(x, u, \dot{u}, \ldots, u^{(\alpha)}) \\ \cdots\cdots\cdots\cdots\cdots\cdots \\ \dot{x}_n = a_n(x, u, \dot{u}, \ldots, u^{(\alpha)}) \\ y_1 \ = b_1 (x, u, \dot{u}, \ldots, u^{(\alpha)}) \\ \cdots\cdots\cdots\cdots\cdots\cdots \\ y_p \ = b_p (x, u, \dot{u}, \ldots, u^{(\alpha)}) \end{cases}$$

III.4. **Remark**. Take two *minimal* states $x = (x_1, \ldots, x_n)$, $\bar{x} = (\bar{x}_1, \ldots, \bar{x}_n)$, i.e., two transcendence bases of k<u,y>/k<u>. The components of $\bar{x}$ (resp. x) are k<u>-algebraically dependent on the components of x (resp. $\bar{x}$):

$$\begin{cases} P_1 (\bar{x}_1, x, u, \dot{u}, \ldots, u^{(\alpha)}) = 0 \\ \text{------------------------} \\ P_n (\bar{x}_n, x, u, \dot{u}, \ldots, u^{(\alpha)}) = 0 \ . \end{cases}$$

As before, this relation is implicit and can only "locally" be put in an explicit form. But the main feature concerns the dependence on the control variables and on a finite number of their derivatives. In the actual realization theory, only changes of coordinates have been considered. Note, however, that the control dependence has already been taken into account in some more applied works (see, e.g., Williamson [32], Zeitz and some of his students [35, 22]).

III.5. The preceding constructions can most easily be adapted for linear systems by taking the components of $x = (x_1, \ldots, x_n)$ in [u,y] such that their image in the quotient [u,y]/[u] is a basis of the latter space. The state-variable representation then reads:

$$
(L)\begin{cases}
\dfrac{d}{dt}\begin{pmatrix} x_1 \\ \vdots \\ x_n \end{pmatrix} = A \begin{pmatrix} x_1 \\ \vdots \\ x_n \end{pmatrix} + \sum_{\text{finite}} B_\mu \dfrac{d^\mu}{dt^\mu} \begin{pmatrix} u_1 \\ \vdots \\ u_m \end{pmatrix} \\
\begin{pmatrix} y_1 \\ \vdots \\ y_p \end{pmatrix} = C \begin{pmatrix} x_1 \\ \vdots \\ x_n \end{pmatrix} + \sum_{\text{finite}} D_\mu \dfrac{d^\mu}{dt^\mu} \begin{pmatrix} u_1 \\ \vdots \\ u_m \end{pmatrix}
\end{cases}
$$

The various matrices are of appropriate sizes. Their entries lie in k.

Two minimal states $x = (x_1, \ldots, x_n)$, $\bar{x} = (\bar{x}_1, \ldots, \bar{x}_n)$ are related by a linear formula which is control dependent:

$$
(T) \qquad \begin{pmatrix} \bar{x}_1 \\ \vdots \\ \bar{x}_n \end{pmatrix} = P \begin{pmatrix} x_1 \\ \vdots \\ x_n \end{pmatrix} + \sum_{\text{finite}} Q_\mu \frac{d^\mu}{dt^\mu} \begin{pmatrix} u_1 \\ \vdots \\ u_m \end{pmatrix}.
$$

P is a square invertible matrix.

III.6. The transformation (T) shows that it is possible to eliminate the derivatives of the control variables in the dynamics of (L). We thus get a Kalman-type description:

$$
\begin{cases}
\dfrac{d}{dt}\begin{pmatrix} \bar{x}_1 \\ \vdots \\ \bar{x}_n \end{pmatrix} = A \begin{pmatrix} \bar{x}_1 \\ \vdots \\ \bar{x}_n \end{pmatrix} + B \begin{pmatrix} u_1 \\ \vdots \\ u_m \end{pmatrix} \\
\begin{pmatrix} y_1 \\ \vdots \\ y_p \end{pmatrix} = C \begin{pmatrix} \bar{x}_1 \\ \vdots \\ \bar{x}_n \end{pmatrix} + \sum_{\text{finite}} D'_\mu \dfrac{d^\mu}{dt^\mu} \begin{pmatrix} u_1 \\ \vdots \\ u_m \end{pmatrix}
\end{cases}
$$

It is straightforward to check that this procedure cannot be extended to general nonlinear systems.

III.7. **Remark**. When compared to Rosenbrock's pseudo-state [27], we do believe that our state-variable representation is of some interest as it behaves more like the classic Kalman state.

III.8. With a nonlinear dynamics K/k<u>, the state-variable description of III.1 and III.2 can be applied by taking a transcendence basis of K/k<u> and forgetting about the output.

With a linear dynamics V, we just take elements in V such that their image in V/[u] is a basis of that space.

IV. Controllability and observability

IV.1. It follows from III.6 that a linear dynamics V possesses the following state-variable representation *à la* Kalman:

$$\text{(K)} \qquad \dot{x} = Ax + Bu.$$

Assume that the ground field k is a field of constants, $\mathbb{R}$ for instance. The dynamics V is said to be *controllable* if, and only if, (K) is controllable according to the usual sense (see [11] for more details). The classic Kalman decomposition into the controllable part and the uncontrollable part shows that the elements of the uncontrollable part are characterized as being differentially k-linearly dependent on zero.

IV.2. Before giving the field theoretic translation of IV.1, we shall recall that the *differentially algebraic closure* of the ground field k in the dynamics K is the differential field $\bar{k}$, $k \subseteq \bar{k} \subseteq K$, consisting of the elements of K which are differentially algebraic over k. The field k is said to be differentially algebraically closed if, and only if, $\bar{k} = k$.

Definition(3). The input-output system $k\langle u,y\rangle/k\langle u\rangle$ (resp. the dynamics $K/k\langle u\rangle$) is said to be controllable if, and only if, the ground field k is differentially algebraically closed in $k\langle u,y\rangle$ (resp. K).

The word controllability is still well chosen in this new context as it now means that all elements in $k\langle u,y\rangle$ or K are indeed influenced by the input.

IV.3. *Unobservability* of a constant linear dynamics V with respect to an output y means that V strictly contains [u,y] (see [11]). Here is a field theoretical translation:

Definition. A dynamics $K/k\langle u\rangle$ is said to be observable with respect to an output y, the components of which belong to K, if, and only if, the extension $K/k\langle u,y\rangle$ is (non-differentially) algebraic.

This means that any element of K is solution of an algebraic equation with coefficients in $k\langle u,y\rangle$.

---

(3) This definition can be found in Pommaret's book [26] which unfortunately contains many highly controversial parts.

V. Controller canonical forms and feedback linearization [7,10,12]

V.1. Take a nonlinear dynamics K/k<u>. Assume that k<u> is not a field of constants. This is for instance the case when the set u of control variables is non-empty and is independent. From I.9, we know the existence of a differential primitive element ξ for the extension K/k<u>. If the (non-differential) transcendence degree of K/k<u> is equal to n, it is quite easy to show that $\xi, \dot{\xi}, \ldots, \zeta^{(n-1)})$ is a (non-differential) transcendence basis. Set $x_1 = \zeta$, $x_2 = \dot{\xi}$, ..., $x_n = \xi^{(n-1)}$; $x = (x_1, \ldots, x_n)$ is a minimal state. This yields the *global implicit controller canonical form*:

$$\begin{cases} \dot{x}_1 = x_2 \\ \text{------------} \\ \dot{x}_{n-1} = x_n \\ C(\dot{x}_n, x, u, \dot{u}, \ldots, u^{(\alpha)}) = 0, \end{cases}$$

and a *local explicit controller canonical form*:

$$\begin{cases} \dot{x}_1 = x_2 \\ \text{------------} \\ \dot{x}_{n-1} = x_n \\ \dot{x}_n = c(x, u, \dot{u}, \ldots, u^{(\alpha)}) \ . \end{cases}$$

We refer to Sommer [29], Zeitz [35] and Krener [24] for other attempts of deriving the controller form.

V.2. Take now a linear dynamics where the ground field k is not a field of constants. From I.14, we know the existence of a cyclic element $\gamma \in V$ such that $V = [u] + [\gamma]$. If the (non-differential) dimension of the quotient V/[u] is equal to n, then, as in V.1, $x_1 = \gamma$, $x_2 = \dot{\gamma}$, ..., $x_n = \gamma^{(n-1)}$ is a minimal state which yields the *generalized local controller canonical form*:

$$
\begin{cases}
\dot{x}_1 = x_2 \\
\text{------------} \\
\dot{x}_{n-1} = x_n \\
\dot{x}_n = a_1 x_1 + \ldots + a_n x_n + \sum\limits_{\text{finite}} b_{i\nu}\, u_i^{(\nu)}
\end{cases}
$$

$(a_{1,} \ldots, a_n,)b_{i\nu} \in k)$

Section III.6 shows that it is possible to obtain a controller form without derivatives of the control variables.

V.3. **Remark.** The preceding formulae in V.1 and V.2 are obvious generalizations of the controller form of a constant linear controllable single-input system. It can be shown [12] that our controller for is not valid in two most important particular cases:

(i) multivariable constant linear system;

(ii) autonomous differential systems, i.e., differential equations with constant coefficients and no input.

This fact has far-reaching implications which will be explored in future works dealing with variable structure systems and/or stabilization techniques.

V.4. The problem of exactly linearizing a nonlinear dynamics via a static dynamic feedback is now an important and classic topic. The first significant results were obtained on the one hand by Jakubczyk and Respondek [19] and on the other by Hunt, Su and Meyer [15]. We refer to Claude [3] and Isidori [16] for surveys. Recently Charlet, Levine and Marino [2] studied linearization with a dynamic feedback. They were able, among other things, to prove that, for a single-input, dynamic and static feedback conditions coincide.

Take the generalized local controller canonical form of V.1. Equate c with a linear homogeneous polynomial ℓ with coefficients in k, in the variables $x = (x_1, \ldots, x_n)$, $v = (v_1, \ldots, v_{m'})$, which is a *new* input, and a finite number of derivatives of v:

$$c(x, u, \dot{u}, \ldots, u^{(\alpha)}) = \ell(x, v, \dot{v}, \ldots, v^{(\beta)}).$$

This obviously defines a linearizing dynamic feedback for any dynamics where u is non-empty.

The apparent contradiction with [2] is due to a different viewpoint on feedback since in [2] the feedback dynamics must also be linearized. Let us stress that this difference should be pinpointed to our realization where the change of state is control-dependent.

References

[1] H. Blomberg and R. Ylinen. Algebraic Theory of Multivariable Linear Systems, Academic Press, London, 1983.

[2] B. Charlet, J. Levine, and R. Marino, Two sufficient conditions for dynamic feedback linearization of nonlinear systems, in "Analysis and Optimization of Systems", A. Bensoussan and J.L. Lions eds., Lect. Notes Control Inform. Sci. 111, pp.181-192, Springer-Verlag, Berlin, 1988.

[3] D. Claude, Everything you always wanted to know about linearization, in "Algebraic and Geometric Methods in Nonlinear Control Theory", M. Fliess and M. Hazewinkel eds., pp.181-220, Reidel, Dordrecht, 1986.

[4] P. Deligne, Equations différentielles à points singuliers réguliers, Lect. Notes Math. 163, Springer-Verlag, Berlin, 1970.

[5] M. Fliess, Réalisation locale des systèmes non linéaires, algèbres de Lie filtrées transitives et séries génératrices non commutatives, Invent. Math., 71, pp.521-533, 1983.

[6] M. Fliess, A note on the invertibility of nonlinear input-output systems, Systems Control Lett., 8, pp.147-151, 1986.

[7] M. Fliess, Nonlinear control theory and differential algebra, in Modelling and Adaptive Control, Ch.I. Byrnes and A. Kurzhanski eds., Lect. Notes Control Inform. Sci. 105, pp.134-145, Springer-Verlag, Berlin, 1988.

[8] M. Fliess, Automatique et corps différentiels, Forum Math., vol.1, 1989.

[9] M. Fliess, Generalized linear systems with lumped or distributed parameters, Int. J. Control, 1989.

[10] M. Fliess, Généralisation non linéaire de la forme canonique de commande et linéarisation par bouclage, C.R. Acad. Sci. Paris, I-308, pp.377-379, 1989.

[11] M. Fliess, Commandabilité, matrices de transfert et modes cachés, C.R. Acad. Sci. Paris, to appear.

[12] M. Fliess, Generalized controller canonical forms for linear and nonlinear dynamics, to appear.

[13] M. Hasler and J. Neirynck, Circuits non linéaires, Presses Polytechniques Romandes, Lausanne, 1985.

[14] R. Hermann, On the accessibility problem in control theory, in "Int. Symp. Nonlinear Diff. Eq. and Nonlinear Mech.", J.P. LaSalle and S. Lefschetz eds., pp.325-332, Academic Press, New York, 1963.

[15] L.R. Hunt, R. Su and G. Meyer, Global transformations of nonlinear systems, IEEE Trans. Autom. Control, AC-28, pp.24-31, 1983.

[16] A. Isidori, Nonlinear Control Systems: An Introduction, Lect. Notes Control Inform. Sci. 72, Springer-Verlag, Berlin, 1985.

[17] B. Jakubczyck, Existence and uniqueness of realizations of nonlinear systems, SIAM J. Control Optimiz., 18, pp.455-471, 1980.

[18] B. Jakubczyck, Realization theory for nonlinear systems; three approaches, in "Algebraic and Geometric Methods in Nonlinear Control Theory", M. Fliess and M. Hazewinkel eds., pp.3-31, Reidel, Dordrecht, 1986.

[19] B. Jakubczyck and W. Respondek, On linearization of control systems, Bull. Acad. Polon. Sci. Sér. Math., vol.28, pp.517-522, 1980.

[20] R.E. Kalman, Mathematical description of linear systems, SIAM J. Control, 1, pp.152-192, 1963.

[21] R.E. Kalman, P.L. Falb and M.A. Arbib, Topics in Mathematical System Theory, McGraw-Hill, New York, 1969.

[22] H. Keller and H. Fritz, Design of nonlinear observers by a two-step transformation, in "Algebraic and Geometric Methods in Nonlinear Control Theory", M. Fliess and M. Hazewinkel eds., pp.89-98, Reidel, Dordrecht, 1986.

[23] E.R. Kolchin, Differential Algebra and Algebraic Groups, Academic Press, New York, 1973.

[24] A.J. Krener, Normal forms for linear and nonlinear systems, Contemp. Math., vol.68, pp.157-189, 1987.

[25] C. Lobry, Contrôlabilité des systèmes non linéaires, SIAM J. Control, 8, pp.573-605, 1970.

[26] J.F. Pommaret, Lie Pseudogroups and Mechanics, Gordon and Breach, New York, 1988.

[27] H.H. Rosenbrock, State Space and Multivariable Theory, Nelson, London, 1970.

[28] L.A. Rubel, An elimination theorem for systems of algebraic differential equations, Houston J. Math., 8, pp.289-295, 1982.

[29] R. Sommer, Control design for multivariable nonlinear time-varying systems, Int. J. Control, 31, pp.883-891, 1980.

[30] H.J. Sussmann, Existence and uniqueness of minimal realizations of nonlinear systems, Math. Systems Theory, 10, pp. 263-284, 1977.

[31] H.J. Sussmann and V. Jurdjevic, Controllability of nonlinear systems, J. Diff. Equations, 12, pp.95-116, 1972.

[32] D. Williamson, Observation of bilinear systems with application to biological control, Automatica, 13, pp.243-254, 1977.

[33] L.A. Zadeh and C.A. Desoer, Linear System Theory: The State Space Approach, McGraw-Hill, New York, 1963.

[34] O. Zariski and P. Samuel, Commutative Algebra, vol.1, Van Nostrand, Princeton, N.J., 1958.

[35] M. Zeitz, Canonical forms for nonlinear systems, in Geometric Theory of Nonlinear Control Systems, B. Jakubczyck, W. Respondek and K. Tchoń eds, pp.255-278, Widawnictwo Politechniki Wrocławskiej, Wrocław, 1985.

Laboratoire des Signaux et Systèmes
CNRS-ESE
Plateau du Moulon
91192 Gif-sur-Yvette Cedex
France

LINEAR-QUADRATIC PROBLEMS AND THE RICCATI EQUATION

A.H.W. Geerts & M.L.J. Hautus
Dept. of Math. & Comp.Sci.
Eindhoven University of technology

0. Introduction

Linear-Quadratic (LQ) control problems have been investigated intensively since the fundamental and seminal paper of R.E. Kalman in 1960 ([KA]). In that paper, it was shown that the Riccati Equation plays an important role for the LQ-Problem. It is the purpose of the present paper to give an overview of the results relating the Riccati Equation with the LQ-Problem. LQ problems are also treated using the Hamiltonian matrix instead of the Riccati Equation. This will not be discussed in this paper. Neither will we deal with the use of Riccati equations outside of the LQ-problem context, e.g. in differential games and the H^∞-optimization problem.

The LQ-problem can be formulated as follows:

L Q - P R O B L E M. *Given matrices* A, B, Q, S, R *of suitable dimensions, a vector* $x_0 \in \mathbb{R}^n$ *and* T *satisfying* $0 < T \le \infty$, *determine a control* u *such that for the system*

$$\dot{x} = Ax + Bu, \quad x(0) = x_0 \tag{0.1}$$

the quantity

$$J(x_0, u, T) := \int_0^T (x'Qx + 2u'Sx + u'Ru)\,dt \tag{0.2}$$

is minimized.

The integrand can be written as

$$\omega(x,u) = [x', u']N\begin{bmatrix} x \\ u \end{bmatrix}, \tag{0.3}$$

where

$$N := \begin{bmatrix} Q & S' \\ S & R \end{bmatrix}. \tag{0.4}$$

It is natural to assume that Q and R are symmetric matrices. Also, we may assume that

$$\operatorname{rank} \begin{bmatrix} B \\ S' \\ R \end{bmatrix} = m, \tag{0.5}$$

where m denotes the number of inputs. Otherwise one could restrict the inputs to a suitable subspace.

The results obtained in literature mostly depend on additional assumptions. We will refer to them as "cases". The following cases have obtained names:

Finite-Horizon Problem:	$T < \infty$.
Infinite-Horizon Problem:	$T = \infty$.
Semidefinite Problem:	$N \geq 0$
Indefinite Problem:	$\neg\, N \geq 0$
Regular case:	R is invertible
Singular case:	R is not invertible

In the ∞-horizon problem, convergence problems play a role. In particular, two cases are distinguished: Problems in which *a priori*, the state variable is required to go to zero for $t \to \infty$ (so-called **zero-endpoint problems**) and problems in which no requirement is made about the asymptotic behavior of the state (**free-endpoint problems**). This distinction was introduced in [Wi]. Recently, interest has arisen in the intermediate situation, where x is required to tend to zero modulo some prescribed subspace. ([G-4, Ch.3])

In the semidefinite case the matrix N in (0.4) is positive semidefinite. Therefore, it can be factorized as

$$N = \begin{bmatrix} C' \\ D' \end{bmatrix} [C \ D], \tag{0.6}$$

where C and D are matrices of suitable dimensions. We will choose C and D such that [C,D] has full row rank. Next one can introduce an artificial output variable y by

$$y = Cx + Du. \tag{0.7}$$

Then the optimization problem can be reformulated as:

0.8 PROBLEM *Minimize the integral*

$$J(x_0,u,T) = \int_0^T \|y(t)\|^2 dt. \tag{0.9}$$

subject to the system

Σ: $\dot{x} = Ax + Bu,\ x(0) = x_0,\quad y = Cx + Du.$

We will identify the system Σ with the matrix quadruple (A,B,C,D). Because of the fact that [C,D] has full row rank, condition (0.5) is equivalent to rank [B',D'] = m. The problem is regular iff D is injective (left invertible).

1. The regular semidefinite case.

A complete solution of the regular semidefinite finite-horizon problem is given by:

1.1 THEOREM (**Regular semidefinite finite-horizon case**). *Assume that the problem is regular and semidefinite, and use the problem formulation as in* Problem 0.8, *in which* D *is injective.*

i) *The* **Riccati Differential Equation** *(=:* RDE*)*

$$\dot{P} = C'C + A'P + PA - (PB + C'D)(D'D)^{-1}(B'P + D'C),\ P(0)=0 \tag{1.2}$$

has a unique solution P(t) *for* $t \geq 0$. *This solution is symmetric, positive semidefinite and depends monotonically on* t *(i.e.,* $t_1 < t_2 \Longrightarrow P(t_1) \leq P(t_2)$*).*

ii) *Define the time-variable feedback matrix*

(1.3) $$F(t) := -(D'D)^{-1}(B'P(T-t)+D'C) \qquad (0 \le t \le T)$$

Then we have for every function u:

(1.4) $$J(x_0,u,T) = x_0'P(T)x_0 + \int_0^T \|D(u(t) - F(t)x(t))\|^2 dt$$

iii) *There exists a unique optimal control* $u^*:[0,T] \to \mathbb{R}^m$. *This is given by the time-variable feedback*

(1.5) $$u^*(t) = F(t)x(t)$$

iv) *The minimal value of* $J(x_0,u,T)$ *equals*

(1.6) $$J^*(x_0,T) = J(x_0,u^*,T) = x_0'P(T)x_0.$$

The statements iii) and iv), which constitute the solution of the LQ-problem, are obvious consequences of ii). The simplest proof of ii) is based on completion of squares. The monotonicity statement in i) is an immediate consequence of iv).

For the free-endpoint infinite-horizon problem, a condition is needed for the existence of an optimal control. This condition can be given in several equivalent forms (See [G&H, Theorem 2.2]).

1.7 THEOREM *The following statements are equivalent:*

i) *For every* $x_0 \in \mathbb{R}^n$, *there exists* u *such that* $\int_0^\infty \|y\|^2 dt < \infty$.

ii) *For every* $x_0 \in \mathbb{R}^n$, *there exists* u *such that* $y(t) \to 0$ $(t \to \infty)$.

iii) *There exists a feedback* $u = Fx$ *such that for the resulting system, we have* $y(t) \to 0$ $(t \to \infty)$ *for every* x_0.

iv) $$\langle A|\mathrm{im}B\rangle + \mathcal{X}^-(A) + \mathcal{V}(\Sigma) = \mathbb{R}^n$$

where

$$\langle A|\mathrm{im}B\rangle := \{\, x_0 \in \mathbb{R}^n | \exists_{T>0} \exists_u \; x(T,x_0,u) = 0 \,\}$$

$$\mathcal{X}^-(A) := \{\, x_0 \in \mathbb{R}^n | e^{tA}x_0 = x(t,x_0,0) \to 0 \quad (t \to \infty)\}$$

$$\mathcal{V}(\Sigma) := \{\, x_0 \in \mathbb{R}^n | \; y(t,x_0,u) = 0 \text{ for all } t \ge 0\}$$

are the **controllable**, **stable** *and* **weakly unobservable subspace**, *respectively.*

v) *The free endpoint problem has a solution.*

If either of these conditions is met, we call the system **output stabilizable**. Sufficient for the output stabilizability of the system is its (state) stabilizability, which is equivalent to $\langle A|\mathrm{im}B\rangle + \mathcal{X}^-(A) = \mathbb{R}^n$. Note that conditions for property iii) were studied in [Wo, Section 4.4] (for the case D = 0).

The following result gives the solution of the infinite-horizon problem (Compare [G-1]):

1.8 THEOREM *Let* $\Sigma = (A,B,C,D)$ *be output stabilizable and* D *injective. Then*

i) *The* **Algebraic Riccati Equation** (=: ARE)

(1.9) $$C'C + A'P + PA - (PB + C'D)(D'D)^{-1}(B'P + D'C) = 0$$

has a smallest positive semidefinite solution P^-. *(Every solution* $P \geq 0$ *satisfies* $P \geq P^-$.)

ii) *Define the time-invariant feedback matrix*

(1.10) $$F := -(D'D)^{-1}(B'P^- + D'C).$$

Then for every input u, *one has*

(1.11) $$J(x_0,u) := \int_0^\infty \|y\|^2 dt = x_0'P^-x_0 + \int_0^\infty \|D(u(t)-Fx(t))\|^2 dt.$$

In particular, $J(x_0,u) < \infty$ *iff* $\int_0^\infty \|D(u(t)-Fx(t))\|^2 dt < \infty$

iii) *The free-endpoint problem has exactly one solution, given by the static state feedback*

(1.12) $$u^*(t) = Fx(t)$$

iv) *The minimal value* $J^*(x_0)$ *of* $J(x_0,u)$ *is*

(1.13) $$J^*(x_0) = J(x_0,u^*) = x_0'P^-x_0.$$

In addition to the above, the existence of a positive semidefinite solution of the ARE implies the output stabilizability of Σ. Hence, this condition may be added to the series of equivalent statements im Theorem 1.7. Formula (1.11) is a consequence of the following formula crucial in this problem area:

(1.14) $$\int_0^T \|y\|^2 dt = x_0'Px_0 - x'(T)Px(T) + \int_0^T \|D(u(t) - Fx(t))\|^2 dt,$$

which holds for any input u, $T > 0$, and any solution P of the ARE. Here $F := -(D'D)^{-1}(B'P + D'C)$. The important step in the transition from (1.14) to (1.11) is the property $P^- x(t) \to 0$ $(t \to \infty)$ if $J(x_0,u) < \infty$, which is true for the particular solution P^- but not for arbitrary solutions $P \geq 0$ of the ARE (see [G-4, Cor. 3.28])

In the **zero-endpoint problem**, one is asked to minimize $J(x_0,u)$ subject to the constraint $x(t) \to 0 \quad (t \to \infty)$. Obviously, for such a control to exist, it is necessary that Σ be stabilizable. In addition, one needs a condition on the zeros of Σ:

1.15 THEOREM *Assume* Σ *to be stabilizable and* D *left invertible.*

i) $J_0^*(x_0) := \inf\{ J(x_0,u) \mid x(t,x_0,u) \to 0 \quad (t \to \infty) \} = x_0' P^+ x_0$, *where* P^+ *is the largest (positive semidefinite) solution of the* ARE.

ii) *The zero endpoint problem has a solution iff there are no zeros on the imaginary axis, i.e.,*

(1.16) $$\operatorname{rank}\begin{bmatrix} sI-A & -B \\ C & D \end{bmatrix} = n + m \quad (\mathrm{Re}\ s = 0).$$

iii) *Formula* (1.11), *with* P^- *replaced by* P^+ *is valid for every input* u *for which the state* $x(t) \to 0 \quad (t \to \infty)$.

iv) *If condition* (1.16) *is satisfied, the optimal control is unique and determined by the static state feedback*

(1.17) $$u^*(t) = -(D'D)^{-1}(B'P^+ + D'C)x(t).$$

1.18 REMARK If the system is strongly detectable, i.e., has all its zeros in $\mathbb{C}^- := \{ s \in \mathbb{C} \mid \mathrm{Re}\ s < 0 \}$, condition (1.16) is certainly satisfied. In this case, the optimal trajectory of the free-endpoint problem converges to zero. Consequently, the solutions of the free and zero endpoint problem coincide, and in particular, $P^- = P^+$. Therefore, the ARE has only one positive semidefinite solution.

In case all zeros of the system are in the *closed* halfplane with at least one of them on the imaginary axis, we still have $P^- = P^+$, but this time, the trajectory $x(t)$, corresponding to the control given in (1.17), need not converge to zero, and hence u^* solves the free endpoint problem, but it is not the solution of the zero-endpoint problem. The latter problem does not have a solution.

2. The singular semidefinite case.

In the singular case, one cannot expect a conventional ($\mathcal{L}_2$) input solving the problem to exist. The following simple example illustrates this: Consider e.g. the system

$$\dot{x} = u, \quad y = x, \quad x(0) = 1$$

The quantity $J_T := \int_0^T x^2(t)\,dt$ can be made arbitrarily small by a suitable choice of u, but it is clear that it cannot be made equal to zero.

This example suggests the use of distributions as allowed inputs. One could e.g. take $u = -\delta(t)$ (the Dirac delta function), because then $x(t) = 0$ for $t > 0$. A rigorous setup based on distributions (of L. Schwartz) was introduced in [H&S]. In order to keep the emphasis of the treatment on the algebraic aspects of the problem, it is convenient to restrict the inputs to the class $\mathcal{C}_{imp}$ consisting of all distributions of the form $u = u_1 + u_2$, where u_1 is impulsive, i.e., of the form $\Sigma a_i \delta^{(i)}$, and u_2 is smooth on $[0,\infty)$ (zero on $(-\infty,0)$). This class has the property that it is closed under convolution, and hence under differentiation and integration.

It is fairly obvious that not all distributions in $\mathcal{C}_{imp}$ can be allowed as inputs. In particular, in the above example, we cannot possibly take $u = \dot{\delta}$, the derivative of the delta function, because then we would have $x = 1 + \delta$. Consequently, $x^2(t)$ and hence J would not be defined. The class of **admissible** inputs (which is system dependent) can be defined as

(2.1) $$\mathcal{U}_\Sigma := \{\ u \in \mathcal{C}^m_{imp} \mid y \text{ is regular }\}.$$

For these inputs, $J := \int_0^\infty \|y(t)\|^2 dt$ is well defined.

For singular systems the space

$$W := \{ x_0 | \exists_{u \in \mathcal{U}_\Sigma} \; x_u(0+,x_0) = 0 \}, \tag{2.2}$$

plays an important role. Here $x_u(t,x_0)$ denotes the value of the state at time t, corresponding to initial state x_0 and input u. This space is called the **strongly controllable** (or **strongly reachable**) space. It is the space of points that can instantly be sent to the origin, by an admissible control. Obviously, $W = 0$ in the regular case. The relevance of this space for the singular case will become clear in the next paragraph. Here we only remark that for $x_0 \in W$, the minimal cost for the LQ problem equals zero, as is obvious from the definition.

Basically, the solution of the singular *free*-endpoint LQ problem consists of two steps. First a series of transformations is performed (the so-called **dual structure algorithm**) resulting in a system of the following form:

$$\dot{\xi} = A\xi + B_0\upsilon_0 + B_1\upsilon_1, \quad y = C\xi + D_0\upsilon_0, \tag{2.3}$$

where D_0 is left invertible. Notice that the input and the state are replaced by new variables. The original input is related to $\upsilon = [\upsilon_0', \upsilon_1']'$ via $u = H(p)\upsilon$, where p stands for d/dt (or $\dot{\delta}$) and H(s) is a nonsingular polynomial matrix. The original and the new state variables satisfy the relation $x - \xi \in W$ for $t > 0$. The output, however, is the same as before. Consequently, the optimality criterion is not affected by these transformation. Secondly, the preliminary feedback $F_0 := -(D_0'D_0)^{-1}D_0C$ is applied to system (2.3)., i.e., $\upsilon_0 = F_0\xi + \omega_0$, $\omega_1 = \upsilon_1$, which yields

$$\bar{A} := A + B_0F_0, \quad \bar{C} := C + D_0F_0$$

Thus one obtains the following system.

$$\dot{\xi} = \bar{A}\xi + B_0\omega_0 + B_1\omega_1, \quad y = \bar{C}\xi + D_0\omega_0, \tag{2.5}$$

with the following additional properties:

(2.6) $\quad \mathrm{im}\, B_1 \subseteq \mathcal{W}, \quad \bar{A}\mathcal{W} \subseteq \mathcal{W}, \quad \bar{C} \subseteq \ker \mathcal{W}.$

For the details of these computations we refer to [G-2, Section 4] and [G-4, Section 2.2]. The relations (2.6) imply that the transfer matrix from the input ω_1 to the output y is zero, so that y is independent of ω_1. Hence, the part ω_1 of the input represents the nonuniqueness of the optimal control. If we choose, in particular, $\omega_1 = 0$, (2.5) reduces to a regular LQ problem, which can be solved as in the previous section. In this way, we find a regular function $[\omega_0^{*\prime},0]'$ as an optimal control for the problem corresponding to (2.5). The corresponding optimal control for system (2.3) is $\upsilon^* := [(F_0\xi + \omega_0^*)',0]'$, which is also a regular function. The optimal control of the original problem, in terms of the original variables is $u^* := H(p)\upsilon^*$, and hence an input which is in $\mathcal{C}^m_{imp}$, but in general not regular. We conclude that finding an optimal control amounts to solving a suitable ARE, specifically:

(2.7) $$\bar{\Phi}(P) := \bar{C}'\bar{C} + \bar{A}'P + P\bar{A} - PB_0(D_0'D_0)^{-1}B_0'P = 0$$

Note that in the case the system Σ is left invertible (i.e. if the transfer function is left invertible), the ω_1 term in (2.3) is not present and the optimal control of the original problem is unique. This is the case studied in [H&S]. In the regular case, (where the system is certainly left invertible) no transformation at all is needed. The general procedure to construct the solution, in particular the details of the method sketched above can be found in [G-2] and [G-4]. We can conclude that the dual structure algorithm basically reduces the singular case to the regular case. In particular, it follows that the free endpoint problem has a solution iff the system is output stabilizable. Hence Theorem 1.7 remains valid in the singular case.

For the zero-endpoint problem a similar method can be given, but the transformations have to be performed more carefully. The difficulty lies in the fact that the stability of the original system is not equivalent to the stability of (2.3). But it turns out that the final result is similar to the regular case, i.e., there exists a solution iff the system is (state) stabilizable and has no zeros on the imaginary axis. (see [WKS] and [G-4, Section 3.2])

3. The dissipation inequality

In the previous section, we saw how the general singular problem could be solved, using the dual structure algorithm. In this section we give a method for expressing the solution in terms of the original variables. This method is based on what is termed the dissipation inequality.

If $x(t)$ is a trajectory, i.e., solution of the differential equation corresponding to some input $u(t)$, then $J^*(x(t))$ is a decreasing function of t, and constant if $x(t)$ is an optimal trajectory. This is the **principle of optimality** for this problem. Taking a differential formulation of this observation and using that $J^*(x)$ is of the form $x'Px$, one obtains the dissipation inequality (see [Po]). For a general (not necessarily semidefinite) LQ-problem, we define the matrix-valued function:

$$(3.1) \qquad F(P) := \begin{bmatrix} A'P+PA+Q & PB+S' \\ B'P+S & R \end{bmatrix}.$$

We call this the **dissipation** matrix. For a semidefinite problem, this formula can be written as

$$(3.2) \qquad F(P) := \begin{bmatrix} A'P+PA+C'C & PB+C'D \\ B'P+D'C & D'D \end{bmatrix}$$

We apply this map exclusively to symmetric matrices P. The result will be an $(n+m)\times(n+m)$-dimensional symmetric matrix. Notice that $F(0) = N$ (See (0.4)). The inequality $F(P) \geq 0$ will play an important role. It is called the **dissipation inequality** (=: DI). Based on the remarks of the previous paragraph, we can express the solution of the free and zero endpoint problems in terms of $F(P)$, even in the singular case:

3.3 THEOREM *Let* $T(s) = C(sI - A)^{-1}B + D$ *be the* **transfer function** *of* Σ *and let* rank T *denote the normal (or global) rank of* T(s). *Then*

i) *For all* $P \geq 0$, *we have* rank $F(P) \geq$ rank T.

ii) *There exists a* $P \geq 0$ *satisfying* $F(P) \geq 0$ *and* rank $F(P)$ = rank T *iff* Σ *is output stabilizable. Let us denote by* Γ^{+}_{min} *the set of such matrices* P, *i.e.,*

(3.4) $\Gamma^{+}_{min} := \{P \in \mathbb{R}^{n\times n} | P' = P,\ P \geq 0,\ F(P) \geq 0,\ \text{rank}\ F(P) = \text{rank}\ T\}$

iii) *If* Σ *is output stabilizable,* Γ^{+}_{min} *has a smallest element* P^{-}. *Furthermore,* $J^{*}(x_0) = x_0' P^{-} x_0$.

iv) *If* Σ *is (state) stabilizable,* Γ^{+}_{min} *has a largest element* P^{+}. *Furthermore,* $J_0^{*}(x_0) = x_0' P^{+} x_0$. *If* Σ *has no zeros on the imaginary axis, the infimum in* $J_0^{*}(x_0)$ *is attained for some input* u *that results in a state trajectory that converges to zero as* $t \to \infty$.

Solutions of the DI, satisfying the additional condition rank F(P) = rank T, are called **rank-minimizing solutions** of the DI. In the regular case, these are exactly the solutions of the ARE. This is a consequence of **Schur's lemma**:

If U,V,W *and* X *are matrices of suitable dimensions and X is nonsingular then*

(3.5) $$\text{rank} \begin{bmatrix} U & V \\ W & X \end{bmatrix} = \text{rank}\ X + \text{rank}(U - VX^{-1}W)$$

Note that D'D is invertible in the regular case. In the general case, it can be shown that ([G-2, Section 6], [G-4, Prop. 2.18])

(3.6) $$\Gamma^{+}_{min} = \{P \in \mathbb{R}^{n\times n} | P = P', P \geq 0,\ \mathcal{W} \subseteq \ker P,\ \overline{\Phi}(P) = 0\},$$

where $\overline{\Phi}$ is defined by (2.7). Note that $F(P) \geq 0$ implies that $P\mathcal{W} = 0$. This condition, together with the equation $\overline{\Phi}(P) = 0$, might suggest that there are more equations than variables, so that the problem is unlikely to have a solution. It can be shown however, that for every symmetric P such that $P\mathcal{W} = 0$, we have $\mathcal{W} \subseteq \ker \overline{\Phi}(P)$, so that the number of equations reduces corresponding to the number of added equations.

3.7 REMARK In Theorem 3.3, P^+ is characterized as the largest element in Γ^+_{min}. Actually, P^+ is the largest of all solutions of the DI. A similar but less well-known general characterization can be given for P^-. First note that it is easily seen from the definition of $\mathcal{V}$ that $P^-\mathcal{V} = 0$. In [G-4, Prop. 3.20], it is shown that P^- is the largest solution P of the DI satisfying $P\mathcal{V} = 0$. □

The conclusion of this section is that the DI together with the rank-minimization condition can be viewed as a generalization of the Riccati equation, which remains valid in the singular case. Note that in the semidefinite case, a positive semidefinite solution of the DI always exists, viz. P = 0. It is the rank-minimization condition which makes the solvability of the DI nontrivial (see Theorem 3.3 ii)).

3.8 REMARK The dissipation inequality was introduced in [Po], and its relevance for the LQ problem became particularly clear in [Wi]. The result of Theorem 3.3, i) was given in [Sc] and ii) appeared in [G&H] and [G-4, Prop.3.62]. Finally we refer to [WKS] for the treatment of the singular zero-endpoint problem from a geometrical point of view. □

4. The indefinite case.

We restrict our attention to the zero-endpoint problem. As of now, the theory for the free-endpoint problem is by no means complete.

As before, it is clear that the stabilizability is a necessary condition for the existence of an optimal control. A first difficulty of the treatment of the problem is the *exact definition* of the nondefinite infinite-horizon optimal control problem. What exactly do we mean by $\inf_u J(x_0,u)$? One possibility we have is to restrict ourselves to situations where the integral $\int_0^\infty \omega(x(t),u(t))dt$ converges to a finite or infinite limit. On the other hand, it may be more natural to infimize the expression $\limsup_{T\to\infty} \int_0^T \omega(x(t),u(t))dt$ without requiring the limit to exist. It is not clear whether this will lead to a different answer.

The crucial question however will be, whether the infimum is $> -\infty$. (Because of our stabilizability assumption, it is obvious that the infimum is less than ∞.) Because of the linear-quadratic nature of the problem, is is to be expected that, if the infimum J_0^* is bounded, it is bounded between two quadratic functions of the initial state. It was shown in [Mo] that in this case, J_0^* must be a quadratic function itself. Once it is known that J_0^* is of the form $x_0'Px_0$, it is easy to prove that P must satisfy the DI. Hence the existence of a solution of the DI is close to necessary for the existence of an optimal control.

Based on these considerations our standing assumption will be:

4.1 ASSUMPTION i) (A,B) *is stabilizable*

ii) *There exists a solution* $\overline{P}$ *of the* DI.

Note that we do not require $\overline{P}$ to be rank minimizing or semidefinite. Also, it is clear that conditon ii) is always satisfied in the semidefinite problem (take $\overline{P} = 0$). Notice that assumption ii) implies that $R \geq 0$.

Similar to (0.6), we may factorize the positive semidefinite matrix $F(\overline{P})$ as

$$F(\overline{P}) = \begin{bmatrix} \overline{C} \\ \overline{D} \end{bmatrix} [\overline{C} \quad \overline{D}] \tag{4.2}$$

and introduce the output

$$\overline{y}(t) := \overline{C}x(t) + \overline{D}u(t) \tag{4.3}$$

The following formula (which can be derived by straightforward calculations) is crucial (see [Wi]):

$$\int_0^T \omega(x,u)\,dt = \int_0^T \|\overline{y}\|^2 dt + x_0'\overline{P}x_0 - x'(T)\overline{P}x(T). \tag{4.4}$$

If u is an admissible control and if $x(t) \to 0$ $(t \to \infty)$, the right-hand side of (4.4) tends to $\int_0^\infty \|\bar{y}\|^2 dt$ for $T \to \infty$. It follows that $\int_0^T \omega(x,u)dt$ is always bounded from below and that $\int_0^\infty \omega(x,u)dt$ always exists. This integral is finite iff $\int_0^\infty \|\bar{y}\|^2 dt < \infty$. In this case we have,

$$(4.5) \qquad \int_0^\infty \omega(x,u)dt = \int_0^\infty \|\bar{y}\|^2 dt + x_0'\bar{P}x_0.$$

Consequently, the original indefinite LQ problem is equivalent to the semidefinite problem corresponding to the system $\bar{\Sigma} := (A,B,\bar{C},\bar{D})$. Exactly the same controls will be optimal, and if we denote the performance criterion of the original system by $J_0^*(x_0)$, we have $J_0^*(x_0) = J_{\bar{\Sigma},0}^*(x_0) + x_0'\bar{P}x_0$. Thus we have reduced the indefinite problem to an equivalent semidefinite one, and the results in the previous sections about the zero-endpoint problem carry over to the indefinite case. The relation is made explicit by the observation that $\bar{F}(\hat{P}) = F(\bar{P} + \hat{P})$, where $\bar{F}$ denotes the dissipation matrix of $\bar{\Sigma}$. It follows that if $\hat{P}^+$ is the largest (rank-minimizing positive semidefinite) solution of the DI for $\bar{\Sigma}$, then $P^+ := \bar{P} + \hat{P}^+$ is the largest solution of of the DI of the original problem and conversely. As a consequence, the existence of a largest solution of the DI follows from assumption 4.1, and $J_0^*(x_0)$, the infimal value of $J(x_0)$, equals $x_0'P^+x_0$. An optimal control (possibly a distribution) exists iff $\bar{\Sigma}$ has no zeros on the imaginary axis. It follows from this statement that the existence of zeros on the imaginary axis is independent of the choice of $\bar{P}$ (See [G-4, Prop.2.37]) for a direct proof of this).

In particular, in the regular case (notice that $\bar{\Sigma}$ is regular iff the original problem is regular), if the system $\bar{\Sigma}$ has no zeros on the imaginary axis, the state feedback based on the largest positive semidefinite solution $\hat{P}^+$ of the ARE of $\bar{\Sigma}$, i.e.

$$u^*(t) = -(\bar{D}'\bar{D})^{-1}(B'\hat{P}^+ + \bar{D}'\bar{C})x(t),$$

stabilizes $\bar{\Sigma}$, and minimizes $\bar{J}(x)$. Using the definition of $\bar{C}$ and $\bar{D}$, we find the optimal feedback equals

$$u^*(t) = -R^{-1}(B'P^+ + S)x(t).$$

In the singular case, the optimal control can be calculated again via the structure algorithm.

If one tries to apply a similar method to the Free-Endpoint Problem, a serious difficulty arises. Specifically, it is not clear whether the term $x'(T)\bar{P}x(T)$ in the right-hand of (4.4) converges to zero as $T \to \infty$. As a consequence, we do not have formula (4.5), which accomplishes the reduction of the indefinite case to the semidefinite case. It is conjectured that under fairly general conditions this term does actually converge to zero, so that the reduction to the semidefinite case is possible again (see [G-4, Section 2.1]).

5. Conclusion

In this paper, it was demonstrated that a fairly general way of solving the LQ-problem can be obtained using the Riccati Equation or its generalization, the Dissipation Inequality. The singular case can be reduced to the regular case via the generalized dual structure algorithm, which exhibits clearly the nonuniqueness of the solution of the LQ-problem. The Dissipation Inequality has the merit of expressing the solution directly in terms of the original data. Also, it enables us to reduce the indefinite case to the semidefinite case under fairly natural assumptions, at least for the Zero-Endpoint Problem. For the Free-Endpoint problem, it is not clear yet whether this method leads to a solution.

References

[G&H] A.H.W. Geerts & M.L.J. Hautus, "The output-stabilizable subspace and linear optimal control", *Proc. Internat. Symp. Math. Th. of Netw. & Syst.*, vol. 8, pp. - , 1990.

[G-1] Ton Geerts, "A necessary and sufficient condition for solvability of the linear-quadratic control problem without stability", *Syst. & Contr. Lett.*, vol. 11, pp. 47 - 51, 1988.

[G-2] Ton Geerts, "All optimal controls for the singular linear-quadratic problem without stability; a new interpretation of the optimal cost", *Lin. Alg. & Appl.*, vol. 116, pp. 135 - 181, 1989.

[G-3] Ton Geerts, "The computation of optimal controls for the singular linear-quadratic problem that yield internal stability", EUT Report 89-WSK-03, Eindhoven University of Technology, 1989.

[G-4] Ton Geerts, Structure of Linear-Quadratic Control, Ph.D. Thesis, Eindhoven, 1989.

[H&S] M.L.J. Hautus & L.M. Silverman, "System structure and singular control", *Lin. Alg. & Appl.*, vol. 50, pp. 369 - 402, 1983.

[Ka] R.E. Kalman, "Contributions to the theory of optimal control", *Bol. Soc. Mat. Mex.*, vol. 5, pp. 102 - 199, 1960.

[Mo] B.P. Molinari, "The time-invariant linear-quadratic optimal control problem", *Automatica*, vol. 13, pp. 347 - 357, 1977.

[Po] V.M. Popov, "Hyperstability and optimality of automatic systems with several control functions", *Rev. Roumaine Sci. Tech. Ser. Electrotech. Energet.*, vol. 9, pp. 629 - 690, 1964.

[Sc] J.M. Schumacher, "The role of the dissipation matrix in singular optimal control", *Syst. & Contr. Lett.*, vol. 2, pp. 262 - 266, 1983.

[Wi] J.C. Willems, "Least squares stationary optimal control and the algebraic Riccati equation", *IEEE Trans. Automat. Contr.*, vol. AC-16, pp. 621 - 634, 1971.

[WKS] J.C. Willems, A. Kitapçi & L.M. Silverman, "Singular

optimal control: A geometric approach", *SIAM J. Contr. & Opt.*, vol., 24, pp. 323 - 337, 1986.

[Wo] W.M. Wonham, **Linear Multivariable Control: A Geometric Approach**, Springer, New York, 1979.

Robustness measures for linear state space systems under complex and real parameter perturbations

D. Hinrichsen A. J. Pritchard

Abstract

In this paper we study the effect of perturbations of a system matrix on its spectrum. We consider perturbations of the form $A \to A + BDC$ where B, C are given matrices. Robustness measures with respect to the location of the spectrum of A (in arbitrary open domains $\mathbb{C}_g \subset \mathbb{C}$) are introduced and characterized. Complex and real perturbations are considered separately. The results are used to analyse the stability of state space systems when subjected to time-varying, nonlinear and dynamic perturbations. Here we observe interesting differences between the complex and real cases.

1 Introduction

No mathematical model can exactly represent the dynamics of a physical plant, so controllers designed on the basis of a nominal model may not achieve the required performance criteria (e.g. stability) when applied to the real system. Roughly speaking, a controller is said to be *robust* if it works well for a large class of perturbed models. If this class reflects the uncertain or neglected features of the plant then one can have more confidence in the behaviour of the controlled system.

In order to make the notion of robustness more precise we need to define robustness measures for a particular performance criterion with respect to a given class of perturbations. Most of the work in this area is based on frequency response techniques. In contrast we adopt a state space approach which we initiated in [12], [13]. Here we present some new results, mainly concerning the real stability radius, but in order to make the paper self contained we also briefly review our previous work.

It is assumed that the required performance of a linear time-invariant finite dimensional system can be expressed in terms of pole locations. Let

$$\mathbb{C} = \mathbb{C}_g \dot{\cup} \mathbb{C}_b \tag{1}$$

be any partition of the complex plane into a "good" and a "bad" region where $\mathbb{C}_g$ is always supposed to be open and hence $\mathbb{C}_b$ is closed. Since we are interested in the effects of real and of complex perturbations we have to consider both fields $\mathbb{K} = \mathbb{R}$ and $\mathbb{K} = \mathbb{C}$ of real and complex numbers. Throughout the paper we assume that the nominal system matrix (representing e.g. the regulated plant) has a spectrum in the good part of the complex plane:

$$A \in \mathbb{K}^{n \times n}, \quad \sigma(A) \subset \mathbb{C}_g \tag{2}$$

Moreover we suppose that the perturbed system matrix has the form $A + BDC$, where $B \in \mathbb{K}^{n \times m}$ and $C \in \mathbb{K}^{p \times n}$ are given matrices defining the structure of the perturbations and $D \in \mathbb{K}^{m \times p}$ is an unknown disturbance matrix. The perturbed system

$$\dot{x}(t) = Ax(t) + BDCx(t),\ t \in \mathbb{R}_+ \quad \text{or} \quad x(t+1) = Ax(t) + BDCx(t),\ t \in \mathbb{N}$$

may be interpreted as a closed loop system with unknown static linear output feedback (Fig. 1). Note, however, that in applications B and C are chosen to reflect the

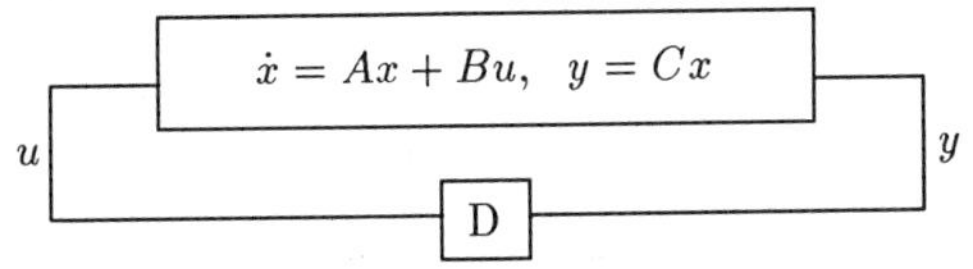

Figure 1: Feedback interpretation of the perturbed system

structure of the model uncertainty, they do not represent input or output matrices in our context and hence controllability or observability assumptions for the pairs (A, B) or (A, C) are not appropriate.

The entries of the perturbed matrix $A + BDC$ depend affinely on the unknown entries d_{ij} of D and provide a flexible mechanism for analysing different robustness properties. For example if $B = C = I_n$ (unstructured case) then all the elements of A are subjected to independent perturbations whereas by a suitable choice of B, C the effect of perturbations of individual elements, rows or columns of A can be studied.

Given a matrix norm $\|\cdot\|$ on $\mathbb{K}^{m \times p}$, various more or less conservative bounds $\delta = \delta(A, B, C) > 0$ have been proposed in the literature that guarantee $\sigma(A + BDC) \subset \mathbb{C}_g$ for all $D \in \mathbb{K}^{m \times p}$ with norm $\|D\| < \delta$. Most of these results are for the unstructured case with $\mathbb{C}_g = \mathbb{C}_-$ (the open left half plane) or $\mathbb{C}_g = \mathbb{C}_1$ (the open unit disk in $\mathbb{C}$), see [1], [27], [23], [24] [15]. The question arises which of these bounds are tight. The smallest norm of a "destabilizing" disturbance matrix $D \in \mathbb{K}^{m \times p}$ (see 1) is called the *stability radius* with respect to the perturbation structure (B, C). In the unstructured case the stability radius is the distance of A from the set of $\mathbb{C}_g$-unstable matrices. The distance from instability has been analysed in [17], [12] and the structured *complex* stability radius in [13]. The complex case is easier to handle and (with respect to Euclidean norms) it has a well developed theory [13], [10], see also [19], [20]. It also extends to infinite dimensional [22] and time-varying systems [8]. There are only a few results concerning the distance from instability in the space of *real* $n \times n$ matrices, see [17], [12]. Here we derive, for the first time, general formulae for the structured real stability radius. Moreover, we prove results which shed some light on the relationship between complex and real stability radii.

We proceed as follows. After a brief review of the basic problem the complex case is considered in Section 3. Specializing to Euclidean norms we describe an algorithm for computing the complex stability radius (see [9]). General formulae for the *real* stability radius are derived in Section 4, and we use them in Section 5 to obtain explicit computable formulae for the case $m = 1$ or $p = 1$. Choosing A to be in companion form and $B = [0, \cdots, 0, 1]^{\mathsf{T}}$, robustness properties of *polynomials* with respect to arbitrary affine perturbations can be deduced (see [14]). Finally in Section 6 we analyse the stability of a linear time-invariant stable state space system when subjected to both real and complex time-varying, nonlinear and dynamic perturbations. We show that the *complex* stability radius is — contrary to the real one — invariant with respect to these extensions of the class of perturbations and this motivates its use for *real* problems.

2 Definitions and elementary properties

We suppose that $\|\cdot\|_{\mathbb{K}^m}$ and $\|\cdot\|_{\mathbb{K}^p}$ are given norms on $\mathbb{K}^m$ resp. $\mathbb{K}^p$ and measure the size of perturbation matrices $D \in \mathbb{K}^{m\times p}$ by the corresponding operator norm

$$\|D\| = \max\{\|Dy\|_{\mathbb{K}^m}; y \in \mathbb{K}^p, \|y\|_{\mathbb{K}^p} \le 1\} \tag{3}$$

Definition 2.1 Given a partition (1), the *stability radius* of A with respect to perturbations of the structure (B,C) is defined by

$$r_{\mathbb{K}} = r_{\mathbb{K}}(A; B, C; \mathbb{C}_b) = \inf\left\{\|D\|; D \in \mathbb{K}^{m\times p}, \sigma(A + BDC) \cap \mathbb{C}_b \neq \emptyset\right\} \tag{4}$$

If A, B, C are real, we obtain two stability radii, $r_{\mathbb{R}}$ or $r_{\mathbb{C}}$, according to whether real ($\mathbb{K} = \mathbb{R}$) or complex ($\mathbb{K} = \mathbb{C}$) perturbations are considered in (4). They are, respectively, called the *real* or the *complex* stability radii of A and are clearly related by

$$r_{\mathbb{R}}(A; B, C; \mathbb{C}_b) \ge r_{\mathbb{C}}(A; B, C; \mathbb{C}_b) \ge 0 \tag{5}$$

Remark 2.2 Note that not every affine perturbation of A can be represented in the form $A + BDC$. In fact, if $(\Delta A)_{ij} = (BDC)_{ij} = 0$ for all $D \in \mathbb{K}^{m\times p}$ then necessarily the i-th row vector of B or j-th column vector of C is zero. Hence if the (i, j)-th element of $A + BDC$ does not depend on D then all the elements of the i-th row or j-th column remain unchanged under the perturbation. In particular it is not possible to represent in this form affine perturbations of A which affect exclusively the diagonal elements of A . □

In the unstructured case where $m = p = n$ and $B = C = I_n$, $r_{\mathbb{K}}$ is the distance, in the normed space $(\mathbb{K}^{n\times n}, \|\cdot\|)$, between A and the set of matrices in $\mathbb{K}^{n\times n}$ having at least one eigenvalue in $\mathbb{C}_b$.

By definition, the infimum of the empty set is infinite so that $r_{\mathbb{K}} = r_{\mathbb{K}}(A; B, C; \mathbb{C}_b) = \infty$ iff there does not exist $D \in \mathbb{K}^{m\times p}$ with $\sigma(A + BDC) \cap \mathbb{C}_b \neq \emptyset$. On the other hand if $r_{\mathbb{K}} < \infty$ an easy compactness argument shows that there exists a destabilizing D with $\|D\| = r_{\mathbb{K}}$. So in this case we may replace the *infimum* in (4) by a *minimum*.

It is immediate from Definition 2.1 that the stability radius $r_{\mathbb{K}}$ is invariant under similarity transformations

$$r_{\mathbb{K}}(A; B, C; \mathbb{C}_b) = r_{\mathbb{K}}(TAT^{-1}; TB, CT^{-1}; \mathbb{C}_b) \tag{6}$$

Moreover if $\sigma(A + BDC) \cap \operatorname{int} \mathbb{C}_b \neq \emptyset$ where $\operatorname{int} \mathbb{C}_b$ denotes the set of interior points of $\mathbb{C}_b$, then by continuity the same is true for all matrices in a small neighbourhood of $D \in \mathbb{K}^{m \times p}$. Hence

$$r_{\mathbb{K}}(A; B, C; \mathbb{C}_b) = r_{\mathbb{K}}(A; B, C; \partial\mathbb{C}_b) \tag{7}$$

where $\partial\mathbb{C}_b = \bar{\mathbb{C}}_g \cap \mathbb{C}_b$ is the boundary of $\mathbb{C}_b$.

As an illustration consider the unstructured case $(B = C = I_n)$ where $\|\cdot\| = \|\cdot\|_2$ is the spectral norm on $\mathbb{K}^{n \times n}$ and $\mathbb{C}_g = \mathbb{C}_-$, $\mathbb{C}_b = \mathbb{C}_+ = \mathbb{C} \setminus \mathbb{C}_-$. If $A \in \mathbb{R}^{n \times n}$ is normal it has been shown in [12] that the distances $d_{\mathbb{K}}^{-}(A) = r_{\mathbb{K}}(A; I, I; \mathbb{C}_+)$, $\mathbb{K} = \mathbb{R}$, $\mathbb{C}$ of A from the set of unstable real or complex matrices are both equal to the distance of the spectrum $\sigma(A)$ from the imaginary axis

$$d_{\mathbb{R}}^{-}(A) = d_{\mathbb{C}}^{-}(A) = \operatorname{dist}(\sigma(A), i\mathbb{R}) \quad \text{if} \quad AA^{\top} = A^{\top}A. \tag{8}$$

For non-normal matrices $\operatorname{dist}(\sigma(A), i\mathbb{R})$ may be a very poor indicator of robustness as the following example shows.

Example 2.3 Consider the matrices

$$A_k = \begin{bmatrix} -k & k^3 \\ 0 & -k \end{bmatrix}, \qquad k \in \mathbb{N}$$

If $D_k = \begin{bmatrix} 0 & 0 \\ k^{-1} & 0 \end{bmatrix}$, then $A_k + D_k$ is singular and hence $d_{\mathbb{R}}^{-}(A_k) \leq \|D_k\| = k^{-1} \to 0$ as $k \to \infty$. On the other hand $\operatorname{dist}(\sigma(A_k), i\mathbb{R}) \to \infty$ as $k \to \infty$. □

More generally it has been shown in [12] that for every stable matrix $A \in \mathbb{R}^{n \times n}$ which is not a multiple of the identity matrix and any $\varepsilon > 0$ there exist matrices $A_i = T_i A T_i^{-1}$, $i = 1, 2$ similar to A such that $d_{\mathbb{R}}^{-}(A_1) < \varepsilon$ and $d_{\mathbb{R}}^{-}(A_2) > \operatorname{dist}(\sigma(A), i\mathbb{R}) - \varepsilon$.

3 Complex stability radius

In this section we assume $\mathbb{K} = \mathbb{C}$. The following formula for the complex stability radius is a straightforward generalization of Theorem 2.1 in [13], see also [10].

Proposition 3.1 *If $G(s) = C(sI - A)^{-1}B$ is the transfer matrix associated with (A, B, C), then*

$$r_{\mathbb{C}}(A; B, C; \mathbb{C}_b) = \left[\max_{s \in \partial\mathbb{C}_b} \|G(s)\|\right]^{-1} \tag{9}$$

where $\|G(s)\|$ denotes the operator norm of $G(s) \in \mathbb{C}^{p \times m}$ with respect to the norms $\|\cdot\|_{\mathbb{C}^m}$ and $\|\cdot\|_{\mathbb{C}^p}$ and, by definition, $0^{-1} = \infty$.

Proof: For any $D \in \mathbb{C}^{m \times p}$, $\sigma(A + BDC) \cap \partial\mathbb{C}_b \neq \emptyset$ iff there exists $x \in \mathbb{C}^n$, $x \neq 0$ and $s \in \partial\mathbb{C}_b$ such that

$$(A + BDC)x = sx. \tag{10}$$

Since $\sigma(A) \subset \mathbb{C}_g$, $(sI - A)$ is invertible for $s \in \partial\mathbb{C}_b$ and (10) is equivalent to

$$x = (sI - A)^{-1}BDCx$$

where $DCx \neq 0$. Set $u = DCx$, then

$$u = DG(s)u \neq 0 \tag{11}$$

and hence

$$\|D\|\,\|G(s)\| \geq \|DG(s)\| \geq 1.$$

This proves the inequality $\geq$ in (9). On the other hand, suppose that $\|G(s_0)\| = \max_{s\in\partial\mathbb{C}_b}\|G(s)\|$ and choose $u \in \mathbb{C}^m, \|u\|_{\mathbb{C}^m} = 1$ so that $\|G(s_0)u\| = \|G(s_0)\|$. If $G(s_0) \neq 0$ there exists (by Hahn-Banach) a linear form $y^* \in (\mathbb{C}^p)^*$ (the dual space of $\mathbb{C}^p$) of dual norm $\|y^*\|_{(\mathbb{C}^p)^*} = 1$ which is aligned with $G(s_0)u$ so that

$$y^*G(s_0)u = \|G(s_0)u\|_{\mathbb{C}^p} \tag{12}$$

It is easy to check that D defined by

$$D = r_{\mathbb{C}}\, uy^* \tag{13}$$

has the norm $\|D\| = r_{\mathbb{C}}$ and satisfies (11) with $s = s_0$. To see that D "destabilizes" A, let $x = (s_0I - A)^{-1}Bu$; then $u = DCx$, hence $x \neq 0$ and $(s_0I - A)x = BDCx$, i.e. $s_0 \in \sigma(A + BDC) \cap \partial\mathbb{C}_b$.
□

Remark 3.2 The stability radius $r_{\mathbb{C}}$ determines the magnitude of the smallest perturbation which destroys the $\mathbb{C}_g$-stability of the system. In applications it may be of interest to know not only the *critical magnitude* but also the *critical directions* of system perturbations. If $r_{\mathbb{C}} < \infty$ and s_0, u, y^* are chosen as in the above proof, then D defined by (13) is a rank one minimum norm destabilizing perturbation. If $\mathbb{C}^m$, $\mathbb{C}^p$ are provided with their usual Hilbert norms, it is easy to see that all minimum norm destabilizing perturbations of rank 1 are given by

$$D = r_{\mathbb{C}}^2\, u(\lambda)u(\lambda)^*G(\lambda)^* \tag{14}$$

where $\lambda \in \partial\mathbb{C}_b$ satisfies $\|G(\lambda)\| = r_{\mathbb{C}}^{-1}$ and $u(\lambda) \in \mathbb{C}^m$ satisfies $\|u(\lambda)\| = 1$ and $\|G(\lambda)u(\lambda)\| = \|G(\lambda)\|$.
□

The computation of the RHS of (9) requires the *global maximization* of the real valued function $s \longrightarrow \|G(s)\|$ on $\partial\mathbb{C}_b$, which may have many local minima, spikes etc., see [10], [21],[9]. In the following we describe the algorithms STABRAD I, II developed in [9] which solve the computational problem for the standard case where $\mathbb{C}_g = \mathbb{C}_-$ and $\mathbb{C}^m$, $\mathbb{C}^p$ are provided with their usual Hilbert norms $\|\cdot\|_2$. These algorithms are based on the following characterization of $r_{\mathbb{C}}^-(A;B,C) = r_{\mathbb{C}}(A;B,C;\mathbb{C}_+)$ [9], see Fig. 2.

Proposition 3.3 *If H_ρ denotes the Hamiltonian matrix associated with (A, B, C)*

$$H_\rho = \begin{bmatrix} A & -BB^* \\ \rho C^*C & -A^* \end{bmatrix} \tag{15}$$

then

$$\rho < r_{\mathbb{C}}^-(A;B,C)^2 \quad \textit{iff} \quad \sigma(H_\rho) \cap i\mathbb{R} = \emptyset. \tag{16}$$

Moreover,

$$i\omega_0 \in \sigma\left(H_{(r_{\mathbb{C}}^-)^2}\right) \quad \textit{iff} \quad \|G(i\omega_0)\| = \max_{\omega\in\mathbb{R}}\|G(i\omega)\|. \tag{17}$$

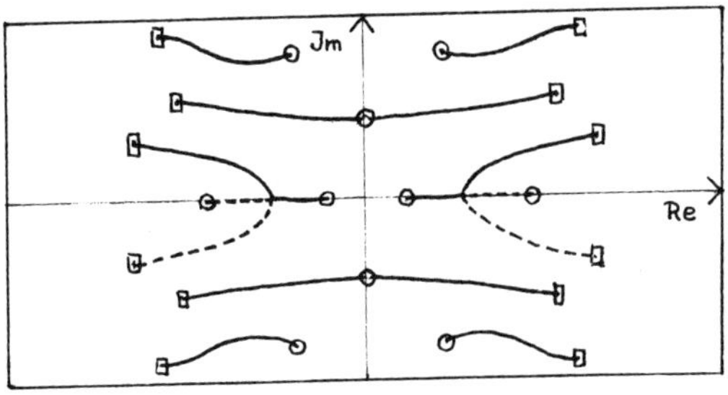

Figure 2: Eigenloci of H_ρ as $\rho \to (r_{\mathbb{C}}^-)^2$

A proof of this proposition can be found in [9].

The Hamiltonian characterization of $r_{\mathbb{C}}$ can be used as a basis for computing $r_{\mathbb{C}}^-$. Starting from some lower and upper estimates

$$\rho_0^- < (r_{\mathbb{C}}^-)^2 < \rho_0^+$$

successively better estimates ρ_k^-, ρ_k^+, $k = 0, 1, 2, \cdots$ are obtained by the following steps:
Take ρ_k such that $\rho_k^- < \rho_k < \rho_k^+$.
If H_{ρ_k} has eigenvalues on the imaginary axis set $\rho_{k+1}^- := \rho_k^-$ and $\rho_{k+1}^+ := \rho_k$. Otherwise set $\rho_{k+1}^- := \rho_k$ and $\rho_{k+1}^+ := \rho_k^+$.

In STABRAD I the strategy for the choice of ρ_k is that of bisection $\rho_k = (\rho_k^- + \rho_k^+)/2$. Bisectional algorithms for the computation of the em unstructured stability radius $r_{\mathbb{C}}^- = d_{\mathbb{C}}^-$ have also been proposed and analyzed in [3], [10], [21]. STABRAD I and these more special bisectional algorithms are reliable, but convergence is slow. An acceleration can be achieved by using an extrapolation technique (STABRAD II) which is motivated by the observation that the graph of the function $\rho \to \operatorname{dist}(\sigma(H_\rho), i\mathbb{R})$ has in most examples an almost parabolic shape [9].

Both STABRAD algorithms have been implemented on a SUN workstation 3/50. For random triples (A, B, C) with $m = p = 3$ and $n = 20$, STABRAD II results in an average speed up by a factor 3 over that of STABRAD I, see [9]. In all applications both algorithms proved to be remarkably reliable, their results coinciding up to 6 digits corresponding to the required relative accuracy ($\varepsilon = 10^{-6}$). Via (17) both versions also yield all ω_0 which maximize the function $\omega \to \|G(i\omega)\|$. This in turn enables us to calculate (from (14)) all minimum norm rank 1 destabilizing perturbations D, see Remark 3.2.

Associated with the Hamiltonian (15) is a parametrized algebraic Riccati equation. The relation between Riccati equations and Hamiltonian matrices has been carefully studied in the literature, see [6], [26]. Utilizing this relationship or by a more direct approach based on the work of *Brockett* [2] and *Willems* [25] , we have a further characterization of $r_{\mathbb{C}}^-$, which was first derived in [13].

Theorem 3.4 *There exist Hermitian solutions of the algebraic Riccati equation*

$$PA + A^*P - \rho C^*C - PBB^*P = 0 \tag{18}$$

iff $\rho \leq (r_{\mathbb{C}}^-)^2$.

The above proposition can be used to construct a Liapunov function of "maximal robustness". Indeed if P is any solution of (18) with $\rho = (r_{\mathbf{C}}^{-})^2$, then it can be shown (see [13]) that $V(x) = -\langle x, Px\rangle$ is a joint Liapunov function for the set of systems

$$\Sigma_D: \qquad \dot{x} = (A + BDC)x, \quad \|D\| < r_{\mathbf{C}}^{-}.$$

Clearly there does not exist a joint Liapunov function for a larger ball of systems Σ_D, $\|D\| < \alpha$ where $\alpha > r_{\mathbf{C}}^{-}$.

This Liapunov function can also be used to consider nonlinear and time-varying perturbations, see Section 6.

4 Real stability radius

We now assume $\mathbf{K} = \mathbf{R}$ so that A, B, C are real matrices. In this case we will, in general, have to consider perturbations of rank 2. Given u_1, $u_2 \in \mathbf{R}^m$, y_1, $y_2 \in \mathbf{R}^p$ we denote by $\mu(y_1, y_2; u_1, u_2)$ the smallest operator norm of all linear maps $D: \mathbf{R}^p \to \mathbf{R}^m$ which take y_1, y_2 into u_1, u_2:

$$\mu = \mu(y_1, y_2; u_1, u_2) = \inf\left\{\|D\|;\ D \in \mathbf{R}^{m\times p}, Dy_1 = u_1 \text{ and } Dy_2 = u_2\right\} \qquad (19)$$

Note that $\mu = \infty$ iff there is no such D and this will be the case iff $\alpha y_1 + \beta y_2 = 0$ and $\alpha u_1 + \beta u_2 \neq 0$ for some $\alpha, \beta \in \mathbf{R}$. If $\mu < \infty$ the infimum in (19) is attained.

The following theorem holds for arbitrary pairs of norms on $\mathbf{R}^p$, $\mathbf{R}^m$ and the corresponding operator norm on $\mathbf{R}^{m\times p}$.

Theorem 4.1 *If $G_R(s)$ and $G_I(s)$ are the real and imaginary parts of $G(s) = C(sI - A)^{-1}B$, then*

$$r_{\mathbf{R}}(A; B, C; \mathbf{C}_b) = \min_{\substack{s\in\partial\mathbf{C}_b \\ (u_1,u_2)\neq(0,0)}} \mu(G_R(s)u_1 - G_I(s)u_2, G_R(s)u_2 + G_I(s)u_1; u_1, u_2) \quad (20)$$

Proof: We use the notation of the proof of Proposition 3.1. If $\sigma(A+BDC)\cap\partial\mathbf{C}_b \neq \emptyset$ for any $D \in \mathbf{R}^{m\times p}$, then $u = DG(s)u$ for some $s \in \partial\mathbf{C}_b$, $u \in \mathbf{C}^m$, $u \neq 0$ (see (11)). Let $u = u_1 + iu_2$, u_1, $u_2 \in \mathbf{R}^m$, then $u = DG(s)u$ can be rewritten in real terms

$$\begin{aligned} u_1 &= D(G_R(s)u_1 - G_I(s)u_2) \\ u_2 &= D(G_R(s)u_2 + G_I(s)u_1) \end{aligned} \qquad (21)$$

Hence the inequality $\geq$ holds in (20). Conversely, the existence of $(D, u_1, u_2, s) \in \mathbf{R}^{m\times p} \times \mathbf{R}^m \times \mathbf{R}^m \times \partial\mathbf{C}_b$, $(u_1, u_2) \neq (0,0)$ satisfying (21) ensures via

$$(u_1 + iu_2) = DG(s)(u_1 + iu_2) \quad \text{and} \quad x = (sI - A)^{-1}B(u_1 + iu_2)$$

the existence of $x \in \mathbf{C}^n$, $x \neq 0$ such that $(A + BDC)x = sx$ and hence $s \in \sigma(A + BDC) \cap \partial\mathbf{C}_b$. This proves the inequality $\leq$ in (20). Finally note that there exists a minimal norm destabilizing disturbance matrix $D \in \mathbf{R}^{m\times p}$ if $r_{\mathbf{R}} < \infty$ so that the minimum on the RHS is always attained. □

It is not difficult to show that for $s = \alpha + i\omega$, α, $\omega \in \mathbf{R}$

$$G_R(s) = C[\omega^2 I + (\alpha I - A)^2]^{-1}(\alpha I - A)B,\ G_I(s) = -\omega C[\omega^2 I + (\alpha I - A)^2]^{-1}B \quad (22)$$

From now on we assume that $\mathbf{R}^p$ and $\mathbf{R}^m$ are both provided with the usual Euclidean norm $\|\cdot\| = \|\cdot\|_2$. In this case the function $\mu(y_1, y_2; u_1, u_2)$ can be determined explicitely (see [12]). This is trivial if y_1, y_2 are linearly dependent. Otherwise one has the following

Lemma 4.2 *If* $y_1, y_2 \in \mathbb{R}^p$ *are linearly independent, then* $\mu^2 = \mu(y_1, y_2; u_1, u_2)^2$ *is the largest of the two real solutions of the quadratic equation in* ρ:

$$(\|y_1\|^2\|y_2\|^2 - \langle y_1, y_2\rangle^2)\rho^2 - (\|y_1\|^2\|u_2\|^2 + \|y_2\|^2\|u_1\|^2 - 2\langle y_1, y_2\rangle\langle u_1, u_2\rangle)\rho + \|u_1\|^2\|u_2\|^2 - \langle u_1, u_2\rangle^2 = 0 \tag{23}$$

The proof is based on Weierstrass' Theorem on quadratic forms and can be found in [12].

To construct a minimum norm destabilizing perturbation in the case $r_{\mathbb{R}} < \infty$ we note that by Theorem 4.1 there exist $s \in \partial\mathbb{C}_b$ and $u_1, u_2 \in \mathbb{R}^m$, $(u_1, u_2) \neq (0, 0)$ such that

$$y_1 = G_R(s)u_1 - G_I(s)u_2, \quad y_2 = G_R(s)u_2 + G_I(s)u_1$$

satisfy

$$r_{\mathbb{R}}(A; B, C; \mathbb{C}_b) = \mu(y_1, y_2; u_1, u_2).$$

It follows by straight forward calculations that

$$D = \begin{cases} u_1 y_1^\top / \|y_1\|^2 & \text{if } y_1, y_2 \text{ are linearly dependent, } y_1 \neq 0 \\ [u_1\ u_2] \begin{bmatrix} \|y_1\|^2 & \langle y_1, y_2\rangle \\ \langle y_1, y_2\rangle & \|y_2\|^2 \end{bmatrix}^{-1} \begin{bmatrix} y_1^\top \\ y_2^\top \end{bmatrix} & \text{if } y_1, y_2 \text{ are linearly independent} \end{cases} \tag{24}$$

is a minimum norm destabilizing disturbance matrix $D \in \mathbb{R}^{m \times p}$.

We see that for an analysis of the real stability radius $r_{\mathbb{R}}$ it suffices to consider rank 2 matrix perturbations. Note, however, that in general there will not exist a minimum norm perturbation D of rank 1. This is in contrast with the complex case (as we have seen in (14)) and indicates why it is easier to analyze than the real one. In fact we will see in the next section that there is a considerable reduction in the complexity of the real stability radius if only rank 1 perturbations need to be considered.

5 Rank 1 perturbations and the stability radius of polynomials

Throughout this section we assume $m = 1$. In order to obtain simplified formulae for this case, we denote by $d(y, v\mathbb{R})$ the distance in the normed space $(\mathbb{R}^p, \|\cdot\|_{\mathbb{R}^p})$ of the point $y \in \mathbb{R}^p$ from the linear subspace $v\mathbb{R} = \{\alpha v; \alpha \in \mathbb{R}\}$ spanned by $v \in \mathbb{R}^p$:

$$d(y, v\mathbb{R}) = \min_{\alpha \in \mathbb{R}} \|y - \alpha v\|_{\mathbb{R}^p} \tag{25}$$

It is easy to see that in the Euclidean case ($\|\cdot\|_{\mathbb{R}^p} = \|\cdot\|_2$)

$$d^2(y, v\mathbb{R}) = \begin{cases} \|y\|_2^2 - \langle y, v\rangle^2/\|v\|_2^2 & , \quad v \neq 0 \\ \|y\|_2^2 & , \quad v = 0 \end{cases}, \tag{26}$$

see Fig. 3. The following result holds for arbitrary norms on $\mathbb{R}^p$. Note that the disturbances D are *linear forms* on $\mathbb{R}^p$ in the present context ($m = 1$) and thus are represented by *row* vectors, i.e. $D \in (\mathbb{R}^p)^* = \mathbb{R}^{1 \times p}$. The corresponding operator norm is simply the dual norm of $\|\cdot\|_{\mathbb{R}^p}$ on $\mathbb{R}^{1 \times p}$:

$$\|D\| = \max\{|Dy|; y \in \mathbb{R}^p, \|y\|_{\mathbb{R}^p} = 1\}$$

Thus $\|\cdot\|_{\mathbb{R}^p}$ and $\|\cdot\|$ form a dual pair of norms on $\mathbb{R}^p = \mathbb{R}^{p \times 1}$ and $\mathbb{R}^{1 \times p} = (\mathbb{R}^p)^*$, respectively.

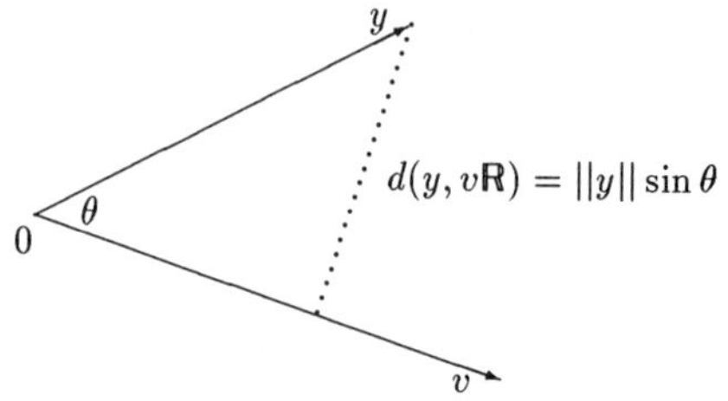

Figure 3: The distance $d(y, v\mathbf{R})$ in $(\mathbf{R}^p, \|\cdot\|_2)$ if $v \neq 0$

Proposition 5.1 *If* $m = 1$ *then*

$$r_{\mathbf{R}}(A; B, C; \mathbf{C}_b) = \left[\max_{s \in \partial \mathbf{C}_b} d(G_R(s), G_I(s)\,\mathbf{R})\right]^{-1}. \tag{27}$$

Proof: To apply formula (20) we have to determine, for any pair $(u_1, u_2) \in \mathbf{R}^2$, $(u_1, u_2) \neq (0,0)$ and $s \in \partial \mathbf{C}_b$, a minimum norm row vector $D \in \mathbf{R}^{1\times p}$ satisfying (21). But (21) is now equivalent to

$$\begin{aligned} (1 - DG_R(s))u_1 + DG_I(s)u_2 &= 0 \\ -DG_I(s)u_1 + (1 - DG_R(s))u_2 &= 0. \end{aligned} \tag{28}$$

Since $(u_1, u_2) \neq (0,0)$, $D \in \mathbf{R}^{1\times p}$ satisfies (28) iff

$$DG_I(s) = 0 \quad \text{and} \quad DG_R(s) = 1. \tag{29}$$

It follows that $\mu(G_R(s)u_1 - G_I(s)u_2, G_R(s)u_2 + G_I(s)u_1; u_1, u_2)$ does not depend upon $(u_1, u_2) \in \mathbf{R}^2$, $(u_1, u_2) \neq (0,0)$ and equals the minimum norm of all $D \in \mathbf{R}^{1\times p}$ satisfying (29). Since an elementary geometric consideration shows that this minimum norm is $[d(G_R(s), G_I(s)\,\mathbf{R})]^{-1}$, see [18], formula (27) is proved. □

Remark 5.2 Of particular interest are the 1, 2, ∞ norms on $\mathbf{R}^p$ and the corresponding dual norms on $\mathbf{R}^{1\times p} = (\mathbf{R}^p)^*$:

$$\|y\|_{\mathbf{R}^p} = \|y\|_\infty = \max_{i \in \underline{p}} |y_i| \quad , \quad \|D\| = \|D\|_1 = \sum_{i=1}^{p} |D_i|$$

$$\|y\|_{\mathbf{R}^p} = \|y\|_2 = \left(\sum_{i=1}^{p} y_i^2\right)^{1/2} \quad , \quad \|D\| = \|D\|_2 = \left(\sum_{i=1}^{p} D_i^2\right)^{1/2}$$

$$\|y\|_{\mathbf{R}^p} = \|y\|_1 = \sum_{i=1}^{p} |y_i| \quad , \quad \|D\| = \|D\|_\infty = \max_{i \in \underline{p}} |D_i|$$

□

In general the map $s \mapsto d(G_R(s), G_I(s)\,\mathbf{R})$ is not continuous at the zeros of $G_I(\cdot)$ and this may create problems in computing the RHS of (26). In order to get around this problem, let

$$\Omega = \{s \in \partial \mathbf{C}_b; G_I(s) = 0\} \tag{30}$$

By (27) we obtain

$$r_{\mathbb{R}}(A;B,C;\mathbb{C}_b) = \min\left\{\left[\max_{s\in\Omega}\|G_R(s)\|_{\mathbb{R}^p}\right]^{-1}, \left[\sup_{s\in\partial\mathbb{C}_b,\, s\notin\Omega} d(G_R(s), G_I(s)\,\mathbb{R})\right]^{-1}\right\} \tag{31}$$

Remark 5.3 (i) Similar formulae to (27), (31) hold if $p = 1$.

(ii) In the case $p = m = 1$ the second term on the RHS of (31) is always infinite and so

$$r_{\mathbb{R}}(A;B,C;\mathbb{C}_b) = \left[\max_{s\in\Omega}|G_R(s)|\right]^{-1} \tag{32}$$

(iii) All of the above formulae may be specialized to $\mathbb{C}_g = \mathbb{C}_-$ or $\mathbb{C}_1$. In particular, (32) becomes

$$r_{\mathbb{R}}^{-}(A;B,C) = \left[\max_{\omega\in\Omega}|G_R(s)|\right]^{-1}$$

where $\Omega = \{\omega \in \mathbb{R}; G_I(i\omega) = 0\}$. A result which could have been easily deduced from the Nyquist criterion, see Fig. 4.

□

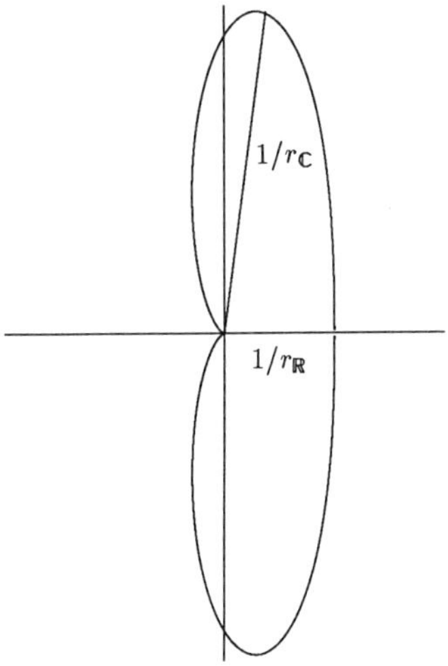

Figure 4: The real and complex stability radii in a Nyquist plot

Equations (31) gives rise to a computable formula for the real stability radius. In fact if $\mathbb{C}_g = \mathbb{C}_-$, so that $\partial\mathbb{C}_b = i\mathbb{R}$, then

$$\Omega = \left\{\omega \in \mathbb{R}; \omega = 0 \text{ or } C(\omega^2 I + A^2)^{-1}B = 0\right\}$$

is the zero set of a real rational function in ω^2 and hence can be effectively computed. With respect to the Hilbert norm we have

$$d^2(G_R(i\omega), G_I(i\omega)\,\mathbb{R}) = f(\omega^2) := \|G_R(i\omega)\|_2^2 - \frac{\langle G_R(i\omega), G_I(i\omega)\rangle^2}{\|G_I(i\omega)\|_2^2}, \quad \omega \notin \Omega$$

where

$$G_R(i\omega) = -C(\omega^2 I + A^2)^{-1}AB, \quad G_I(i\omega) = -\omega C(\omega^2 I + A^2)^{-1}B.$$

So $f(\omega^2)$ is a strictly proper real rational function of ω^2. The maximum of this function can be approximately determined via a computer plot of the function $f(\cdot)$ or, more precisely, by some global optimization procedure for real functions on $\mathbb{R}$.

An alternative method consists of the following three steps:

1. Compute a Cholesky factorization of $f(\omega^2)$, i.e. a strictly proper real rational function $g(s)$ which has its poles and zeros in $\mathbb{C}_-$ such that

$$f(\omega^2) = g(-i\omega)g(i\omega) = |g(i\omega)|^2 \tag{33}$$

2. Determine a realization $(\hat{A}, \hat{b}, \hat{c})$ of $g(s)$.

3. Compute $r_{\bar{\mathbb{C}}}(\hat{A}, \hat{b}, \hat{c})$ by means of the algorithm STABRAD II.

Then

$$r_{\bar{\mathbb{R}}}(A; B, C) = \min\left\{\left[\max_{\omega\in\Omega} \|G_R(i\omega)\|_2\right]^{-1}, r_{\bar{\mathbb{C}}}(\hat{A}, \hat{b}, \hat{c})\right\}.$$

These rank 1 results can be used to obtain explicit formulae for stability radii of polynomials. Let

$$p(s, a) = s^n + a_{n-1}s^{n-1} + \ldots + a_0 \tag{34}$$

where $a = [a_0, \ldots, a_{n-1}] \in \mathbb{K}^{1\times n}$ (row vector). Now assume that $p(s, a)$ is a nominal polynomial with all its roots in $\mathbb{C}_g$ and the coefficients are perturbed to

$$a_{j-1}(D) = a_{j-1} - \sum_{i=1}^{p} D_i c_{ij}\,, \quad j \in \underline{n} \tag{35}$$

where $C = (c_{ij}) \in \mathbb{K}^{p\times n}$ is a given matrix. The coordinates of the unknown *disturbance vector* $D = [D_1, D_2, \ldots, D_p] \in \mathbb{K}^{1\times n}$ represent the deviations of the polynomial parameters from their nominal values.

Definition 5.4 Given a partition (1) of the complex plane, an arbitrary norm $\|\cdot\|$ on $\mathbb{K}^{1\times p}$ and perturbations of the form (35) the stability radius of the polynomial (34) is defined to be

$$r_{\mathbb{K}}(a, C; \mathbb{C}_b) = \inf\left\{\|D\|; D \in \mathbb{K}^{1\times p}, \exists\lambda \in \mathbb{C}_b : p(\lambda, a(D)) = 0\right\} \tag{36}$$

In this definition, $\|\cdot\|$ is not a priori an operator norm on $\mathbb{K}^{1\times p}$. However, let $\|\cdot\|_{\mathbb{K}^p} = \|\cdot\|_*$ be the dual norm on $\mathbb{K}^p$, i.e.

$$\|y\|_* = \max\{|Dy|; D \in \mathbb{K}^{1\times p}, \|D\| \leq 1\} \tag{37}$$

Then $\|\cdot\|$ is conversely the dual norm of $\|\cdot\|_*$. For the rest of this section we always choose the norm $\|\cdot\|_{\mathbb{R}^p} = \|\cdot\|_*$ on $\mathbb{R}^p$ and write '$\|\cdot\|_*$' instead of '$\|\cdot\|_{\mathbb{R}^p}$'.

Proposition 5.5 *Suppose that* $p(\cdot, a)$, $a \in \mathbf{K}^{1\times p}$ *has all its roots in* $\mathbb{C}_g$ *and let*

$$G(s) = \frac{1}{p(s,a)}[c_1(s), \ldots, c_p(s)]^\top, \; c_i(s) = \sum_{j=1}^{p} c_{ij} s^{j-1}, \; i \in \underline{p} \tag{38}$$

Then

$$r_{\mathbb{C}}(a, C; \mathbb{C}_b) = \left[\max_{s\in\partial\mathbb{C}_b} \|G(s)\|_*\right]^{-1} \tag{39}$$

$$r_{\mathbb{R}}(a, C; \mathbb{C}_b) = \left[\max_{s\in\partial\mathbb{C}_b} d_*(G_R(s), G_I(s)\,\mathbb{R})\right]^{-1} \tag{40}$$

where $d_*(y, v\mathbb{R}) = \min_{\alpha\in\mathbb{R}} \|y - \alpha v\|_*$ *and* $G_R(s)$, $G_I(s)$ *are the real and imaginary parts of* $G(s)$.

Proof: It is easy to see that

$$r_{\mathbf{K}}(a, C; \mathbb{C}_b) = r_{\mathbf{K}}(A; B, C; \mathbb{C}_b)$$

where

$$A = \begin{bmatrix} 0 & 1 & 0 & \cdots & 0 \\ 0 & 0 & 1 & \cdots & 0 \\ \vdots & \vdots & \vdots & \ddots & \vdots \\ 0 & 0 & 0 & \cdots & 1 \\ -a_0 & -a_1 & -a_2 & \cdots & -a_{n-1} \end{bmatrix}, \quad B = \begin{bmatrix} 0 \\ 0 \\ \vdots \\ 0 \\ 1 \end{bmatrix}$$

So (39) follows from (9) and (40) from (27). □

Many results which follow from the above proposition can be found in [14]. As an illustration let $[\underline{a}, \bar{a}]$ denote the closed n-dimensional interval between $\underline{a}$ and $\bar{a}$:

$$[\underline{a}, \bar{a}] = \{p(\cdot, a); \underline{a}_i \leq a_i \leq \bar{a}_i, \; i = 0, \ldots, n-1\} \tag{41}$$

Kharitonov [16] has given necessary and sufficient conditions for the interval $[\underline{a}, \bar{a}]$ to consist of Hurwitz polynomials. Rather surprisingly (at first sight) these conditions only require that *four* special polynomials (which can easily be constructed from the vectors $\underline{a}, \bar{a}$) are Hurwitz. Here we determine conditions in terms of stability radii for the more general requirement that the roots of all the polynomials lie in $\mathbb{C}_g$.

Proposition 5.6 *The interval* $[\underline{a}, \bar{a}]$ *consists of polynomials which have all their roots in* $\mathbb{C}_g$ *iff the following two conditions are satisfied*

(a) $p(\cdot, a)$ *has all its roots in* $\mathbb{C}_g$

(b) with respect to the ∞ *norm* $r_{\mathbb{R}}(a, C; \mathbb{C}_b) > 1$, *where*

$$\begin{aligned} a &= 1/2\,[\underline{a}_0 + \bar{a}_0, \underline{a}_1 + \bar{a}_1, \ldots, \underline{a}_{n-1} + \bar{a}_{n-1}] \qquad (42) \\ C &= 1/2\,diag\,[\underline{a}_0 - \bar{a}_0, \underline{a}_1 - \bar{a}_1, \ldots, \underline{a}_{n-1} - \bar{a}_{n-1}] \qquad (43) \end{aligned}$$

Proof: The coefficient vectors $a(D)$ (35) of the perturbed polynomials are

$$a(D) = 1/2[(1 + D_1)\underline{a}_0 + (1 - D_1)\bar{a}_0, \ldots, (1 + D_n)\underline{a}_{n-1} + (1 - D_n)\bar{a}_{n-1}]$$

Thus

$$\{a(D); \|D\|_\infty \leq 1\} = [\underline{a}, \bar{a}].$$

□

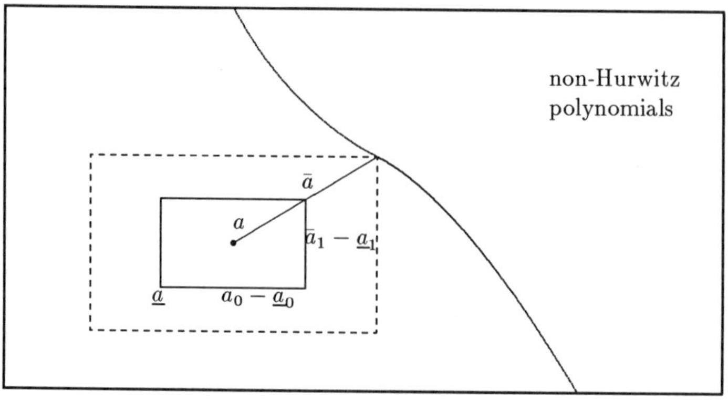

Figure 5: Optimal blowing up of $[\underline{a}, \bar{a}]$

In fact $r_{\mathbb{R}}(a, C; \mathbb{C}_b)$ is the minimal factor by which the interval must be blown up before one of the polynomials has a root in $\mathbb{C}_b$, see Fig. 5. The Kharitonov result can be used to considerably simplify the calculation of $r_{\mathbb{R}}^{-}(a, C) = r_{\mathbb{R}}(a, C; \mathbb{C}_b)$ for the Hurwitz case $\mathbb{C}_g = \mathbb{C}_-$.

Proposition 5.7 *Let a, C be as in (42), (43) and*

$$\begin{aligned} \hat{c} &= [\underline{a}_0 - \bar{a}_0, \underline{a}_1 - \bar{a}_1, -\underline{a}_2 + \bar{a}_2, -\underline{a}_3 + \bar{a}_3, \underline{a}_4 - \bar{a}_4, \ldots] \\ \tilde{c} &= [\underline{a}_0 - \bar{a}_0, -\underline{a}_1 + \bar{a}_1, -\underline{a}_2 + \bar{a}_2, \underline{a}_3 - \bar{a}_3, \underline{a}_4 - \bar{a}_4, \ldots]. \end{aligned} \tag{44}$$

Then

$$r_{\mathbb{R}}^{-}(a, C) = \min\left\{r_{\mathbb{R}}^{-}(a, \hat{c}),\ r_{\mathbb{R}}^{-}(a, \tilde{c})\right\} \tag{45}$$

Proof: The coefficient vectors $a - D\hat{c}$, $a - D\tilde{c}$ yield the four optimal Kharitonov polynomials for $D = \pm 1$. □

Remark 5.8 When $p = m = 1$ stability radii can be calculated via an optimisation over a finite set (see Remark 5.3 (iii), for details cf. [14]). Hence the formula (45) yields an efficient method for determining $r_{\mathbb{R}}(a, C; \mathbb{C}_+)$ with respect to the ∞ norm, when C is diagonal. □

6 Other perturbation classes

If A is a real matrix it would seem that we should restrict considerations to real perturbations and real stability radii. However, with the exception of the cases where $m = 1$ or $p = 1$, we have seen that the formulae which characterize these radii are far more complicated than the corresponding complex ones. Also the formula of *Van Loan* [17] for the unstructured real stability radius (with respect to the Frobenius norm) involves a constrained global minimization problem of similar degree of difficulty. In contrast the complex stability radius, $r_{\mathbb{C}}$, can be computed via STABRAD II and has control theoretic associations with Riccati equations and Hamiltonian matrices. It is useful therefore to explore possible motives for using the complex stability radius in real problems. Clearly it is a lower bound for the real radius and it is natural to ask how conservative is this bound. Unfortunately it can be very bad as we now illustrate by considering the distances from instability $d_{\mathbb{R}}^{-}(A)$, $d_{\mathbb{C}}^{-}(A)$ for $A \in \mathbb{R}^{2\times 2}$, see [10].

Proposition 6.1 *Let $A \in \mathbb{R}^{2\times 2}$ be stable, then with respect to the Euclidean norm on $\mathbb{R}^2$*

$$d_{\mathbb{R}}^{-}(A) = \min \{s_2(A), dist(\sigma(A), i\mathbb{R})\}$$

where $s_2(A)$ is the lowest singular value of A.

Proof: If $D \in \mathbb{R}^{2\times 2}$ is a destabilizing perturbation of minimal norm then either $\det(A+D) = 0$ or $i\omega \in \sigma(A+D)$ for some $\omega \neq 0$. In the first case $d_{\mathbb{R}}^{-}(A) = s_2(A)$ whereas in the second case $\mathrm{trace}(A+D) = 0$, so that $d_{\mathbb{R}}^{-}(A) = |\mathrm{trace}\, A|/2 = dist(\sigma(A), i\mathbb{R})$. □

Using the above proposition one can prove the following, see [10].

Proposition 6.2 *For every $q \in (0,1]$ there exists a stable matrix $A \in \mathbb{R}^{2\times 2}$ such that $d_{\mathbb{C}}^{-}(A)/d_{\mathbb{R}}^{-}(A) = q$ (with respect to the Euclidean norm).*

For a proof, see [10].

This rather negative result indicates that the complex stability radius is not in general a good estimate for the real radius. Nevertheless we will show in this section that $r_{\mathbb{C}}^{-}$ has some remarkable properties which justify its use as a robustness measure also of real systems. We only consider the Hilbert norm case.

We begin by applying the characterization of $r_{\mathbb{C}}$ via the algebraic Riccati equation (see Proposition 3.4) to prove that a generalized multivariable version of *Aizerman's conjecture* is true over the field of complex numbers (contrary to the real case). A local version of this result may be found in [13].

Proposition 6.3 *The linear systems*

$$\dot{x} = Ax + BDCx, \quad D \in \mathbb{C}^{m\times p},\ \|D\|^2 < \rho$$

are asymptotically stable iff the origin of the nonlinear system

$$\dot{x} = Ax + BN(Cx, t) \tag{46}$$

is globally asymptotically stable for all differentiable $N : \mathbb{C}^p \times \mathbb{R}_+ \to \mathbb{C}^m$ satisfying $N(0,t) = 0$ for all $t \in \mathbb{R}_+$ and

$$\|N(y,t)\|_2^2 < \rho\|y\|_2^2, \quad y \in \mathbb{C}^p,\ y \neq 0,\ t \in \mathbb{R}_+ \tag{47}$$

Proof: The 'if' part is trivial. Conversely, suppose that the linear system is asymptotically stable for all $D \in \mathbb{C}^{m\times p}$, $\|D\|^2 < \rho$. Then necessarily $\rho \leq \left(r_{\mathbb{C}}^{-}\right)^2$. By Proposition 3.4 there exists an Hermitian solution $P \prec 0$ of (18). Multiplying (18) from the left by x^* and from the right by $x \in \mathbb{R}^n$ we see that $\ker P \subset \ker C$.

Consider $V(x) = -\langle x, Px\rangle$. Computing $\dot{V}(x)$ along arbitrary solutions of (46) we obtain by (18)

$$\begin{aligned}
\dot{V}(x) &= -\langle Ax(t) + BN(y(t),t), Px(t)\rangle - \langle x(t), P\,(Ax(t) + BN(y(t),t))\rangle \\
&= -\rho\|y(t)\|^2 - \|B^*Px(t)\|^2 - 2\,\mathrm{Re}\langle B^*Px(t), N(y(t),t)\rangle \\
&= -\|B^*Px(t) + N(y(t),t)\|^2 - \left[\rho\|y(t)\|^2 - \|N(y(t),t)\|^2\right]
\end{aligned}$$

where $y(t) = Cx(t)$. Hence $\dot{V}(x) \leq 0$, $x \in \mathbb{C}^n$ and by (47) $\dot{V}(x) = 0$ only if $y = Cx = 0$.
Consider the decomposition

$$x(t) = x_1(t) + x_2(t)\,,\; x_1(t) \in (\ker P)^{\perp}\;,\; x_2(t) \in \ker P.$$

Since

$$\langle x(t), Px(t)\rangle = \langle x_1(t), Px_1(t)\rangle \leq \langle x_1(t_0), Px_1(t_0)\rangle, \quad t \geq t_0 \tag{48}$$

$x_1(t)$ and $y(t) = Cx_1(t)$, $t \geq t_0$ are bounded. Moreover, by (48), there exists a constant $\alpha > 0$ such that

$$\|y\|_{L^\infty} \leq \alpha \|x_1(t_0)\|. \tag{49}$$

But

$$x(t) = e^{A(t-t_0)}x_0 + \int_{t_0}^{t} e^{A(t-s)}BN(y(s),s)\,ds$$

and $\|e^{At}\| < Me^{-\omega t}$, $t \geq 0$ for suitable M, $\omega > 0$. It follows therefore from (47) that there exists a constant $c > 0$

$$\|x(t)\| \leq Me^{-\omega(t-t_0)}\|x_0\| + c\|y\|_{L^\infty(t_0,\infty;\mathbb{C}^p)}\,, \quad t \geq t_0. \tag{50}$$

Hence, by (49), the origin of the nonlinear system (47) is stable.
Finally, applying *LaSalle's* theorem we conclude that $x(t) \to S$ as $t \to \infty$ where S is the largest invariant set in $\dot{V}^{-1}(0)$. Since $\dot{V}^{-1}(0) \subset \ker C$ we get $y(t) \to 0$ as $t \to \infty$ and hence, replacing t_0 by a sufficiently large t_0' in (50), $x(t) \to 0$ as $t \to \infty$. □

The notion of stability radius can be extended to include perturbations which are

- linear but time-varying,
- time-varying and nonlinear,
- time-varying, nonlinear and dynamic.

We introduce these extensions for both the complex and the real case. The perturbed system equations are

$$\begin{aligned} \Sigma_{D(\cdot)} &: \quad \dot{x}(t) = Ax(t) + BD(t)Cx(t) \\ \Sigma_N &: \quad \dot{x}(t) = Ax(t) + BN(Cx(t),t) \\ \Sigma_{\mathcal{N}} &: \quad y = G(I + \mathcal{N}G)^{-1}u \end{aligned}$$

$\Sigma_{\mathcal{N}}$ is the input–output representation of the feedback system obtained by combining the (L^2-stable) input–output operator

$$\begin{gathered} G : L^2(0,\infty;\mathbb{K}^m) \longrightarrow L^2(0,\infty;\mathbb{K}^p)\,, \quad u(\cdot) \mapsto y(\cdot) \\ y(t) = \textstyle\int_0^t Ce^{A(t-s)}Bu(s)\,ds\,, \quad t \geq 0 \end{gathered}$$

with an arbitrary L^2-stable feedback operator

$$\mathcal{N} : L^2(0,\infty;\mathbb{K}^p) \longrightarrow L^2(0,\infty;\mathbb{K}^m).$$

As perturbation norms we choose

- in the time-varying case:
  $\|D\| = \text{ess sup}_{t\in\mathbf{R}_+} \|D(t)\|, \quad D(\cdot) \in L^\infty(\mathbf{R}_+, \mathbf{K}^{m\times p})$
- in the nonlinear case:
  $\|N\| = \inf\{\gamma \in \mathbf{R}_+; \|N(y,t)\|_2 \leq \gamma\|y\|_2\}, \quad N : \mathbf{K}^p \times \mathbf{R}_+ \to \mathbf{K}^m,\ N(0,t) \equiv 0$
- in the operator case:
  $\|\mathcal{N}\| = \inf\{\gamma > 0; \|\mathcal{N}(y)\|_{L^2} \leq \gamma\|y\|_{L^2}\}, \quad \mathcal{N} : L^2(0,\infty;\mathbf{K}^p) \to L^2(0,\infty;\mathbf{K}^m)$

The corresponding stability radii are, by definition, the infima of the norms of each of the above perturbations which yield a non asymptotically stable system $\Sigma_{D(\cdot)}$, Σ_N or a non L^2–stable system $\Sigma_{\mathcal{N}}$. They are denoted by $r_{\mathbf{K},tv}(A;B,C)$, $r_{\mathbf{K},nl}(A;B,C)$, $r_{\mathbf{K},dyn}(A;B,C)$, respectively. It is easily seen that

$$r_{\mathbf{K}}^{-} \geq r_{\mathbf{K},tv} \geq r_{\mathbf{K},nl} \geq r_{\mathbf{K},dyn} \tag{51}$$

The *complex* stability radius $r_{\mathbf{C}}^{-}$ has the following very convenient property.

Proposition 6.4

$$r_{\mathbf{C}}^{-} = r_{\mathbf{C},tv} = r_{\mathbf{C},nl} = r_{\mathbf{C},dyn} \tag{52}$$

Proof: The first two equations are a direct consequence of Proposition 6.3. The last equality follows, for example, by applying the small gain theorem. □

Remark 6.5 Note that the stability results which are contained in the first two equalities of the above proposition are more restrictive than the result expressed in Proposition 6.3 when $\rho^{1/2} = r_{\mathbf{C}}^{-}$. For example, $r_{\mathbf{C}}^{-} = r_{\mathbf{C},nl}$ implies that the nonlinear system (46) will be globally asymptotically stable provided there exists $\gamma < r_{\mathbf{C}}^{-}$ such that

$$\|N(y,t)\| \leq \gamma\|y\|, \quad y \in \mathbf{C}^p,\ t \in \mathbf{R}_+.$$

Whereas Proposition 6.3 yields the same conclusion provided

$$\|N(y,t)\| < r_{\mathbf{C}}^{-}\|y\|, \quad y \in \mathbf{C}^p,\ y \neq 0,\ t \in \mathbf{R}_+.$$

□

The real stability radius $r_{\mathbf{R}}^{-}$ does not have the same remarkable robustness to a change in the *class* of perturbations. Indeed it is well known (see e.g. [7]) that the *Aizerman conjecture* does not hold over the reals. Hence one cannot expect that $r_{\mathbf{R},nl}$ will be equal to $r_{\mathbf{R}}^{-}$. In general, we have not been able to characterize $r_{\mathbf{R},nl}$, but we do have the following interesting result.

Proposition 6.6 *For any real triple (A,B,C), $\sigma(A) \subset \mathbf{C}_-$ and $\varepsilon > 0$ there exists a stable real proper rational perturbation matrix $D(s) \in \mathbf{R}^{m\times p}(s)$ with H^∞ norm less than $r_{\mathbf{C}}^{-}(A,B,C) + \varepsilon$ which destabilizes the system. In particular,*

$$r_{\mathbf{R},dyn}(A,B,C) = r_{\mathbf{C}}^{-}(A,B,C) \tag{53}$$

This proposition can be proved by applying similar techniques as in [4]. We omit the proof, see [11].

As a consequence of Proposition 6.6 one should take $r_{\mathbf{C}}^{-}$ rather than $r_{\mathbf{R}}^{-}$ as a robustness measure whenever neglected dynamics play an important role in the model uncertainty.

The situation for nondynamic but time-varying linear perturbations is much more complicated. In fact, the following instructive 2-dimensional example illustrates that $r_{\mathbf{R},tv}$ may lie anywhere between $r_{\mathbf{C}}^{-}$ and $r_{\mathbf{R}}^{-}$.

Example 6.7 Consider the linear oscillator

$$\dot{x} = \begin{bmatrix} 0 & 1 \\ -1 & -2b \end{bmatrix} x + d(t) \begin{bmatrix} 0 & 0 \\ 1 & 0 \end{bmatrix} x \tag{54}$$

where the restoring force coefficient is subjected to time-varying perturbations $d(t)$. For the corresponding triple

$$A = \begin{bmatrix} 0 & 1 \\ -1 & -2b \end{bmatrix}, \quad B = \begin{bmatrix} 0 \\ 1 \end{bmatrix}, \quad C = \begin{bmatrix} 1 & 0 \end{bmatrix}$$

it is easy to show that $r_{\mathbb{R}}^- = 1$ for all $b \in (0,2)$, whereas $r_{\mathbb{C}}^- = 1$ for $b \in (0.708, 2)$ and $r_{\mathbb{C}}^- < r_{\mathbb{R}}^-$ for $b \in (0, 0.708)$. Moreover, $r_{\mathbb{C}}^- \to 0$ as $b \to 0$, see Fig. 6. For time-varying perturbations $d(t)$, *Colonius and Kliemann* [5] determined analytically the maximal achievable Liapunov exponent for (54) when $d(t)$ is restricted to a given interval $[\alpha, \beta] \subset \mathbb{R}$. Their analysis involves a separate discussion of many different cases even for this simple example. $r_{\mathbb{R},tv}$ is the smallest positive real ρ for which there exists a perturbation $d : \mathbb{R}_+ \to [-\rho, \rho]$ such that the maximum Liapunov exponent of (54) is greater than or equal zero. As a consequence of the analysis in [5] we obtain the dependence of $r_{\mathbb{R},tv}$ on b as shown in Fig. 6. Because of (51) and (52), $r_{\mathbb{C}}^- \leq r_{\mathbb{R},tv} \leq r_{\mathbb{R}}^-$. We find $r_{\mathbb{R},tv} = 1$ for $b \in (0.26, 2)$, but more interestingly, we have this equality also for parameter values $b \in (0.26, 0.708)$ where $r_{\mathbb{C}}^- < r_{\mathbb{R}}^-$. Moreover Fig. 6 shows that as parameter b runs from 0 to 1, $r_{\mathbb{R},tv}$ moves in the varying interval $(r_{\mathbb{C}}^-, r_{\mathbb{R}}^-]$ from one extreme to the other.

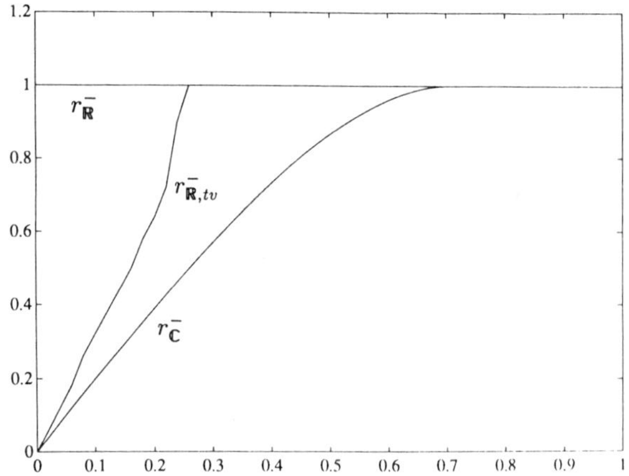

Figure 6: Stability radii $r_{\mathbb{C}}^-$, $r_{\mathbb{R},tv}$ and $r_{\mathbb{R}}^-$ of the linear oscillator (54)

□

Acknowledgement
We would like to thank Mr. Bernd Kelb for performing the necessary computations for Figure 6. This work was partly supported by the EEC Program "Stimulation Action" under grant ST2J-0066-2.

References

[1] M.F. Barret. Conservatism with robustness test for linear feedback control systems. In *Proc. of 25th IEEE CDC, Athens*, pages 751–755, 1980.

[2] R. Brockett. *Finite Dimensional Linear Systems.* J. Wiley, 1970.

[3] R. Byers. A bisection method for measuring the distance of a stable matrix to the unstable matrices. *SIAM J. Sci. Stat. Comput.*, 9:875–881, 1988.

[4] M.J. Chen and C. A. Desoer. Necessary and sufficient condition for robust stability of linear distributed feedback systems. *International Journal of Control*, 35:255–267, 1982.

[5] F. Colonius and W. Kliemann. Exact exponential stabilization and destabilization of bilinear control systems. Report 116, Institut für Mathematik, Universität Augsburg, 1989.

[6] I. Gohberg, P. Lancaster, and L. Rodman. On the Hermitian solutions of the symmetric algebraic Riccati equation. *SIAM J. Control and Opt.*, 24:1323–1334, 1986.

[7] W. Hahn. *Theory and Application of Liapunov's Direct Method.* Prentice-Hall, 1963.

[8] D. Hinrichsen, A. Ilchmann, and A.J. Pritchard. Robustness of stability of time-varying linear systems. Report 161, Institut für Dynamische Systeme, Universität Bremen, 1987. To appear in J. Diff. Equ.

[9] D. Hinrichsen, B. Kelb, and A. Linnemann. An algorithm for the computation of the structured stability radius with applications. Report 182, Institut für Dynamische Systeme, Universität Bremen, 1987. To appear in Automatica.

[10] D. Hinrichsen and M. Motscha. Optimization problems in the robustness analysis of linear state space systems. In *Approximation and Optimization*, LN Mathematics 1354, Berlin-Heidelberg-New York, 1988. Springer Verlag.

[11] D. Hinrichsen and A. J. Pritchard. Destabilization by output feedback. Technical report, Institut für Dynamische Systeme, Universität Bremen, 1989.

[12] D. Hinrichsen and A.J. Pritchard. Stability radii of linear systems. *Systems & Control Letters*, 7:1–10, 1986.

[13] D. Hinrichsen and A.J. Pritchard. Stability radius for structured perturbations and the algebraic Riccati equation. *Systems & Control Letters*, 8:105–113, 1986.

[14] D. Hinrichsen and A.J. Pritchard. An application of state space methods to obtain explicit formulae for robustness measures of polynomials. In *Robustness in Identification and Control*, Torino, 1988. To appear.

[15] D.C. Hyland and D.S. Bernstein. The majorant Lyapunov equation: A nonnegative matrix equation for robust stability and performance of large scale systems. *IEEE Transactions on Automatic Control*, AC-32(11):1005–1013, 1987.

[16] V.L. Kharitonov. Asymptotic stability of an equilibrium position of a family of systems of linear differential equations. *Diff. Uravn.*, 14:2086–2088, 1979.

[17] C. Van Loan. How near is a stable matrix to an unstable matrix? *Contemporary Mathematics*, 47:465–478, 1985.

[18] D. G. Luenberger. *Optimization by Vector Space Methods.* J. Wiley, New York, 1968.

[19] J.M. Martin. State space measures for stability robustness. *IEEE Transactions on Automatic Control*, AC-32:509–512, 1987.

[20] J.M. Martin and G.A. Hewer. Smallest destabilizing perturbations for linear systems. *International Journal of Control*, 45:1495–1504, 1987.

[21] M. Motscha. An algorithm to compute the complex stability radius. *International Journal of Control*, 48:2417–2428, 1988.

[22] A.J. Pritchard and S. Townley. Robustness of linear systems. *Journal of Differential Equations*, 77(2):254–286, 1989.

[23] L. Qiu and E.J. Davison. New perturbation bounds for the robust stability of linear state space models. In *Proc. 25th Conference on Decision and Control*, pages 751–755, Athens, 1986.

[24] K.H. Wei and R.K. Yedavalli. Invariance of strict Hurwitz property for uncertain polynomials with dependent coefficients. *IEEE Transactions on Automatic Control*, AC-32:907–909, 1987.

[25] J.C. Willems. Least squares optimal control and the algebraic Riccati equation. *IEEE Transactions on Automatic Control*, AC-16:621–634, 1971.

[26] H.K. Wimmer. Monotonicity of maximal solutions of algebraic Riccati equations. *Systems and Control Letters*, 5:317–319, 1985.

[27] R.K. Yedavalli. Improved measures of stability robustness for linear state space models. *IEEE Transactions on Automatic Control*, AC-30:577–579, 1985.

Singular Optimal Problems

V. Jurdjevic

Introduction.

Much of the content for this paper is motivated by the following variational problem. Suppose that the state space M is a real, analytic manifold of dimension m, and that $X_0, \ldots, X_c$ are real analytic vector fields on M. Consider the following differential system on M defined by this data:

$$(1) \quad \frac{dx(t)}{dt} = X_0(x(t)) + \sum_{i=1}^{c} u_i X_i(x(t)) .$$

The functions $u_1, \ldots, u_c$, to be regarded as the control functions, will be assumed to be locally integrable on the compact intervals of $[0, \infty)$. Such control functions will be referred to as *admissible.*

Each admissible control $u = (u_1, \ldots, u_c)$ defines an integral curve x of (1) through each point $a \in M$. That is, x is an absolutely continuous curve in M, defined on some interval $[0, T], T > 0$, which satisfies (1) for almost all t in $[0, T]$. The pair (x, u) will be called a *trajectory.*

In addition to the differential system (1), it will be assumed that there is a functional $f_0 : M \times R^c \to R$ which is analytic, and quadratic in the variables $u_1, \ldots, u_c$ for each $x \in M$. f_0 will be expressed as:

$$(2) \quad f_0(x, u) = \frac{1}{2} \sum_{i,j=1}^{c} P_{ij}(x) u_i u_j + \sum_{i=1}^{c} Q_i(x) u_i + R(x) .$$

The basic concern is to discuss the nature of the extremals associated with the question of minimizing $\int_0^T f_0(x(t), u(t)) dt$ over all the trajectories of (1) which satisfy the prescribed boundary conditions $x(0) = a$, and $x(T) = b$. The focus will be on the following two situations:

(A) The Linear-Quadratic Problem.

The state space M is equal to R^m, X_0 is a linear vector field, while each $X_1, \ldots, X_c$ is a constant vector field. The cost functional f_0 is quadratic jointly on $R^m \times R^c$. In this context, a characterization of the optimal trajectories, for the free and point problem, in terms of the associated Riccati equation, has been known since the early days of modern control theory [K]. The assumptions under which this study was made were that $f_0(x, u) \geq 0$ for each $(x, u) \in R^m \times R^c$, that (1) is controllable, and that the Hessian matrix $\left(\frac{\partial^2 f_0}{\partial u_i \partial u_j}\right)$ is strictly positive definite. With the exception of a few isoslated papers [W], much of the subsequent literature on this subject is done under the assumption that $f_0(x,u) \geq 0$ for all (x,u). When the Hessian matrix $\left(\frac{\partial^2 f_0}{\partial u_i \partial u_j}\right)$ is only semi positive definite, the problem is known as a singular problem. In contrast to the regular problem, which seems to be well understood, the singular problem still continues to be of interest ([HS]).

In a joint work with Kupka ([JK3]), the singular quadratic case is completely resolved under the following assumptions:

(i) It is assumed that $\text{Inf} \int_0^T f_0(x(t), u(t))dt > -\infty$ over all the trajectories (x,u) which satisfy the given boundary conditions $x(0) = a$, $x(T) = b$, and moreover it is assumed that this condition holds for any a, b in R^m, and any $T > 0$.

(ii) It easily follows from (i) that the zero trajectory is optimal for $a = b = 0$ and any $T > 0$. The second assumption is that the zero trajectory is the only optimal trajectory for $a = b = 0$ and any $T > 0$.

(iii) Finally it is assumed that (1) is controllable.

It is easy to find examples for which f_0 is not positive, nor even convex, which satisfy all of the above assumptions.

The main result of [JK3], called the Generalized Optimal Synthesis is described in terms of jump fields, and the generalized trajectories of the system. The space of jump fields J, naturally expressed through the notion of the Lie Saturate of the system, can also be defined in more immediate terms as follows: let $v(a,b,T) = \text{Inf}\{\int_0^T f_0(x,u)dt : (x,u) \text{ trajectory of } (1)$ with $x(0) = a$ and $x(T) = b\}$. Then the space of jump directions J consists of all vectors d with the property that if $b - a = \alpha d$ for some $\alpha \in R$ then $\liminf_{(c,T)\to(b,0)} v(a,c,T) < \infty$. Since it is well known that $\lim_{T\to 0} v(a,b,T) = \infty$ for any a, b with $a \neq b$, in the regular case, it follows that J is a non-empty vector subspace of M whenever the problem is singular. If the criterion of $\lim_{T\to 0} v(a,b,T) = \infty$ is interpreted as the symbol of "high price for high speed", then the jump directions may be described as the "cheap" directions. The controls which correspond to the jump direction are sometimes known as the cheap controls.

Associated with each jump direction there is a jump cost. A generalized trajectory consists of concatenations of the regular trajectories and the jumps. The cost of such a trajectory, called the generalized cost, is equal to the sum of the regular cost, and the jump costs. In the Hautus-Silverman study [HS], the jump directions are called strongly reachable states. In that context the jump cost is equal to zero because $f_0 \geq 0$, and the generalized trajectories correspond to the trajectories given by the distributional controls [HS].

The Generalized Optimal Synthesis is as follows:

For any a, b in M and any $T > 0$, there exists a unique optimal regular trajectory $x(t)$ which initiates at $a + J$ and which satisfies $x(T) \in b + J$. If $x(0) = \bar{a}$ and $x(T) = \bar{b}$, then both $b - \bar{b}$ and $a - \bar{a}$ belong to J, and therefore the above data defines a unique generalized trajectory consisting of the initial jump, followed by a regular trajectory, followed by the final jump. Hence in the singular case, for any a, b and $T > 0$, there is a unique generalized trajectory which connects a to b in T units of time, with exactly two jumps, one at the beginning, and the other at the end of a regular trajectory. The second situation which will be discussed here is as follows.

(B) The problems of elastica and the total squared curvature.

This class of variational problems arizes from modelling an elastic rod in equilibrium, and naturally leads to a differential system on the Euclidean group of motions in R^3. Each rod configuration corresponds to a curve in R^3 with its elastic strain determined by the curvature and torsion. The elastic equilibria correspond to the minima of the total energy represented by an integral of quadratic form in the curvature and torsion ([T]).

A geometric analogue of the elastica problem extended to two dimensional surfaces has been recently studied by [BG] and [LS]. Both of these papers are concerned with the nature of the closed extremals associated with minimizing $\frac{1}{2}\int_0^T k^2 dt$. The differential equations for the extrema in both of these papers are obtained through Lagrangian formalism based on the use of Lagrange multipliers.

The basic setting considered in this paper is the following differential system:

$$(3) \quad \begin{cases} \frac{d\sigma}{dt} = v_1 \\ \frac{dv_i}{dt} = k_i v_{i+1} - k_{i-1} v_{i-1} \quad \text{for} \quad i = 1, \ldots, n-1, \quad \text{and} \\ \frac{dv_n}{dt} = -k_{n-1} v_{n-1} \end{cases}$$

σ is a curve in $R^n, v_1, \ldots, v_n$ are orthonormal vectors in R^n, and $k_1, \ldots, k_{n-1}$ are non-negative numbers except for the last one which may be also negative.

Differential system (3) can be naturally viewed as a left-invariant system on G_0, the semi-direct product of R^n with the orthogonal group $O_n(R)$. That is, elements of G_0 are ordered pairs (σ, R) with $\sigma \in R^n$ and $R \in O_n(R)$ with the group multiplication $(\sigma_1, R_1)(\sigma_2, R_2) = (\sigma_1 + R_1\sigma_2, R_1R_2)$. Each pair (σ, R) may be represented as an element of $GL_{n+1}(R)$ via the matrix $\begin{pmatrix} 1 & 0 \\ \sigma & R \end{pmatrix}$ with σ an n-column vector. Since G_0 is not connected, it is convenient to restrict to the connected component of G_0 through the identity. That is, G is equal to the semi-direct product of R^n with $SO_n(R)$.

G is a Lie group whose Lie algebra $L(G)$ consists of all pairs (a, A) with $a \in R^n$, and A skew symmetric. As an element of $(n+1)\times(n+1)$ matrices, (a, A) is equal to $L = \begin{pmatrix} 0 & 0 \\ a & A \end{pmatrix}$. For each $L = (a, A)$, the mapping $g \to gL$ defines a left-invariant vector field whose value at the identity is equal to L. In particular, let $L_0 = \begin{pmatrix} 0 & 0 \\ e_1 & 0 \end{pmatrix}$, and $L_i = \begin{pmatrix} 0 & 0 \\ 0 & A_i \end{pmatrix}$ $1 \le i \le n-1$ where $A_i = e_i \otimes e_{i+1} - e_{i+1} \otimes e_i$. In this notation, $e_1, \ldots, e_n$ is the standard orthonormal basis. Let $X_0, \ldots, X_{n-1}$ be the left invariant vector fields corresponding to $L_0, \ldots, L_{n-1}$. Then equations (3) become:

$$(4) \quad \frac{dx}{dt} = X_0(x) + \sum_{i=1}^{n-1} k_i X_i(x)$$

where $x = (\sigma, g)$ with $\sigma \in R^n$ and g the element of $SO_n(R)$ whose columns are $v_1, \ldots, v_n$.

Among a number of interesting variational problems that can be associated with system (4), we will consider the one studied by [Gr], and subsequently by [BG] and [LS], namely that of minimizing $\frac{1}{2}\int_0^T k_1^2(s)ds$. In particular, when $n = 2$, Bryant and Griffiths have shown that each extremal curvature must satisfy the following equation

$$(5) \quad \left(\frac{dk}{dt}\right)^2 + \left(c_1 - \frac{1}{2}k^2\right)^2 = c_2^2$$

for some constants c_1 and c_2. Since equation (5) is integrable by means of elliptic functions, all extremal curvatures are determined in terms of the constants c_1 and c_2.

For $n = 3$, let $k_1 = k$ and $k_2 = \tau$. k is called the curvature, and τ is the torsion associated with the curve σ. In this setting, Griffiths ([Gr]) has shown that each extremal curvature and torsion conform to the following formulas

$$(6) \quad k^2\tau = \text{constant and} \quad 2\frac{d^2k}{dt^2} + k^3 - 2kc - 2\tau^2 k = 0$$

for some constant c. He has also shown that for $n > 3$, k_1 and k_2 are given by the same equations as (6) while each $k_i = 0$ for $i > 2$.

The principal purpose of this paper is to discuss all of these problems through a common perspective provided by the Maximum Principle, and the associated Hamiltonian formalism with a particular focus on the singular case given by the degenerate Hessian $\left(\frac{\partial^2 f_0}{\partial u_i \partial u_j}\right)$.

This framework is also of sufficient generality to encompass the geodesic problems in sub-Riemannian geometry, ([S]), also recognized by Brockett some time ago ([Br]). In this setting, the drift X_0 is equal to zero, the associated system (1) is assumed controllable, and $f_0(x,u) \geq 0$ for all $(x,u) \in M \times R^c$. Then, f_0 may be used to define geodesics through the infimum of $\int_0^T \sqrt{f_0\big(x(t),u(t)\big)}dt$ over the trajectories of (1). Such a metric is also known as the Carathéodory-Carnot metric (see [S] for further references).

I. The Maximum Principle and Its Consequences.

Let T^*M denote the cotangent bundle of M. T^*M has a natural symplectic structure which sets up a correspondence between functions on T^*M and the vector fields on T^*M. If H is any function on T^*M, then its corresponding vector field called the Hamiltonian vector field of H, will be denoted by $\vec{H}$. In terms of the canonical coordinates $x_1, \ldots, x_n, p_1, \ldots, p_n$, the integral curves of $\vec{H}$ are defined by the well known system of differential equations,

$$\begin{aligned} \frac{dx_i}{dt} &= \frac{\partial H}{\partial p_i}\big(x(t),p(t)\big), \qquad \text{and} \\ \frac{dp_i}{dt} &= -\frac{\partial H}{\partial x_i}\big(x(t),p(t)\big), \quad 1 \leq i \leq n\,. \end{aligned}$$

In particular each vector field X on M has a natural lifting to a Hamiltonian vector field via the function $H_X(x,p) = \langle X(x),p\rangle$. $\langle X(x),p\rangle$ denotes the evaluation of p on $X(x)$. H_X is called the Hamiltonian of X, and $\vec{H}_X$ is called the Hamiltonian lifting of X.

The Maximum Principle is naturally expressed through the liftings of the extended system on $R \times M$ defined by:

$$(7) \quad \frac{dx^0}{dt} = f_0(x,u), \qquad \frac{dx}{dt} = X_0(x) + \sum_{i=1}^{c} u_i X_i(x)$$

$\tilde{x} = (x^0, x)$ will be used to denote the points of the extended state space $\tilde{M} = R \times M$. Each choice of the control function, determines a time varying field F_u through (7). Let $H\big(\tilde{x},\tilde{p},u(t)\big)$ be the corresponding Hamiltonian on $\tilde{M}$. Then the Maximum Principle goes as follows:

Suppose that (x,u) is optimal, in the sense that $\int_0^T f_0(x,u)dt$ is minimal among all trajectories which satisfy $x(0) = a$ and $x(T) = b$. Let $(\tilde{x},u)$ be the extended trajectory with $x^0(t) = \int_0^t f_0\big(x(s),u(s)\big)ds$ for each $t \in [0,T]$. Then $\tilde{x}$ is the projection of an integral curve $\big(\tilde{x}(t),\tilde{p}(t)\big)$ of $\vec{H}\big(\tilde{x},\tilde{p},u(t)\big)$ defined on the interval $[0,T]$ for which $\tilde{p}(t)$ is not identically zero, such that

(i) $H(\tilde{x}(t), \tilde{p}(t), v) \le H(\tilde{x}(t), \tilde{p}(t), u(t))$ for any $v \in R^c$, and almost all t in $[0,T]$, and

(ii) if $\tilde{p}(t) = (p^0(t), p(t))$, then $p^0(T) \le 0$.

Since the vector fields defined by (7) do not depend explicitly on x^0, $p^0(t)$ is constant along the integral curves of $\vec{H}$. Therefore, it may be assumed, after rescaling, that p_0 is either equal to zero, or equal to -1. The case $p^0 = 0$ corresponds to the Hamiltonian $H_0(x,p,u) = \langle X_0(x), p\rangle + \sum_{i=1}^{c} u_i \langle X_i(x), p\rangle$, while $p^0 = -1$ corresponds to the Hamiltonian $H_{-1}(x,p,u) = -f_0(x,u) + \langle X_0(x),\ p\rangle + \sum_{i=1}^{c} u_i\langle X_i(x), p\rangle$.

The triples (x,p,u) which are the integral curves of either $\vec{H}_0$ or $\vec{H}_{-1}$, along which (i) holds are called *extremals*. The extremals of $\vec{H}_0$ are called exceptional, and those of $\vec{H}_{-1}$ regular. In either case, along each extremal, $\frac{\partial H_\lambda}{\partial u_i}(x(t), p(t), u(t)) = 0$ for almost all t as a consequence of (i). ($\lambda = 0$, or $\lambda = -1$, depending on the type of extremal).

Let (x,p,u) be an extremal triple. In the exceptional case, $\frac{\partial H_0}{\partial u_i}(x(t), p(t), u(t)) = 0$ for almost all t in the domain of definition of (x,p,u), and for each $i = 1, 2, \ldots, c$. Therefore,

$$(8) \quad \langle X_i(x(t), p(t))\rangle = 0 \qquad \text{for} \quad i = 1, 2, \ldots, c\,.$$

In the regular case, $\frac{\partial H_{-1}}{\partial u_i}(x(t), p(t), u(t)) = 0$ for $i = 1, 2, \ldots, c$, and almost all t for which the extremal is defined. Therefore,

$$(9) \quad -\frac{\partial f_0}{\partial u_i}(x(t), p(t), u(t)) + \langle X_i(x(t), p(t))\rangle = 0$$

for $i = 1, 2, \ldots, c$, and almost all t. In either case, if (x,p,u) is the extremal which corresponds to an optimal trajectory (x,u), then the Hessian matrix $\left(\frac{\partial H_\lambda}{\partial u_i \partial u_j}(x(t), p(t), u(t))\right)$ is semi-negative definite for almost all t (since H_λ is maximal).

Although in general it may happen that an optimal trajectory lifts only to an exceptional extremal, it will be assumed that such is not the case here. For that reason the remaining analysis pertains only to the regular extremals. It is then convenient to use H to denote H_{-1}. In this situation, $\frac{\partial H}{\partial u_i} = -\sum_{j=1}^{c} P_{ij}(x)u_j - Q_i(x) + \langle X_i(x), p\rangle$, and $\frac{\partial^2 H}{\partial u_i \partial u_j} = -P_{ij}(x)$.

It follows from the preceding remarks that along an extremal (x,p,u) which corresponds to the optimal trajectory (x,u), $P(x(t)) \ge 0$. P is the matrix whose entries are P_{ij}.

Assume that $P(x) \ge 0$ for each $x \in M$. It is convenient to denote by H_i the expression $-Q_i(x) + \langle X_i(x), p\rangle$. H_i is the regular Hamiltonian of the extended vector field $(Q_i(x), X_i(x))$. In terms of this notation equations (9) become

$$-\sum P_{ij}(x)u_j + H_i(x(t), p(t)) = 0 \qquad \text{for each} \quad i = 1, 2, \ldots, c\,.$$

When $P(x) > 0$ the equations

$$(10) \qquad -\sum_{j=1}^{c} P_{ij}(x)u_j + H_i(x,p) \;=\; 0 \qquad i = 1, 2, \ldots, c$$

are uniquely solvable for $u_1, \ldots, u_c$ in terms of x and p. Suppose $u(x,p)$ is such a solution. Let $H_f(x,p) = H_{-1}(x, p, u(x,p))$. Then the the set of integral curves of $\vec{H}_f$ coincides with the set of the regular extremals. This setting is essentially the same as the classical one from the calculus of variations.

In the singular case, $P(x)$ contains null directions. Then, along each regular extremal (x, p, u)

$$(11) \qquad \sum_{i=1}^{c} v_i(t)H_i(x(t), p(t)) \;=\; 0$$

for all t, provided that $(v_1(t), \ldots, v_c(t))$ belongs to $\ker P(x(t))$.

Suppose that the rank of $P(x_0)$ is constant in a neighborhood of a fixed point x_0. Under such an assumption, there is no loss of generality in assuming that $P(x)$ is diagonal with $P(x)e_1 = 0$ for $1 \le i \le r$, and $P(x)e_i = 1$ for $i > r$. This form of P can be obtained by choosing a suitable basis $a_1(x), \ldots, a_c(x)$ in R^c. The equations (10) define $u_{r+1}, \ldots, u_c$, through $u_i(x,p) = H_i(x,p)$ while $u_1, \ldots, u_r$ are arbitrary, provided that $H_i(x,p) = 0$.

Let

$$\hat{H}(x,p) = H_{-1}(x, p, o, \ldots, o, H_{r+1}(x,p), \ldots, H_c(x,p))$$

$$(12) \qquad \text{and } \mathcal{H} = \{\hat{H} + \sum_{i=1}^{r} u_i H_i : \; u_1, \ldots, u_r \text{ arbitrary functions of } x \text{ and } p\}\,.$$

Then the regular extremals satisfy the following criterion:

$$(13) \qquad \begin{aligned} &(x,p,u) \quad \text{is a regular extremal if and only if} \\ &\frac{d}{dt}(x(t), p(t)) \in \vec{\mathcal{H}}(x(t), p(t)) \quad \text{for almost all} \quad t, \quad \text{and} \\ &H_i(x(t), p(t)) = 0 \quad i = 1, 2, \ldots, r. \quad (\vec{\mathcal{H}}(x,p) = \{\vec{H}(x,p) : H \in \mathcal{H}\}.) \end{aligned}$$

This type of a constrained Hamiltonian system is quite similar to the one studied by Dirac in his attempt to provide a mathematical foundation for quantum mechanics ([D]). Its resolution involves various higher order derivatives of the system. These calculations are naturally expressed through the Poisson brackets of various Hamiltonians which occur in the above equations.

For any functions F and G on T^*M, let $\{F,G\}$ be their Poisson bracket. In terms of the canonical coordinates $x_1, \ldots, x_n, p_1, \ldots, p_n$, the Poisson bracket is given by $\sum_{i=1}^{m} \frac{\partial F}{\partial x_i}\frac{\partial G}{\partial p_i} - \frac{\partial F}{\partial p_i}\frac{\partial G}{\partial x_i}$. The essential property of the Poisson bracket is related to the Lie bracket $[\vec{F},\vec{G}]$ of the Hamiltonian vector fields $\vec{F}$ and $\vec{G}$ through the formula: $\overrightarrow{\{F,G\}} = [\vec{F},\vec{G}]$. Another basic fact concerning the Poisson bracket is the following:

Suppose that $z(t)$ is an integral curve of a Hamiltonian vector field $\vec{H}$, and suppose that G is a function on T^*M. Then, $\frac{d}{dt}G(z(t)) = \{H,G\}(z(t))$. In particular, if $G(z(t)) = 0$, then $\{H,G\}(z(t)) = 0$.

This formalism when applied to the extremals in (13) yields that

$$\{\hat{H}, H_i\}(x(t),p(t)) + \sum_{j=1}^{r} u_j\{H_j, H_i\}(x(t),p(t)) = 0$$

for $i = 1,2,\ldots,r$ for any extremal (x,p,u). The matrix $\{H_i,H_j\}$ is antisymmetric, and therefore it is possible for it to be nonsingular in the case r is even. In such a case the above system of equations is uniquely solvable for $u_i, \ldots, u_r$ in terms of x and p. But as Gabasov notes $\{H_i,H_j\}(x(t),p(t)) = 0$ along an extremal whose projection (x,u) is optimal ([G]). His method of proof consists of inventing special variations, along an optimal trajectory which he calls "variation packets" which carry information beyond that contained in the Maximum Principle.

The point of view taken in this paper is quite different. It exploits the symmetry among the optimal trajectories through the notion of the Lie Saturate developed by Jurdjevic-Kupka on the earlier papers concerning controllability of non-linear systems ([JK1] and [JK2]). This approach naturally leads to a Generalized Maximum Principle which properly accounts for the role of the Poisson brackets involving the terms $H_1, \ldots, H_r$.

II. The Lie Saturate and the Jump Fields.

Rather than stating the basic definitions in terms of equations (1) and (2), it is more natural to proceed in terms of arbitrary families of vector fields as is customarily done in studies on non linear control theory.

For any family $\mathcal{F}$ of vector fields on M, $A_{\mathcal{F}}(a,T)$ denotes the set of reachable points of $\mathcal{F}$ from a in exactly T units of time. That is $y \in A_{\mathcal{F}}(a,T)$ if and only if there exists an absolutely continuous curve $x(t)$

defined on $[0,T]$ such that $\frac{dx}{dt}(t) \in \mathcal{F}(x(t))$ for almost all t in $[0,T]$, $x(0) = a$, and $x(T) = y$. $A_{\mathcal{F}}(a, \leq T)$ is equal to the union of $A_{\mathcal{F}}(a,t)$ for all t, $0 \leq t \leq T$. Lie $(\mathcal{F})$ denotes the Lie algebra generated by $\mathcal{F}$, while $\text{Lie}_{\mathbf{x}}(\mathcal{F})$ denotes its evaluation at a fixed point x in M.

Definition 1. *Families of vector fields $\mathcal{F}$ and $\mathcal{G}$ are said to be strongly equivalent if*

(i) $\mathcal{F} \subset \text{Lie}(\mathcal{G})$, $\mathcal{G} \subset \text{Lie}(\mathcal{F})$, *and*

(ii) $c\ell\, A_{\mathcal{F}}(a, \leq T)\ c\ell\, A_{\mathcal{G}}(a, \leq T)$ *for all* $a \in M$, *and all* $T > 0$.

Definition 2. *The Lie Saturate of a family of vector fields $\mathcal{F}$ is the union of families strongly equivalent to $\mathcal{F}$. $LS(\mathcal{F})$ will denote the Lie Saturate of $\mathcal{F}$.*

Evidently, $LS(\mathcal{F}) \subset \text{Lie}(\mathcal{F})$, and $c\ell\, A_{LS(\mathcal{F})}(a, \leq T) = c\ell\, A_{\mathcal{F}}(a, \leq T)$ for all $a \in M$, and $T > 0$. In this notation, $c\ell$ denotes the topological closure.

In order to get some explicit criteria for computing the Lie Saturate, it is necessary to introduce a topology on the space of all vector fields under consideration. Assume that the space of all C^∞ vector fields on M is topologized with its C^∞ topology, and that $\mathcal{F}$ is any family such that Lie $(\mathcal{F})$ is *topologically closed.*

Then, $c\ell\, LS(\mathcal{F})$ is strongly equivalent to $LS(\mathcal{F})$, and therefore, $LS(\mathcal{F}) = c\ell\, LS(\mathcal{F})$. The basic properties of $LS(\mathcal{F})$ are summarized in the following proposition whose proofs can be found in [JK1] and [JK2]:

Proposition 1. *Suppose that $\mathcal{F}$ is such that* Lie $(\mathcal{F})$ *is closed. Then,*

(i) $LS(\mathcal{F})$ *is a closed, convex subset of* Lie $(\mathcal{F})$

(ii) $\lambda\, LS(\mathcal{F}) \subset LS(\mathcal{F})$ *for any* $\lambda \in [0,1]$

(iii) If there exists a vector field X in $LS(\mathcal{F})$ such that $\lambda X \in LS(\mathcal{F})$ for all $\lambda \in R$, then $LS(\mathcal{F})$ is invariant under $\exp \lambda X_{\#}$ for each $\lambda \in R$. $\exp \lambda X_{\#}(Y)$ denotes the infinitesimal generator of $\{\exp \lambda X \exp tY \exp -\lambda V : t \in R\}$

(iv) If $LS(\mathcal{F})$ contains a vector space $\mathcal{V}$, then $\text{Lie}(\mathcal{V}(\subset \text{LS}(\mathcal{F})$.

For the remaining part of the paper, it will be assumed that $\mathcal{F}$ is the family of vector fields defined by (7) on the extended state space $\tilde{M} = R \times M$. That is, each $u \in R^c$ defines an element $F_u(x) = \left(f_0(x,u), X_0(x) + \sum_{i=1}^{c} u_i X_i(x)\right)$ which belongs to $\mathcal{F}$.

Definition 3. *A vector field X is a jump field of $\mathcal{F}$ if $\lambda X \in LS(\mathcal{F})$ for each $\lambda \in R$.*

The set of all jump fields of $\mathcal{F}$ is denoted by $J(\mathcal{F})$. Equivalently, $J(\mathcal{F})$ is a maximal subalgebra of Lie $(\mathcal{F})$ as can be easily seen through Proposition 1.

Let G be the group of diffeomorphisms generated by the union of $\{\exp tX : t \in R\}$ as X varies over $J(\mathcal{F})$. $G(x)$ denotes the orbit of $J(\mathcal{F})$ through x for each x in M. Points of $G(x)$ can be "instantaneously" joined to each other along the trajectories of $J(\mathcal{F})$.

A particular feature of this study is the Generalized Maximum Principle which goes as follows: Let $\mathcal{F}_g = \{\exp V_{\#}(\mathcal{F}) : V \in J(\mathcal{F})\}$. $\mathcal{F}_g$ is called the generalized family of $\mathcal{F}$. Recall that $\exp V_{\#}(X)$ is the infinitesimal generator of $\{\exp V \, \exp tX \, \exp -V : t \in R\}$.

Suppose that (x, u) is an optimal trajectory of (1) on $[0, T]$, and assume that $\mathcal{F}$ is locally controllable along x. Let $\tilde{x} = \big(x^0(t), x(t)\big)$ be the extended trajectory of $\mathcal{F}$. Then $\tilde{x}$ is the projection of an integral curve $(\tilde{x}, \tilde{p})$ of $\vec{H}\big(\tilde{x}, \tilde{p}, u(t)\big)$. for which $\tilde{p}$ is not identically zero, such that

(i) $p_0 f\big(x(t)\big) + \langle X\big(x(t), p(t)\rangle \leq H\big(\tilde{x}(t), \tilde{p}(t), u(t)\big)$ for any $\big(f(x), X(x)\big) \in \mathcal{F}_g(x)$ and almost all t, and

(ii) $\tilde{p} = \big(p^0(t), p(t)\big)$ with $p^0 \leq 0$.

It follows by standard arguments from [JK1] and [JK2] that for each $u(x) \in \ker P(x)$, $\big(f_0\big(x, u(x)\big), \sum_{i=1}^{c} u_i(x) X_i(x)\big)$ is a jump field of (1). In the case that $P(x)$ is diagonal with $P(x)e_i = 0$ for $i \leq r$, it follows that each $\big(Q_i(x),\ X_i(x)\big)$ is a jump field.

But then, the entire Poisson algebra generated by $H_1, \ldots, H_r$ corresponds to the regular Hamiltonians of jump fields, since the space of jump fields is a Lie subalgebra of Lie $\mathcal{F}$ (Proposition 1, (IV)).

(14) The Generalized Maximum Principle implies that if (x, p, u) is a regular extremal such that (x, u) is optimal then $H\big(x(t), p(t)\big) = 0$ for any H which is a regular Hamiltonian of a jump field.

Therefore, $H\big(x(t), p(t)\big) = 0$ for any H in the Poisson algebra generated by $\{H_1, \ldots, H_r\}$. Gabasov's result that $\{H_i, H_j\}\big(x(t), p(t)\big) = 0$ is a particular case of this fact. The proofs of the Generalized Maximum Principle along with its consequences will be done in a separate study and will appear elsewhere. Instead, the scope of the theory will be illustrated through the applications outlined in the introduction.

III. Applications.

This formalism readily applies to the two situations mentioned in the introduction. In either case, the calculations are done in terms of the Poisson brackets. It will be important to not that when F_X and G_X are the Hamiltonians of the right (left) invariant vector fields X and Y on a Lie group G, then their Poisson bracket corresponds to the right (left) invariant vector field given by $[X, Y]$.

We return now to the linear-quadratic problem described in the Introduction (A). An important example which is not positive definite and which yet satisfies assumptions (i), (ii) and (iii) is the following

Example. *Let* $M = R^3$, $x = (x_1, x_2, x_3)$ *and*

$$\frac{dx_1}{dt} = x_2, \qquad \frac{dx_2}{dt} = x_3, \qquad \frac{dx_3}{dt} = u \ .$$

$$f_0(x_1, x_2, x_3) = \tfrac{1}{2}(x_1^2 - x_2^2 + x_3^2) \ .$$

It follows by an easy integration by parts that

$$\int_0^T f_0\big(x_1(t), x_2(t), x_3(t)\big)dt = \tfrac{1}{2}\int_0^T (x_1^2 - x_1x_3 + x_3^2)dt - \tfrac{1}{2}x_1x_2\Big|_{t=0}^{t=T}$$

for any integral curve $x(t)$ of the above system. Hence assumption (i) is satisfied. To verify assumption (ii) assume that $x(t)$ satisfies $x(0) = x(T) = 0$, and is optimal for $a = b = 0$. Then,

$$\tfrac{1}{2}\int_0^T (x_1^2 - x_2^2 + x_3^2)dt = \tfrac{1}{2}\int_0^T \Big[\big(x_1 - \tfrac{x_3}{2}\big)^2 + \tfrac{3}{4}x_3^2\Big]dt = 0 \ .$$

Hence, $x_3(t) = 0$ for all t, and therefore $x_1(t) \equiv 0$. But then $x_2(t) = 0$, and thus (ii) holds. Property (iii) obviously holds, since the system is in Brunovski form.

Since $P = 0$ in this case, every control is in $\ker P$, and therefore the constant vector field $(0, e_3)$ is a jump field. Its Hamiltonian h is equal to $h = p_3$. It turns out that in this example the space of jump fields is one-dimensional, and is equal to the line containing $(0, e_3)$.

Equations (12) and (13) are given by the following data: $\hat{H} = -\frac{1}{2}(x_1^2 - x_2^2 + x_3^2) + p_1x_2 + p_2x_3$, and $\vec{\mathcal{H}} = \{\vec{H} + uh \ : \ u \in R\}$. $h = 0$ along an extremal. Upon differentiation, $\{\hat{H}, h\} = 0$. Since $\{\hat{H}, h\} = x_3 - p_2$, it follows that in addition to $p_3 = 0$, each extremal satisfies $x_3 - p_2 = 0$.

Let $\Omega = \{(x,p) : p_3 = 0 \text{ and } x_3 - p_2 = 0\}$. Ω is a symplectic subspace of $T^*(R^3)$ which contains all the extremals. In terms of the coordinates (x_1, x_2, p_1, p_2) on Ω, each extremal satisfies:

$$\frac{dx_1}{dt} = x_2, \quad \frac{dx_2}{dt} = p_2, \quad \frac{dp_1}{dt} = x_1 \quad \frac{dP_2}{dt} = -x_2 - p_1 \ .$$

It is easily verified that the extremals are the integral curves of the quadratic Hamiltonian

$$H = -\tfrac{1}{2}x_1^2 + \tfrac{1}{2}x_2^2 + \tfrac{1}{2}p_2^2 + p_1x_2 \ .$$

The generalized optimal synthesis for this problem is as follows: Suppose that $a = (a_1, a_2, a_3)$, $b = b_1, x_2, b_3)$ and $T > 0$ are given. There is a unique extremal $x_1(t), x_2(t), p_1(t), p_2(t)$ in Ω which satisfies $x_1(0) = a_1$ $x_1(T) = b_1$, $x_2(0) = a_2$ and $x_2(T) = b_2$. Let $p_2(0) = \bar{a}_3$, and $p_2(T) = \bar{b}_3$.

The optimal way to go from a to b is to jump to $(a_1, a_2, \bar{a}_3)$ with zero cost, follow the above extremal to $(b_1, b_2, \bar{b}_3)$ and then to jump to b with zero cost.

A more detailed analysis fo this problem is treated in [J], while the full details of the general case are done in [JK3]. Return now to the problems of elastica and the total squared curvature as described in (B) of the introduction.

The problem is non-singular only in the case $n = 2$. Consider that case first: The exceptional extremals are given by $k = 0$, i.e., their projections on R^2 are straight lines. For the regular extremals, $\frac{\partial H}{\partial k} = 0$ yields that $k(x,p) = H_1(x,p)$. Then the regular extremals are the integral curves of $H(x,p) = \frac{1}{2}H_1^2 + H_0$.

Suppose that (x,p) is any such an extremal. It defines the curvature $k(t)$ through $H_1(x(t), p(t))$. Then,

$$\frac{dk}{dt} = \{H, H_1\}\big(x(t), p(t)\big) = \{H_0, H_1\}\big(x(t), p(t)\big) ,$$

and

$$\frac{d^2k}{dt^2} = H_1\{H_1, \{H_0, H_1\}\}\big(x(t), p(t)\big) + \{H_0, \{H_0, H_1\}\}\big(x(t), p(t)\big) .$$

It follows that $[L_0, [L_0, L_1]] = 0$, and that $[L_1, [L_0, L_1]] = L_0$. Hence $\{H_0, \{H_0, H_1\}\} = 0$ and $H_1\{H_1, \{H_0, H_1\}\} = H_0$. Therefore,

$$\frac{d^2k}{dt^2} = kH_0\big(x(t), p(t)\big)$$

Along the extremal $\big(x(t), p(t)\big)$, $H\big(x(t), p(t)\big) = \text{constant}$. Therefore, $H_0\big(x(t), p(t)\big) = (H - \frac{1}{2}k^2)$ and hence $\frac{d^2k}{dt^2} = k(H - \frac{1}{2}k^2)$. Furthermore,

$$\frac{d}{dt}\frac{1}{2}\left(\frac{dk}{dt}\right)^2 = \frac{dk}{dt}\frac{d^2k}{dt^2} = (H - \tfrac{1}{2}k^2)k\frac{dk}{dt} ,$$

and consequently

$$\left(\frac{dk}{dt}\right)^2 + (H - \tfrac{1}{2}k^2)^2 = \text{constant} .$$

The last equation is the same as the result of Bryant-Griffiths (Equation (5)).

For $n > 2$, the problem is singular. It follows from the remarks on page 14 that not only each of the extended vector fields (o, X_2), ..., (o, X_{n-1}) is a jump field, but that the Lie algebra generated by them is also contained in the space of jump fields.

Let $H_0, H_1, \ldots, H_{n-1}$ be the Hamiltonians corresponding to the vector fields $X_0, X_1, \ldots, X_{n-1}$, i.e., $H_i(x,p) = \langle X_i(x), p\rangle$. According to equations (13) $H_i\big(x(t), p(t)\big) = 0$ for $i \geq 2$ along each extremal $\big(x(t), p(t)\big)$. But then it follows from equations (14) that $H\big(x(t), p(t)\big) = 0$ for each H in the Poisson algebra generated by $\{H_2, \ldots, H_{n-1}\}$; for each such H is the Hamiltonian of a jump field.

The preceding remarks allow for an easy derivation of equations (6). The argument is as follows:

Let $z(t) = \big(x(t), p(t)\big)$ be a regular extremal. Then according to equations (12), $k_1(x,p) = H_1$ and $\hat{H} = \frac{1}{2}H_1^2 + H_0$.

Since $H_2\big(z(t)\big) = 0$, it follows by differentiation that $H_1\{H_1, H_2\}\big(z(t)\big) = 0$. If $k \neq 0$, then $\{H_1, H_2\}\big(z(t)\big) = 0$. Upon further differentiation,

$$H_1\big\{H_1, \{H_1, H_2\}\big\} + \big\{H_0, \{H_1, H_2\}\big\} + \tau\big\{H_2, \{H_1, H_2\}\big\}\big(z(t)\big) = 0.$$

Since $\big\{H_1, \{H_1, H_2\}\big\} = H_2$, and $\big\{H_2, \{H_1, H_2\}\big\} = H_1$ it follows that

$$H_1(z)H_2(z) + \big\{H_0, \{H_1, H_2\}\big\}(z) + \tau H_1(z) = 0\ .$$

Moreover, $H_2(z) = 0$, and therefore

$$\tau k = -\big\{H_0, \{H_1, H_2\}\big\}(z) = \big\{H_2, \{H_0, H_1\}\big\}(z)\ .$$

Then

$$\begin{aligned} \frac{d}{dt}k^2\tau = \frac{d}{dt}H_1\big\{H_2, \{H_0, H_1\}\big\} &= \{H_0, H_1\}\big\{H_2, \{H_0, H_1\}\big\} + \\ &\quad H_1\tau\Big\{H_2, \big\{H_2, \{H_0, H_1\}\big\}\Big\} = \\ &= \{H_0, H_1\}\big\{H_2, \{H_0, H_1\}\big\} - H_1\tau\{H_0, H_1\} = 0\ . \end{aligned}$$

Therefore, $k^2\tau$ = constant. This argument proves the first part of (6). To get the differential equation for k,

$$\begin{aligned} \frac{d}{dt}k &= \{H_0, H_1, \}(z), \quad \text{and} \\ \frac{d^2k}{dt^2} &= H_1(z)\big\{H_1, \{H_0, H_1\}\big\}(z) \\ &\quad + \big\{H_0, \{H_0, H_1\}\big\}(z) + \tau\big\{H_2, \{H_0, H_1\}\big\}\ . \end{aligned}$$

Since $\big\{H_1, \{H_0, H_1\}\big\} = H_0$ and $\big\{H_0, \{H_0, H_1\}\big\} = 0$ it follows that $\frac{d^2k}{dt^2} = H_1\big(z(t)\big)H_0\big(\big(z(t)\big) - \tau\big\{H_0, \{H_1, H_2\}\big\}\big(z(t)\big)$.

Along $z(t)$, $\hat{H} = \frac{1}{2}H_1^2 + H_0$ is constant. Therefore, $\frac{d^2k}{dt^2} = k(\hat{H} - \frac{1}{2}k^2) + \tau^2 k$, or $2\frac{d^2k}{dt^2} + k^3 - 2k\hat{H} - 2\tau^2 k = 0$. This equation is the same as the result of Griffiths (equation (6)).

References

[BG] R. Bryant and P. Griffiths, Reduction for constrainted variational problems and $\frac{1}{1}\int k^2 ds$, *American Journal of Mathematics* **108** (1986), 525-570.

[Br] R.W. Brockett, *Control theory and singular Riemannian geometry*, New Directions in Applied Mathematics, (P.J. Hilton and G.S. Young eds.) Springer-Verlag 1981, 11-27.

[D] P.A.M. Dirac, Generalized Hamiltonian dynamics, *Canadian Journal of Mathematics* **2** (1950), 129-148.

[G] R. Gabasov, *Singular Optimal Controls*, Plenum Publishing Corp., New York.

[Gr] P. Griffiths, *Exterior differential systems and the calculus of variations*, Birkhäuser, Boston, 1982.

[HS] M.L.I. Hautus and L.M. Silverman, Systems structure and singular control, *Linear Algebra and Its Applications* **50** (1983), 369-402.

[J] V. Jurdjevic, Linear systems with quadratic cost, to be published in the Proceedings of a Conference on Linear Control, held at Rutgers University, May 1987.

[JK1] V. Jurdjevic and I. Kupka, Control systems on semi-simple Lie groups and their homogenous spaces, *Ann. Inst. Fourier* **31** (1981), 151-179.

[JK2] V. Jurdjevic and I. Kupka, Polynomial control systems, *Math. Ann.* **272** (1985), 361-368.

[JK3] V. Jurdjevic and I. Kupka, Linear systems with singular quadratic cost, to appear.

[K] R. Kalman, contributions to the theory of optimal control, *Bol. Soc. Mat. Mexicana* **5** (1960), 102-119.

[LS] J. Langer and D. Singer, The total squared curvature of closed curves, *J. Differential Geometry* **20** (1984), 1-22.

[S] R.S. Strichartz, Sub-Riemannian geometry, *J. Differential Geometry* **24** (1986), 221-263.

[T] C. Truesdell, The influence of elasticity on analysis: the classical heritage, *Bull. Amer. Math. Soc.* **9** (1983), 293-310.

STOCHASTIC REALIZATION OF STATIONARY PROCESSES: STATE-SPACE, MATRIX FRACTION AND ARMA FORMS

L. KEVICZKY AND J. BOKOR

ABSTRACT This paper discusses results from the stochastic realization theory of second order stochastic process. The forward and backward stochastic state-space representation are derived and transfer relations are given to obtain their associated matrix fraction description and ARMA forms. The correspondence among these realizations are elaborated.

1. INTRODUCTION

The theory and application of signals and systems have gone trough an outstanding development in the past twenty years. Originating mainly from the field of system and control engineering, essential advances have been made e.g. in understanding the algebraic and topological structure of linear dynamic systems. These results were succesfully applied in various fields of system and control engineering, e.g. in modelling and statistically interpreting data in stochastic control, signal processing and fault detection. It can also be observed, that system theory and the theory of stochastic processes and time series approached closer each other resulting what is called "statistical theory of linear systems" as elaborated in the recent monographs of Hannan and Deistler [21], and Caines [9]. From the many exciting problems described e.g. in the above mentioned references, here we review only some results from stochastic realization theory associated with second order stochastic processes.

The stochastic realizations derived in State Space Representation (SSR), Matrix Fraction Description (MFD) and monic AutoRegressive Moving Average (ARMA) forms can be applied in solving system identification, parametrization, prediction, filtering and stochastic control problems [2] [5] [11] [16] [30] [31] [32] [36] [37].

To be more specific this paper considers the following problem. Given a stochastic process, what systems can generate it as its output process. In this discussion a system is always given in a specific representation e.g., in SSR, MFD or ARMA form, and we do not intend to give more abstract and general definitions like given e.g. in Kalman at al [26] or in Williems [39]. The stochastic process considered here can be generated or realized by any of the above representations in a *weak* sense, where only second - order properties or in a *strong* sense when almost all of the sample paths are to be reproduced.

We consider here the realization of discrete time multivariable second order Gaussian stochastic processes only. The continuous time case is treated e.g.in Lindquist and Picci [28,29]. The derivation of stochastic forward and backward SSR-s are elaboratd in more detailed, and the equivalent MFD and ARMA forms are obtained from these SSR-s by deriving appropriate transfer relations. Emphasis will be taken to characterize a complete interrelations among the minimal order forward and backward canonical SSR, MFD and ARMA forms realizing the same process.

The next paragraph gives an overview about the weak state-space realizations obtained from covariance and spectral factorizations. The properties and correspodence among these weak SSR-s will be given.

The 3. paragraph discusses the derivation of strong state-space realizations. Based on the Wold decomposition of second order processes, canonic SSR-s are derived. The correspodence between forward and backard SSR-s having the same state-space and generating the same process is derived. The derivation of forward MFD-s from forward SSR-s is discussed in the 4. paragraph.

Monic ARMA forms are the most frequently used in the aplications mentioned above. It is shown in the 5. paragraph, that monic backward ARMA forms are naturally related to backward SSR-s, and it is referenced, that forward monic ARMA forms can be similarly derived from forward canonic SSR-s by using constructibility concepts (Bokor and Keviczky [7]).

The correspodence among the forward and backward strong SSR-s obtained from the Wold - representation and among their associated MFD and ARMA forms is described in the 6. paragraph.

The discussion of the paper is by far not complete in the sense, that the results are stated, but instead of giving proofs, they are either referenced or the concepts applied to obtain the stated results are only described.

2. Stochastic state-space representations

Consider the p component stationary Gaussian stochastic process $z_t, t \in Z$, $Z = \{0, \pm 1, \pm 2, \dots\}$, with $Ez_t = 0$ and covariance structure $Ez_{t+k}z_t = R_k$ where $E(\cdot)$ denotes the expectation. Denote $F(d\lambda)$ the spectral measure and $\Phi(d\lambda)$ the stochastic spectral measure of the process z_t, i.e.,

$$z_t = \int_{-\pi}^{\pi} e^{i\lambda t}\Phi(d\lambda), R_k = \int_{-\pi}^{\pi} e^{i\lambda k} f(d\lambda). \tag{2.1}$$

and $E\Phi(d\lambda)\Phi^*(d\lambda) = F(d\lambda)$ see Rozanov [35], where Φ^* denotes the transponse conjugate of Φ.

Assume that $f(d\lambda)$ is absolutely continuous with respect to the Lebesque measure, i.e. $F(d\lambda) = f(\lambda)d\lambda$, where $f(\lambda), \lambda \in [\pi, \pi]$ is the spectral density matrix of the process z_t.

Assume also that z_t is full rank regular (or purely linearly nondeterministic) process. It is known e.g. from Wiener and Masani [38] that z_t is regular and full rank, if it has an absolutely continuous spectral distribution $F(\lambda) = f(\lambda)d\lambda$, such that $f(\lambda)$ is integrable and

$$det\Sigma = exp\{(2\pi)^{-1} \int_{-\pi}^{\pi} logdetf(\lambda)d\lambda\}. \tag{2.2}$$

where Σ is the covariance matrix of the one-step-ahead prediction error for z_t.

The stochastic State Space Representations /SSR/ are generally described by the following equations (the primes denote matrix transpositions).

Forward SSR:

$$x_{t+1} = Fx_t + v_t, \qquad y_t = Hx_t + w_t \tag{2.3}$$

driven forwards in time by the white noise process v_t, w_t, and where F, H denote appropriate dimensional matrices, x_t, y_t denote the state and output process respectively,$dim\, X_t = dim\, v_t = n$, $dim\, y_t = p$, and

$$Ev_t = Ew_t = 0, \quad Ev_iv_j' = Q\delta_{ij}, \quad Ew_iw_j' = R\delta_{ij}, \quad Ev_iw_j' = S\delta_{ij} \tag{2.4a}$$

$$Ex_tx_x' = P > 0, \quad Ex_kv_t' = 0, \quad Ex_kw_t' = 0, \quad t \geq k, \quad t,k \in Z \tag{2.4b}$$

In (2.4a), (2.4b) the δ_{ij} denotes the Kronecker symbol.

Backward SSR:

$$x_t^B = F_B x_{t+1}^B + v_t^B, \qquad y_t = H_B x_{t+1}^B + w_t^B \tag{2.5}$$

driven backwards in time by the white noise process v_t^B, w_t^B, F_B, H_B are appropriate dimensional matrices, and

$$Ev_t^B = Ew_t^B = 0, \quad Ev_i^B(v_j^B)' = Q_B\delta_{ij}, \quad Ew_i^B(w_j^B)' = R_B\delta_{ij}, \quad Ev_i^B(w_j^B)' = S_B\delta_{ij}, \tag{2.6a}$$

$$Ex_t^B(x_t^B)' = P_b > 0, \quad Ex_k^B(v_t^B)' = 0, \quad Ex_k^B(w_t^B)' = 0, \quad t \geq k, \quad t,k \in Z. \tag{2.6b}$$

Definition 2.1 Let the covariance function $R_k, k \in Z$ of the process z_t be given. Then the SSR given by (2.3), (2.4) is called a weak (forward) stochastic SSR of z_t, if for the output process y_t generated by (2.3) $Ey_{t+k}y_t' = R_k, \;\; k \in Z$.

The weak backward stochastic SSR-s are defined analogously. An immediate consequence of Definition 2.1 is that weak stochastic realizations can also be defined by using the spectral densities, i.e. (2.3), (2.4) is a weak stochastic SSR of z_t, if the spectral density of the output process y_t is the same as that of z_t.

The definition suggests that the construction of weak stochastic realizations can be performed either from the knowledge of the covariance function R_k, $k \in Z$, or from the spectral density $f(\lambda)$. This construction will be now shown using both covariance and spectral factorization approaches.

Realizations derived from covariance factorization

Let the covariance function $R_k, k \in Z$ (where R_k are $p \times p$ dimensional matrices) be given, and denote by $\mathbf{H}$ the Hankel matrix associated with this covariance function:

$$\mathbf{H} = \begin{bmatrix} R_1 & R_2 & R_3 & \cdots \\ R_2 & R_3 & \cdots & \cdots \\ R_3 & \cdot & & \\ \cdot & \cdot & & \\ \cdot & \cdot & & \\ \cdot & \cdot & & \end{bmatrix}. \tag{2.7}$$

The following results are known from the deterministic realization theory, see e.g. Ho and Kalman [22].

THEOREM 2.1. *Let the covariance function $R_k, k \in Z$ and the associated Hankel matrix* **H** *be given. Then there exists matrices F, H, L of dimensions $n \times n$, $p \times n$, $n \times p$ respectively, such that*

$$R_k = HF^{k-1}L, \qquad k \in Z \tag{2.8}$$

if rank $H = n$.

The matrices F, H, L are unique modulo change of basis, i.e. if $\overline{H}, \overline{F}, \overline{L}$ also satisfy (2.8), then there exists an $n \times n$ dimensional matrix T, such that

$$\overline{F} = TFT^{-1}, \qquad \overline{H} = HT^{-1}, \qquad \overline{L} = TL. \tag{2.9}$$

Since we assume that z_t is purely nondeterministic, thus $R_k \to 0$ if $k \to \infty$ and this implies that F is an asymptotically stable matrix.

Using (2.8), the covariance function can be written as

$$R_k = HF^{k-1}L\delta_k + L'(F')^{-k-1}H'\delta_{-k} + R_0\delta_{0k}. \tag{2.10}$$

where $\delta_1 = 1$ if $k > 0$ and $\delta_k = 0$ if $k < 0$.

Taking two-sided Z-transform, one obtaines the spectrum

$$f(z) = \sum_{k \in Z} R_k z^{-k} = Z_1(z) + Z_2(z) \tag{2.11}$$

where the spectral summands Z_1, Z_2 are given by

$$Z_1(z) = H(zI - F)^{-1}L + R_0/2, \quad Z_2(z) = L'(zI - F')^{-1}H' + R_0/2. \tag{2.12}$$

The form of the spectral summand Z_1 immediately results in a forward SSR:

$$x_{t+1} = Fx_t + v_t, \qquad y_t = Hx_t + w_t \tag{2.13}$$

where v_z, w_t are any white noise processes (with n and p dimensions respectively) such that they satisfy (2.4), and x_t is the n-dimensional, state vector. We note, that the SSR (2.13) with state dimension $dim x_t$ = rank **H** $= n$ is called a minimal realization. Given $(F, H, L, R_0/2)$ one can easily deduce that the unknown P, Q, R, S matrices corresponding to a Markovian SSR in (2.3), (2.4) are obtained as

$$P - FPF' = Q, \qquad L - FPH' = S, \qquad R_0 - HPH' = R \tag{2.14}$$

with the nonlinear constraints

$$\begin{bmatrix} Q & S \\ S' & R \end{bmatrix} \geq 0, \qquad P \geq 0. \tag{2.15}$$

It follows from (2.14) that the matrices Q, S, R are uniquely determined from P, i.e. to any given matrix P there corresponds a Markovian SSR as described by (2.13), (2.14), and conversely.

Following Faurre [14], the set of all Markovian SSR-s associated with a given covariance function R_k, $k \in Z$ is defined as follows.

Definition 2.2 Let $(F, H, L, R_0/2)$ be an n-dimensional (minimal) realization of the spectral summand $Z_1(z)$. Define $\mathcal{P}(F, H, L, R_0)$ as the set of all symmetrical matrices P that satisfy (2.14), (2.15).

The properties of the set $\mathcal{P}(F, H, L, R_0)$ that are invariant with respect to the originally choosen F, H, L was studied e.g. by Faurre [14]. It was shown that $\mathcal{P}(F, H, L, R_0)$ is closed, convex and bounded, and there are extremal elements P_*, P^*, such that

$$P_* \leq P, \qquad P^* \geq P \text{ for all } P \in \mathcal{P}(F, H, L, R_0), \tag{2.16}$$

in the sense that $P^* - P$ is positive definite. The SSR-s associated with P_* and P^* are called minimal and maximal SSR-s respectively. Faurre [14] gives two algorithms to generate P_* and P^* recursively. The extremal realizations can also be related to the forward and backward Kalman-filters as will be indicated in the subsequent paragraphs.

Backward weak stochastic SSR-s can be obtained from the right spectral summands directly:

$$x_t^B = F' x_{t+1}^B + v_t^B, \qquad y_t = L' x_{t+1}^B + w_t^B, \tag{2.17}$$

where v_t^B, w_t^B are any white noise processes (with n and p dimensions respectively, such that they satisfy (2.6). The unknown covariances P_B, Q_B, R_B, S_B are related to the matrices F, H, L, R_0 as follows:

$$P_B - F' P_B F = Q_B, \quad H' - F' P_B L = S_B, \quad R_0 - L' P_B L = R_B, \tag{2.18}$$

where R_B, Q_B, R_B, S_B satisfy the nonlinear constraints

$$\begin{bmatrix} Q_B & S_B \\ S_B' & R_B \end{bmatrix} \geq 0, \qquad P_B \geq 0. \tag{2.19}$$

Similarly to Definition 2.2, it is straightforward to define $\mathcal{P}_B(F, H, L, R_0)$ as the set of all symmetric matrices P_B that satisfy (2.18), (2.19). It can be proved that $\mathcal{P}$ and $\mathcal{P}_B$ correspond to each other by matrix inversion, i.e. $P \in \mathcal{P}$ if $P^{-1} \in \mathcal{P}_B$.

Stochastic realizations obtained from spectral factorization

Recall that the stochastic process z_t was assumed to be a full rank regular process having spectral density $f(\lambda)$, $\lambda \in [-\pi, \pi]$, where $f(\lambda)$ is given by

$$f(\lambda) = K(e^{i\lambda}) K'(e^{-i\lambda}) = \overline{K}(e^{i\lambda}) \Sigma \overline{K}'(e^{-i\lambda}), \tag{2.20}$$

where

$$K(e^{i\lambda}) = \overline{K}(e^{i\lambda}) \Sigma^{1/2}, \quad \overline{K}(e^{i\lambda}) = \Sigma_{k=0}^{\infty} \overline{K}_k e^{i\lambda k}, \quad \overline{K}_0 = I_p. \tag{2.21}$$

The problem of finding $K(e^{i\lambda})$ or $\overline{K}(e^{i\lambda}), \Sigma$ satisfying (2.20) is called spectral factorization, see e.g. Youla [42] and Anderson [4]. Consider the complex $p \times p$ dimensional matrix

$$K(z) = \sum_{k=0}^{\infty} K_k z^k, \qquad z \in C, \tag{2.22}$$

satisfying the boundary condition (2.20), and denote its McMillan degree by $\delta(K) = n$. Then $K(e^{i\lambda})$ is called the left minimal spectral factor of $f(\lambda)$ if $\delta(K)$ is the smallest among the McMillan degree of each $p \times p$ dimensional spectral factors satisfying (2.20).

Minimal degree spectral factors are unique modulo a multiplication by an orthogonal matrix. This leads to the following definition.

Definition 2.3 Two $p \times p$ dimensional minimal spectral factors $K_1(e^{i\lambda})$, $K_2(e^{i\lambda})$ of the full rank spectral density $f(\lambda)$ are called equivalent if there exists a $p \times p$ dimensional orthogonal matrix U such that

$$K_2(e^{i\lambda}) = K_1(e^{i\lambda})U. \tag{2.23}$$

The direct relations between the minimal realization (F, H, L, R_0) of the spectral summand $Z_1(z)$ and the left spectral factor $K(e^{i\lambda})$ can be shown by applying the Positive Real Lemma see e.g. Anderson [3].

This Lemma tells, that given the minimal n-dimensional realization (F, H, L, R_0) of $Z_1(z)$, then $Z_1(z)$ is positive real iff there exists matrices $P > 0$, G, D of dimensions $n \times n$, $n \times p, p \times p$ respectively, such that

$$P - FPF' = GG', \quad P - FPH' = GD', \quad R_0 - HPH' = DD' \tag{2.24}$$

In addition, if $Z_1(z)$ is strictly positive real, then one can choose D such that $DD' > 0$ and $F - GD^{-1}H$ is asymptotically stable. It can easily be deduced, that using (2.24), $K(z)$ can be written as

$$K(z) = H(zI - F)^{-1}G + D$$

which is the transfer function of the SSR given by (F, H, G, D) as

$$x_{t+1} = Fx_t + Gw_t, \qquad y_t = Hx_t + Dw_t \tag{2.26}$$

where w_t is any p-dimensional normalized white noise process, $Ew_t = 0$, $Ew_i w_j' = \delta_{ij} I_p$, $Ex_k w_t' = 0$ if $k < t$, $k, t \in Z$ and $DD' = K_0$. It can be seen, that the matrices G, D in (2.24) are not unique, i.e. there exists an orthogonal matrix U such that $\overline{G} = GU, \overline{D} = DU$ also satisfy (2.24). Thus the SSR (F, G, H, D) in (2.26) is also not unique, and this leads to the following definition.

Definition 2.4 The n-dimensional stochastic SSR-s $(H_1, H_1, G_1 D_1)$ and (F_2, H_2, G_2, D_2) are called equivalent if there exists an $n \times n$ dimensional nonsingular matrix T and a $p \times p$ dimensional orthogonal matrix U such that

$$F_1 = TF_2T^{-1}, \qquad H_1 = HT^{-1}, \qquad G_1 = TG_2U, \qquad D_1 = D_2U. \tag{2.27}$$

Backward SSR-s can also be derived from spectral factorization Lindquist and Picci [28] have shown that to each causal $p \times p$ dimensional minimal spectral factor $K(e^{i\lambda})$ of $f(\lambda)$, there corresponds an anticausal $p \times p$ dimensional minimal spectral factor $K_B(e^{i\lambda})$ satisfying

$$f(\lambda) = K_B(e^{i\lambda})K_B'(e^{-i\lambda}).$$

The associated backward SSR-s can be derived similarly to (2.26), see also Caines [9], pp. 260.

Correspondence among the equivalent week SSR-s.

We have shown, that given either the covariance function R_k, $k \in Z$ or the spectral density $f(\lambda)$ of a wide sense stationary process z_t one can construct a weak SSR which produce an output process y_t, such that its covariance or spectral density function coincides with R_k, $k \in Z$ or $f(\lambda)$ respectively. It was found that there are various equivalent representations, like the elements from the set $\mathcal{P}(F, H, L, R_0)$ or from $\mathcal{S}(F, H, G, D)$ which denotes the equivalence classes of all minimal order SSR-s.

Denote the $\mathcal{K}(\overline{K}, \Sigma)$ the set of equivalent classes of minimal degree spectral factors satisfying $f(\lambda) = \overline{K}(e^{i\lambda})\Sigma\overline{K}(e^{i\lambda}), \Sigma \geq 0$. Following Caines [9], pp. 265, [10] the correspodence among the sets $\mathcal{P}, \mathcal{K}, \mathcal{S}$ can be given as follows.

THEOREM 2.2. *Let y_t be a p-component, full rank, linearly regular wide-sense stationary stochastic process with rational spectral density $f(\lambda)$, where the McMillan degree of the rational matrix $\overline{K}(z)$ associated with the left spectral factor of $f(\lambda)$ is $\delta(\overline{K}) = n$.*

Then the following sets are in one-to-one correspondence:

(i) The set $\mathcal{K}(K, \Sigma)$ of equivalence classes of minimal (n-degree) causal left spectral factors of $f(\lambda)$.

(ii) The set $\mathcal{S}(F, H, G, D)$ of equivalence classes of minimal (n-order) stochastic SSR-s for y_t.

(iii) The set $\mathcal{P}(F, H, L, R_0)$ of all symmetric matrices P satisfying (2.14), (2.15).

3. Construction of Strong Stochastic State-Space and Innovation Representations

Based on the concept of splitting subspaces, the construction of strong state space representations was discussed by Lindquist and Picci [27]. The discussion given here is based on the Wold-representation of second order processes, see Rozanov [35] and is closed to the approach given in Faurre [14].

Definition 3.1 Let y_t, x_t, v_t, w_t in (2.3), (2.4) and z_t be defined on the same probability space $(\Omega, \mathcal{B}, \mathbf{P})$. Then (2.3), (2.4) is called a strong forward stochastic SSR of z_t, if $y_t = z_t$ $\mathbf{P}$ a.s., $t \in Z$.

The strong backward stochastic SSR of z_t is defined analogously replacing (2.3), (2.4) by (2.5), (2.6) respectively.

Consider the following Hilbert-spaces generated by the $y_t^i, i = 1, \cdots, p$ components of y_t, $t \in Z$ as

$$\mathbf{Y} = \{y_t, \quad t \in Z\}, \tag{3.1a}$$
$$\mathbf{Y}_t = \{y_t^i, \quad i = 1, \cdots, p\}, \tag{3.1b}$$
$$\mathbf{Y}_t^+ = \{y_{t+\tau}, \quad \tau \geq 0\} = \{y_t, y_{t+1}, \cdots\}, \tag{3.1c}$$
$$\mathbf{Y}_t^- = \{y_{t+\tau}, \quad \tau < 0\} = \{y_{t-1}, y_{t-2}, \cdots\}, \tag{3.1d}$$

Due to the Gaussian assumption, the conditional expectations and orthogonal projections in corresponding spaces defined above are identical, and we denote these by

$$E\{y_t/\mathbf{Y}_t^-\} = y_{t/t-1}, \qquad E(y_t/\mathbf{Y}_{t+1}^+) = y_{t/t+1}^B \tag{3.2}$$

Denote by e_t, e_t^B the forward and backward innovation processes respectively, given as

$$e_t = y_t - y_{t/t-1}, \qquad E e_t e_{t'} = \Sigma \tag{3.3a}$$
$$e_t^B = y_t - y_{t/t+1}^B, \qquad E e_t^B (e_t^B)_{'} = \Sigma_B \tag{3.3b}$$

Clearly e_t, e_t^B are both Gaussian white noise processes and let us denote the Hilbert spaces spanned by their components

$$\mathbf{E}_t = \{e_t^1, \cdots, e_t^p\}, \qquad \mathbf{E}_t^B = \{e_t^{B,1}, \cdots, e_t^{B,p}\}. \tag{3.4}$$

Define the forward and backward state-spaces of y_t by

$$\mathbf{X}_t = \mathbf{Y}_t^+/\mathbf{Y}_t^-, \qquad \mathbf{X}_t^B = \mathbf{Y}_{t+1}^-/\mathbf{Y}_{t+1}^+, \tag{3.5}$$

respectively. The following results follow easily from the stationarity of the process y_t.

Lemma 3.1. If $dim\mathbf{X}_t = n_t < \infty$ for an arbitrary $t \in Z$, then dim $\mathbf{X}_t = n_t = n$ is constant for all $t \in Z$.

Similar result is true for $dim\mathbf{X}_t^B$.

Lemma 3.2 The forward and backward state-space dimensions are equal, i.e.

$$dim\mathbf{X}_t = dim\mathbf{X}_t^B = n \qquad \text{for all} \quad t \in Z. \tag{3.6}$$

We note, that the stationary stochastic process $y_t, t \in Z$ is called rational iff above property holds. We will consider only rational process $y_t, t \in Z$ in the sequel.

Defining the state spaces $\mathbf{X}_t, \mathbf{X}_t^B$ in (2.5), one can derive forward and backward Markovian state-space representations for $y_t, t \in Z$, see e.g. Faurre [14].

`Proposition 3.1` Let x_t and x_t^B be a basis for $\mathbf{X}_t$ and $\mathbf{X}_t^B$ defined in (3.5) respectively. Then there exists quadruples (F, G, H, Σ) and $(F_B, G_B, H_B, \Sigma_B)$ that realize the rational process $y_t, t \in Z$ in state-space forms:

Forward SSR:

$$x_{t+1} = Fx_t + Ge_t, \qquad y_t = Hx_t + e_t \tag{3.7}$$

$$Ee_t = 0, \quad Ee_ie_j' = \Sigma\delta_{ij}, \quad Ex_s = 0, \quad Ex_sx_s' = P > 0, \tag{3.8a}$$

$$Ex_se_t' = 0, \qquad t \geq s \qquad \text{(forward orthogonality condition).} \tag{3.8b}$$

Backward SSR:

$$x_t^B = F_Bx_{t+1}^B + G_Be_t^B, \qquad y_t = H_Bx_{t+1}^B + e_t^B, \tag{3.9}$$

$$Ee_t^B = 0, \quad Ee_i^B(e_j^B)\prime = \Sigma_B\delta_{ij}, \quad Ex_t^B = 0, \quad Ex_t^B(x_t^B)\prime = P_B > 0, \tag{3.10a}$$

$$Ex_s^B(e_t^B)\prime = 0, \qquad t \leq s \quad \text{(backward orthogonality condition).} \tag{3.10b}$$

Proof. The proof is based on the orthogonal decomposition of $\mathbf{Y}_{t+1}^-, \mathbf{Y}_t^+$. For the forward realization, one can write

$$\mathbf{Y}_{t+1}^- = \mathbf{Y}_t^- \otimes \mathbf{E}_t \tag{3.11}$$

where $\otimes$ denotes direct sum of orthogonal subspaces. Then the projection $\mathbf{Y}_{t+1}^-/\mathbf{Y}_{t+1}^-$ can be written as:

$$\mathbf{Y}_{t+1}^+/\mathbf{Y}_{t+1}^- = \mathbf{Y}_{t+1}^+/\mathbf{Y}_t^- \otimes \mathbf{Y}_{t+1}^+/\mathbf{E}_t. \tag{3.12}$$

Since $\mathbf{Y}_{t+1}^+/\mathbf{Y}_t^- \subset \mathbf{Y}_t^+/\mathbf{Y}_t^-$, $\mathbf{Y}_{t+1}^+/\mathbf{Y}_t^-$ is a subspace in $\mathbf{X}_t$ and also $y_t/\mathbf{Y}_t^- \in \mathbf{X}_t$. This means, that $\mathbf{Y}_{t+1}^+/\mathbf{Y}_t^-$ can be generated by the basis vectors x_t of $\mathbf{X}_t$, and $y_t/\mathbf{Y}_t^-$ can also be obtained from linear combinations of x_t, i.e. there exists matrices F, G, such that the recursive formula between the basis vectors x_{t+1}, x_t has the form $x_{t+1} = Fx_t + Ge_t$. The observational equation $y_t = Hx_t + e_t$ and the properties in (3.8) follow directly from the definition of e_t in (3.3).

Similar reasoning can be applied to obtain the backward realization in (3.9), (3.10)$\triangle$.

Remark When comparing the SSR (3.7) to that in (2.4), one can see that the process v_t, w_t are generated in (3.7) by the same orthogonal process e_t, which is the so called innovation process. This explains why the SSR in (3.7) is called innovation SSR. It can be seen that the SSR-s in (2.7), (3.4) are not unique, and the forms of the F, G, H matrices depend on the particular choice of basis in X_t and X_t^B as it is known from deterministic realization theory. Since our interest is in specifying stochastic MFD and ARMA forms via transfer relations, there is a need for deriving unique canonical forms for the equivalent minimal (n-dimensional) SSR-s. These will be derived from the Wold-representations as follows.

It is known e.g. from Rozanov [35] that every purely linearly nondeterministic process y_t can be written in the infinite moving average (MA) form:

$$y_t = \sum_{k=0}^{\infty} K_k\varepsilon_{t-k}, \qquad \sum_{k=0}^{\infty} \| K_k \|^2 < \infty \tag{3.13a}$$

$$y_t = \sum_{k=1}^{\infty} K_k^B\varepsilon_{t+k}^B, \qquad \sum_{k=1}^{\infty} \| K_k^B \|^2 < \infty \tag{3.13b}$$

called forward and backward Wold-representations respectively, where $\varepsilon_t, \varepsilon_t^B$ are the normalized forward and backward innovations

$$\varepsilon_t = \Sigma^{-1/2} e_t, \qquad \varepsilon_t^B = \Sigma_B^{-1/2} e_t^B, \qquad t \in Z \tag{3.14}$$

having the properties that $Y_t = \mathbf{E}_t = \mathbf{E}_t^B$. i.e. the Hilbert-spaces spanned by $y_t, \varepsilon_t, \varepsilon_t^B$ are identical. Using the Wold representations the predictions $y_{t+k/t-1} = y_{t+k}/\mathbf{Y}_t^-$ $k \geq 0$ needed to construct the state space are very easy to calculate.

We will use the following notations:

$$\begin{aligned} E_t =& [e_{t-1}, e_{t-2}, \cdots]', \\ X_t =& [y_{t/t-1}, y_{t+1/t-1}, \cdots]' \\ X_{t+1/t} =& [y_{t+1/t-1}, y_{t+2/t-1}, \cdots]' \end{aligned} \tag{3.15}$$

i.e. the state spaces $\mathbf{X}_t, \mathbf{X}_{t+1}$ are spanned by the components of X_t, X_{t+1}, respectively, and the Hilbert space $\mathbf{Y}_{t+1}^+/\mathbf{Y}_t^-$ is spanned by the components of $X_{t+1/t}$.

It follows from (3.13) that the components of X_t are

$$y_{t+k/t-1} = \sum_{i=k+1}^{\infty} K_i e_{t-i+k}, \qquad k = 0, 1, \cdots, \tag{3.16}$$

and one can write

$$X_t = \mathbf{H} E_t, \tag{3.17}$$

where $\mathbf{H}$ is a Hankel-matrix

$$\mathbf{H} = \begin{bmatrix} K_1 & K_2 & K_3 & \dots \\ K_2 & K_3 & \dots & \\ K_3 & \dots & & \\ \dots & & & \end{bmatrix}. \tag{3.18}$$

It also follows that the components of x_{t+1} can be obtained as

$$y_{t+k/t} = y_{t+k/t-1} + K_k e_t, \qquad k = 1, 2, \ldots, \tag{3.19}$$

and the relationship between x_{t+1} and x_t corresdponding to the orthogonal decomposition in (3.12) is given by

$$X_{t+1} = X_{t+1/t} + \tilde{K} e_t, \tag{3.20}$$

where $\tilde{K}$ consists of the first block column of $\mathbf{H}$.

Since $\mathbf{Y}_{t+1}^+/\mathbf{Y}_t^-$ is a subspace of $\mathbf{X}_t$ any basis in $\mathbf{X}_{t+1/t}$ can be expressed by the linear combination of an appropriately choosen basis x_t of $\mathbf{X}_t$.

Before going into details of a particular selection of basis vectors for $\mathbf{X}_t$, we mention some results concerning the properties of the Hankel-matrix $\mathbf{H}$ and from the deterministic realization theory.

It is known, that if rank $\mathbf{H} = n$, then there exists triple (F, G, H) such that $\mathbf{H}$ cause factorized as

$$\mathbf{H} = \mathbf{OC} = \begin{bmatrix} H \\ HF \\ \vdots \end{bmatrix} [G\, FG \cdots], \tag{3.21}$$

where (H, F) and (F, G) are completely observable and controllable pairs respectively (i.e. rank$\mathbf{O}$ =rank$\mathbf{C} = n$), and

$$K_k = HF^{k-1}G, \qquad k = 1, 2 \ldots . \tag{3.22}$$

The $p \times n$, $n \times n$, $n \times p$ dimensional H, F, G matrices represent the n-dimensional minimal SSR of y_t.

The matrices $\mathbf{O}$ the $\mathbf{C}$ are called observability and controllability matrices, respectively.

Coming back to the choice of basis in $\mathbf{X}_t$ the following result will be used.

Lemma 3.1 Assume that the rational process y_t has a finite dimensional realization of minimum order n =rank $\mathbf{H}$. Let the vectors $h_1, \cdots, h_n$ be a basis selected from the rows of the Hankel matrix $\mathbf{H}$ (or equivalently let $v_1, \cdots, v_n$ be a basis of R^n selected from the rows of the observability matrix $\mathbf{O}$). Then the $y^i_{t+k/t-1}$ components of X_t associated with $h_1, \cdots, h_n$ (or $v_1, \cdots, v_n$) also form a basis for $\mathbf{X}_t$.

The proof of this Lemma can be obtained by the direct use of (3.17).

The basis selection stategies in the deterministic realization theory are called selection plans, see Kailath [25].

The basis selection from the rows of observability matrix when applying a selection plan (Scheme II selection plan in Kailath [25]) is performed by choosing the first linearly independent vectors from left to right in the following scheme:

$$\begin{array}{llll} h'_1, & h'_2, & \cdots, & h'_p \\ h'_1F, & h'_2F, & & \\ \cdots & & & \end{array} \tag{3.23}$$

A vector h'_iF^k is included in the selected vectors if it is linearly independent of each previously selected adjacent vectors. It can be deduced that among the selected vectors the following dependency relations hold:

$$h'_iF^{\nu_i} = \sum_{j=1}^{i-1} \sum_{k=0}^{\min(\nu_i, \nu_j - 1)} \alpha_{ijk} h'_jF^k + \sum_{j=i}^{p} \sum_{k=0}^{\min(\nu_i, v_j)-1} \alpha_{ijk} h'_jF^k \quad i = 1, \cdots, p,$$

$$\nu_1 + \nu_2 + \cdots + \nu_p = n = \dim\mathbf{X}_t. \tag{3.24}$$

Theorem 3.1 Let (H, F) be a competely observable pair and H is of full rank. Then $\{\nu_i, \alpha_{ijk}\}$ constitute a complete system of independent invariants for (H, F) with respect to the transformation $\bar{F} = TFT^{-1}, \bar{H} = HT^{-1}$ for arbitrary nonsingular matrix T.

The proof of this theorem can be obtained as the dual of the results in Popov [], Theorem 1. Using the above results, the SSR of y_t can be obtained from (3.20) as follows.

Proposition 3.2 Let the basis vectors selected by Scheme II selection plan from the rows of the observability matrix $\mathbf{O}$ be arranged in the following chain:

$$h_1', \; h_1'F, \cdots, h_1'F^{v_1-1}; \; h_2', \cdots, \quad , \cdots h_p'F^{v_p-1} \tag{3.25}$$

and define the associated basis for $\mathbf{X}_t$ as

$$\bar{x}_t = [y^1_{t/t-1}, \; y^1_{t+1/t-1}, \cdots, y^1_{t+v_1-1/t-1}; \; y^2_{t/t-1}, \cdots, \; y^p_{t+v_p-1/t-1}]' \, . \tag{3.26}$$

In this basis the rational process y_t has the forward Markovian SSR:

$$\overline{x}_{t+1} = F_c\overline{x}_t + G_c e_t \tag{3.27a}$$

$$y_t = H_c\bar{x}_t + e_t, \tag{3.27b}$$

$$F_c = \begin{bmatrix} 0 & 1 & & & \mid\,\mid & 0 & & & \\ \vdots & & & & \mid\,\mid & & & & \\ 0 & & & 1 & \mid\,\mid & & & & \\ \alpha_{110} & \alpha_{111} & \dots & \alpha_{11v_1} & \mid\,\mid & \alpha_{1p0} & \alpha_{1p1} & \dots & \\ -- & -- & -- & -- & \mid\dots\mid & -- & -- & -- & -- \\ & & & & \mid\,\mid & & & & \\ -- & -- & -- & -- & \mid\dots\mid & -- & -- & -- & -- \\ & 0 & & & \mid\,\mid & 0 & 1 & & \\ & & & & \mid\,\mid & \vdots & & & \\ & & & & \mid\,\mid & 0 & & & 1 \\ \alpha_{p10} & \alpha_{p11} & \dots & & \mid\,\mid & \alpha_{pp0} & \alpha_{pp1} & \dots & \alpha_{ppv_p} \end{bmatrix}; \quad G_c = \begin{bmatrix} K_1^1 \\ \dots \\ K_{v1} \\ -- \\ \\ -- \\ K_1^p \\ \dots \\ K_{vp}^p \end{bmatrix}$$

$$H_c = \begin{bmatrix} 1 & 0 & \dots & 0 & \mid & \mid & 0 & 0 & \dots & 0 \\ \vdots & \vdots & & \vdots & \mid & \mid & \vdots & \vdots & & \vdots \\ 0 & 0 & \dots & 0 & \mid & \mid & 1 & 0 & \dots & 0 \end{bmatrix}, \tag{3.28}$$

The backward stochastic SSR-s can be derived from the backward Wold-representation (3.13b) similarly to the procedure described above, but the results do not indicate how a specific forward SSR is related to a backward one. This relationship is provided by the following result.

Proposition 3.3 Let the forward minimal SSR $(F, G, H, \Sigma, \varepsilon_t)$ with $E\varepsilon_t\varepsilon_t' = I_p, Ex_tx_t' = P \geq 0, \; e_t, x_t, t \in Z$ satisfying the forward orthogonality properties, be given. Then the associated backward SSR $(F_B, G_B, H_B, D_B, e_t^B)$ is given by:

$$x_t^B = F'x_{t+1}^B + G_B e_t^B, \qquad y_t = H_B x_{t+1}^B + D_B e_t^B \tag{3.29}$$

where

$$H_B = HPF' + \Sigma^{1/2}G', \qquad G_B = [P^{-1} - F'P^{-1}F]^{1/2},$$
$$D_B = (H - H_BP^{-1})G_B^{-1}. \tag{3.30}$$

and

$$x_t^B = P^{-1}x_t,\ Ex_t^B(x_t^B)' = P^{-1} \geq 0, \quad e_t^B = G_B^{-1}(x_t^B - PF'x_{t+1}^B)$$
$$Ee_i^B(e_j^B)' = \delta_{ij}I_n, \quad Ee_k^B(x_t^B)' = 0, \quad t \geq k, \quad t,k \in Z \tag{3.31}$$

Proof This is based on the orthogonal decompositions $x_t = x_t/\overline{\mathbf{X}}_{t+k} + x_t - x_t/\overline{\mathbf{X}}_{t+k}$, where $\overline{\mathbf{X}}_{t+k} = \{x_{t+k},\ k \geq o\}$, and on the calculation of the projections y_t/v_{t+k}^B, $v_{t+k}^B = P^{-1}(x_t - x_t/\overline{\mathbf{X}}_{t+k})$, $k \geq 1.\triangle$

4. Stochastic matrix fraction descriptions

Matrix fraction descriptions /MFD/ have been introduced to solve various problems in the control literature as an alternative to the state-space approach, see e.g. the monographs of Wolovich [40] and Kailath [25] as representatives.

We consider the left MFD-s given by two polynomial matrices $P(z), Q(z)$ of $p \times p$ dimensions as:

$$y_t = P^{-1}(z)Q(z)e_t + e_t \tag{4.1}$$

where z is the forward shift operator, $Ee_je_j' = \delta_{ij}\Sigma$, $\det P(z) \neq 0$, and $P^{-1}(z)Q(z) = K(z)$ is a strictly proper rational transfer function.

Definition 4.1 The MFD (P, Q, Σ, e_t) in (4.1) is called minimal or irreducible, if

$$\deg \det P = \delta(K) = n, \tag{4.2}$$

where $\delta(K)$ is the Smith McMillan degree of $K(z)$.

It can be deduced, that the MFD in (4.1) is irreducible if P, Q are left coprime, see e.g. Kailath [25]. Irreducible MFD-s are also not unique but form equivalence classes.

Definition 4.2 Two irreducible MFD-s (P_1, Q_1) and (P_2, Q_2) are called equivalent, if there exists an unimodular matrix $U(z)$ such that $P_1 = UP_2$, $Q_1 = UQ_2$.

There are some alternative approaches to construct irreducible MFD-s. Forney's approach [15] is based on selecting a minimal polynomial basis for the right null space of the extended transfer function $[I_p|K(z)]$, Wolovich [40] and Kailath [25] showed the derivation of MFD-s from SSR-s. The canonic forms of MFD-s obtained from SSR-s are discussed e.g. Guidorzi [19,20], pseudocanonic forms are derived by Gevers and Wertz [16,18].

In the approach discussed here the irreducible MFD-s are obtained from stochastic canonic SSR-s via transfer relations. This reduces the problem of obtaining MFD-s for second order processes to an algebraic problem.

Proposition 4.1 Let the n-dimensional (minimal) SSR (F, G, H, Σ, e_t) be given in observable canonic form. Then the associated canonic MFD is given by the transfer relations

$$y_t = P^{-1}(z)Q(z)e_t + e_t, \tag{4.3}$$

where

$$P(z) = \text{diag}(z^{\nu_1}, \dots, z^{\nu_p}) - F_P S_P(z), \quad Q(z) = [S_Q(z) - F_P S(z)]G \tag{4.4}$$

the integers $\nu_1, \dots, \nu_p$ are the observability (left Kronecker) indices, $\Sigma_{i=1}^{p} \nu_i = n$, the elements of F_P are the invariant parameters $\{\alpha_{ijk}\}$,

$$F_P = \begin{bmatrix} \alpha_{110}, & \alpha_{111}, & \dots & \alpha_{11\nu_1-1} & |\dots| & \alpha_{p10}, & \alpha_{p11}, & \dots & \\ \cdot & & & & | \quad | & \cdot & & & \\ \cdot & & & & | \quad | & \cdot & & & \\ \cdot & & & & | \quad | & \cdot & & & \\ \alpha_{p10}, & \alpha_{p11}, & \dots & & |\dots| & \alpha_{pp0}, & \alpha_{pp1}, & \dots & \alpha_{pp\nu_p-1} \end{bmatrix} \tag{4.5}$$

and

$$S_P(z) = \text{blockdiag}\{(1, z, \dots, z^{\nu_1-1})', \dots, (1, z, \dots, z^{\nu_p-1})'\}, \tag{4.6a}$$

$$S_Q(z) = \text{blockdiag}\{(z^{\nu_1-1}, \dots, z, 1), \dots, (z^{\nu_p-1}, \dots, z, 1)'\}, \tag{4.6b}$$

$$S(z) = S + zS^2 + \dots + z^{\nu-2} S^{\nu-1}, \qquad \nu = max\ \nu_i \tag{4.6c}$$

$$S = \text{blockdiag}\{S_{\nu_1}, \dots, S_{\nu_p}\}, \tag{4.6d}$$

S_k is a $k \times k$ dimensional Toeplitz-matrix

$$S_k = \begin{bmatrix} 0 & 0 & . & . & . & 0 \\ 1 & 0 & & & & . \\ . & . & & & & . \\ . & . & . & & & . \\ . & & . & . & & . \\ 0 & & & . & . & 0 \end{bmatrix}.$$

Proof This is constructive, the key point is to find a one-to-one relationship between the state vector x_t of the SSR and a basis $\mathbf{Y}_t^+$ by utilizing the concept of observability of the state from most recent future outputs.△

Given a backward SSR, the associated MFD can be derived analogously.

5. Stochastic autoregressive-moving average /ARMA/ forms

Rational stochastic process are very frequently described in (forward) ARMA forms given by

$$A_0 y_t + A_1 y_{t-1} + \dots + A_p y_{t-p} = C_0 e_t + C_1 e_{t-1} + \dots + C_q e_{t-q}, \tag{5.1}$$

where $Ee_i e_j' = \delta_{ij}\Sigma$, and A_i, C_j are $p \times p$ dimensional constant matrices.

The ARMA form (5.1) is called *monic* if $A_0 = I_p$, and applying the commonly used "stochastic normalization convention", one can choose $C_0 = I_p$. In this case (5.1) can be rewritten as

$$y_t = A^{-1}(d)B(d)e_t + e_t \tag{5.2}$$

where d is the backward shift operator and $A(d), B(d)$ are polynominal matrices, $detA(d) \neq 0$, and the coefficient matrices in $B(d)$ are in an obvious relations with the MA coefficients $C_j, j = 0, \dots, q$.

Definition 5.1 The ARMA form (5.2) is called irreducible if

$$\deg \det A = \delta(K) = n,$$

where $\delta(K)$ is the McMillan-degree of the rational matrix $K(d) = A^{-1}(d)B(d)$.

Definition 5.2 Two irreducible ARMA forms (A_1, B_1) and (A_2, B_2) are called equivalent if there exists an unimodular matrix $U(d)$, such that $A_1 = UA_2, B_1 = UB_2$.

The use of monic ARMA forms is dominant e.g. in system identification, prediction and certain control applications and received a renewed interest, see Gevers and Wertz [16], Deister and Gevers [12]. Here we consider mainly the construction of forward and backward monic ARMA forms.

In earlier papers the deterministic ARMA forms where derived from left MFD-s, see e.g. Wolovich and Elliott [41], Bokor and Keviczky [6]. Gevers [17], however, pointed out that this procedure generally leads to nonmonic ARMA forms. Janssen [23,24] applied Forney's approach [15] to obtain ARMA forms from $K(d)$ and concluded that the resulting ARMA forms are generally not monic. Bokor and Keviczky [7] considered the realization of deterministic ARMA forms from constructibility canonic SSR-s. They showed, that in the case when the matrix F in the (F, G, H) SSR is invertible, monic ARMA forms can always be obtained via transfer relations. This concept can be extended to obtain forward monic ARMA forms from stochastic forward SSR-s.

In addition, it will be shown here, that the backward monic ARMA forms are naturally related to backward stochastic SSR-s via transfer relations. This is formally stated as follows.

Proposition 5.1 Let the n-dimensional (minimal) backward SSR $(F_B, G_B, H_B, \Sigma_b, e_t^B)$ be given in observable canonic form, with F_B invertible. Then the associated canonic backward ARMA form is given by the transfer relations

$$y_t = A_B^{-1}(d)B_B(d)e_t^B + e_t^B,$$

where

$$A_B(d) = I_p - H_B F_B^{-1} S_A(d); \qquad B_B(d) = H_B F_B^{-1} S_B(d) G_B$$

and

$$S_A(d) = \text{blockdiag}\left\{[d, d^2, \ldots, d^{\nu_1}]', \ldots, [d, d^2, \ldots, d^{\nu_p}]'\right\},$$

$$S_B(d) = \sum_{i=1}^{\nu} S^{i-1} d^i, \qquad \nu = \max \nu_i$$

and S is the block Toeplitz matrix given by (4.6 d).

Proof. This is constructive, and it is based on estabiliting a one-to-one correspodence between the state vector x_t^B of the backward SSR and a basis of $\mathbf{Y}_t^-$ by utilizing the concept of observability of the state from most recent past outputs.

$\downarrow y_t$

$\downarrow$		$\downarrow$
{Forward Wold representation}		{Backward Wold representation}
$\downarrow$	$X_t \simeq X_t^B$	$\downarrow$
{ Irreducible forward SSR } in canonic form	$\longleftarrow\text{-}\text{-}\text{-}\longrightarrow$ (F^{-1} exists)	{ Irreducible backward SSR } in canonic form
$\downarrow$		$\downarrow$ F_B^{-1} exists
{Irreducible forward left MFD} canonic form	$\longleftarrow\text{-}\text{-}\text{-}\longrightarrow$	{ Irreducible backward monic } ARMA canonic form

Fig.1

6. Correspondences among the stochastic realizations

The correspondence and duality between the above discussed stochastic realizations can be illustrated by the following diagram (see Fig.1).

Remark. In case of F is invertible, forward monic ARMA canonic forms can be derived from forward SSR using constructibility concepts. Backwards left MFD canonic forms can similarly be derived from backward SSR-s.

Since each strong realization specify also weak stochastic realization, one can establish the above correspondences in terms of extremal weak realizations too.

Assume that the forward SSR (F, H, e_t) is obtained from the Wold-representation as described by Proposition 3.2, and (F, H) are fixed. Assume also that (F_B, H_B, e_t^B) are obtained from (F, H, e_t) using Proposition 3.3. Then the following correspondences can be given.

$$\begin{array}{ccc} Min.FSSR & \xrightarrow{F^{-1} exists} & Max.BSSR \\ \downarrow & & \uparrow \\ Min.FMFD & \longrightarrow & Max.monicBARMA \end{array}$$

where the notations Min.FSSR, Max.BSSR, Min.FMFD and Max.monicBARMA stand for minimal forward SSR, maximal backward SSR, minimal forward MFD and maximal backward monic ARMA representations respectively. On minimal SSR we mean an SSR having minimal state covariance matrix P_* in the sense of (2.16).

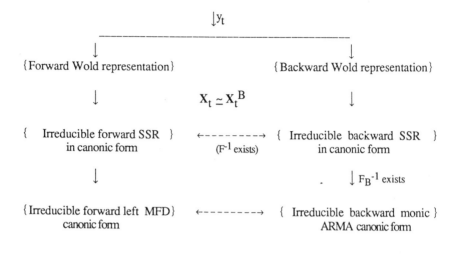

Applying constructibility concepts (F^{-1} exists) the following correspondences can also be given.

$$\begin{array}{ccc} Min.constFSSR & \longrightarrow & Max.BSSR \\ \downarrow & & \uparrow \\ Min.monicFARMA & \longrightarrow & Max.BMFD \end{array}$$

where Min.constFSSR, Min.monicFARMA and Max.BMFD denotes minimal forward SSR in constructibility canonical form, minimal forward monic ARMA and maximal backward MFD forms respectively.

7. Conclusions

The construction of weak and strong realizations of wide sense stationary multivariable stochastic processes was discussed. The state-space, matrix fraction and ARMA realizations can be used to solve various identification, prediction, smoothing and control problems associated with stochastic signals and systems. The transfer relations derived can be applied to find a particular form of realizations that is the most conveniently suited to solve the given problem.

References

[1] Akaike, H. (1974), *Stochastic Theory of Minimal Realization*, IEEE Transactions on Automatic Control **AC-19**, pp. 667–674.

[2] Akaike, H. (1975), *Markovian Representation of Stochastic Processes by Canonical Variables*, SIAM Journal of Control **13**, pp. 162–173.

[3] Anderson, B.D.O. (1967), *A System Theory Criterion for Positive Real Matrices*, SIAM Journal of Control **5**, pp. 171–172.

[4] Anderson, B.D.O. (1967), *An Algebraic Solution to the Spectral Factorization Problem*, IEEE Transactions on Automatic Control **AC-12**, pp. 410–414.

[5] Anderson, B.D.O., and T. Kailath (1979), *Forwards, Backwards and Dynamically Reversible Markovian Models of Second-Order Processes*, IEEE Transactions on Circuits and Systems **CAS-26**, pp. 956–965.

[6] Bokor, J. and L. Keviczky (1982), *Structural Properties and Structure Determination of Vector Difference Equations*, International Journal of Control **36**, pp.461–475.

[7] Bokor, J. and L. Keviczky (1987), *ARMA Canonical Forms Obtained from Constructibility Invariants*, International Journal of Control **45**, pp.861–873.

[8] Bokor J., Cs. Bányász and L. Keviczky (1989) Realization of Stochastic Processes in State-Space, MFD and ARMA Forms, in "Proceedings of the IEEE Conference on Control and Applications, ICCON '89," Jerusalem, pp. WC-5-4.

[9] Caines, P.E. (1988), "Linear Stochastic Systems," John Wiley & Sons.

[10] Caines, P.E. and D.Delchamps (1980), *Splitting Subspaces, Spectral Factorization and the Positive Real Equation: Structural Features of the Stochastic Realization Problem*, in "Proceedings of the IEEE Conference on Decision and Control," Albuquerque, NM, pp. 358–362.

[11] Deistler, M. (1983), *The Structure of ARMA Systems and its Relation to Estimation*, in "Geometry and Identification," Caines, P.E. and R. Hermann (Eds.), Mathematical Sciences Press, Brookline, MA, pp. 49–61.

[12] Deistler, M. and M. Gevers (1989), *Properties of the Parametrization of Monic ARMAX Systems*, Automatica **25**, pp.87–96.

[13] Desai, U.B., and Pal, D. (1984), *A Transformation Approach to Stochastic Model Reduction*, IEEE Transactions on Automatic Control **AC-29**, pp.1097–1100.

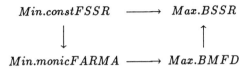

[14] Faurre, P.L. (1976), *Stochastic Realization Algorithms*, in "System Identification: Advances and Case Studies," Mehra, R.K. and D.R. Lainiotis (Eds.), Academic Press, New York, pp. 1–25.

[15] Forney, G.D. (1975), *Minimal Bases of Rational Vector Spaces, with Applications to Multivariable Linear Systems*, SIAM Journal of Control **13**, pp.493–520.

[16] Gevers, M. and V. Wertz (1984), *Uniquely Identifiable State-Space and ARMA Parametrization for Multivariable Linear Systems*, Automatica **20**, pp.333–347.

[17] Gevers, M. (1986), *ARMA Models, Their Kronecker Indices and Their McMillan Degree*, International Journal of Control **43**, pp.1745–1761.

[18] Gevers, M. and V. Wertz (1987), *Techniques for the selection of identifiable parametrizations for multivariable linear systems*, in "Control and Dynamic Systems, Vol. XXVI," C.T. Leondes (Ed.), Academic Press, pp. 35–86.

[19] Guidorzi, R.P. (1975), *Canonical Structures in The Identification of Multivariable Systems*, Automatica **11**, pp.361–374.

[20] Guidorzi, R.P. (1981), *Invariants and Canonical Forms for Systems: Structural and Parametric Identification*, Automatica **17**, pp.117–133.

[21] Hannan, E.J., and M. Deistler (1988), "The Statistical Theory of Linear Systems," John Wiley & Sons, New York.

[22] Ho, L. and R.E. Kalman (1965), *Effective Construction of Linear State Variable Models from Input/Output Data*, in "Proceedings of the 3rd Allerton Conference," pp. 449–459.

[23] Janssen, P. (1987), *MFD Models and Time Delays; Some Consequences for identification*, International Journal of Control **45**, pp.1179–1196.

[24] Janssen, P. (1988), *General Results on the McMillan Degree and Kronecker Indices of ARMA and MFD models*, To appear in the International Journal of Control.

[25] Kailath, T. (1980), "Linear Systems," Prentice-Hall, Englewood Cliffs, NJ.

[26] Kalman, R.E., P. Falb and M. Arbib (1969), "Topics in Mathematical System Theory," McGraw-Hill, New York.

[27] Lindquist, A., and G. Picci (1977), *On the structure of minimal splitting subspaces in Stochastic realization Theory*, in "Proceedings of the Conference on Decision and Control New Orleans, La.," pp. 42–48.

[28] Lindquist, A., and G. Picci (1979), *On the Stochastic Realization Problem*, SIAM Journal of Control Theory **17**, pp.365–389.

[29] Lindquist, A., and G. Picci (1981), *State Space Models for Gaussian Stochastic Processes*, in "Stochastic Systems: The Mathematics of Filtering and Identification and Applications," Hazewinkel, M. and J.C. Willems (Eds.), Reidel, Dordrecht, pp. 169–204.

[30] Lindquist, A., and Pavon, M. (1984), *On the Structure of State-Space Models for Discrete-Time Stochastic Vector Processes*, IEEE Transactions on Automatic Control **AC-29**, pp.418–432.

[31] Ljung, L. and T. Kailath (1976), *Backwards Markovian Models for Second-Order Stochastic Processes*, IEEE Transactions on Information Theory **IT-22**, pp.488–491.

[32] Pavon, M. (1980), *Stochastic Realizations and Invariant Directions of Matrix Riccati Equations*, SIAM Journal of Control and Optimatization **18**, pp.155–180.

[33] Picci, G. (1976), *Stochastic Realization of Gaussian Processes*, Proc. IEEE **64**, pp.112–122.

[34] Popov, V.M. (1972), *Invariant Description of Linear Time Invariant Controllable Systems*, SIAM Journal of Control **10**, pp.252–264.

[35] Rozanov, Yu. A. (1967), "Stacionary Random Processes," Holden-Day, San-Francisco.

[36] Sidhu, G.S., and Desai, U.B. (1976), *New Smoothing Algorithms Based on Reversed-time Lumped Model*, IEEE Transactions on Automatic Control, pp.538–541.

[37] Verghese, G., and Kailath, T.(1979), *A Further Note on Backwards Markovian Models*, IEEE Transactions on Information Theory **IT-25**, 121–124; *Correction to the above paper in ibid*, **IT-25**, p. 50.

[38] Wiener, N. (1958), *The Prediction Theory of Multivariate Stochastic Processes, Part I: The regularity condition*, Acta Mathematica **98**, pp.111–150; *Part II: The linear predictor*, Acta Mathematica **99**, pp.93–137.

[39] Willems, J.C. (1986), *From Time Series to Linear Systems, Part I*, Automatica **22**, pp.561–580; *Part II*, Automatica **22**, pp.675–694; *Part III*, Automatica **23**, pp.87–115.

[40] Wolovich, W.A. (1974), "Linear Multivariable Systems," Applied Mathematical Sciences, No. 11, Springer- Verlag, New York.

[41] Wolovich, W.A. and H. Elliott (1983), *Discrete Models for Linear Multivariable Systems*, International Journal of Control **38**, pp.337–357.

[42] Youla, D.C. (1961), *On the factorization of Rational Matrices*, IEEE Transactions on Information Theory **IT-7**, pp.172–189.

CONTRIBUTIONS TO NONLINEAR INVERSE PROBLEMS ARISING IN PARAMETER ESTIMATION FOR ELLIPTIC SYSTEMS*

K. Kunisch

Abstract

In this paper various aspects of the problem of determining the diffusion coefficient in elliptic equations from knowledge of the state are considered. These include identifiability and stability of the coefficient with respect to the problem data, Tikhonov regularization and numerical techniques for practical determination of the coefficient from given data.

1. Introduction

The purpose of this paper is to discuss recent contributions to the estimation of parameters in elliptic equations from knowledge of the state. We shall focus on the equation

$$(1.1)\qquad \begin{cases} -\operatorname{div}(a\,\operatorname{grad} u) + cu = f \text{ in } \Omega \\ \text{boundary condition,} \end{cases}$$

where Ω is a bounded domain in R^2 or R^3 with sufficiently smooth boundary $\partial\Omega$. The "boundary condition" will only be

*This work was carried out while the author visited the Division of Applied Mathematics at Brown University. Support through the Fonds zur Förderung der wissenschaftlichen Forschung under S3206 and P6005, and through AFOSR-URI-F49620-86-C-0111 is acknowledged.

specified when it is required by the context. If the scalar-valued functions a, c and f are known and if a is uniformly positive, then (1.1) describes a wellknown elliptic equation for the state variable u. In this paper our attention is directed to the inverse problem: given u and only some of the parameters a, c and f, determine the remaining unknown parameters. In applications, as for instance in groundwater flow modeling [Y] or in the geophysical sciences [M], u is not known on all of Ω, but rather only at discrete locations in Ω. In this case we suppose that the pointwise data are interpolated and give rise to a function z defined on all of Ω. This function may also be thought of as an error corrupted perturbation of the solution u of (1.1) evaluated at the "true", but unknown model parameters.

In the case that u, a and c are known, one would address the linear inverse problem of identifying f in (1.1). If, on the other hand, u and f are known with a and/or c unknown, then the inverse problem is nonlinear. In this paper we concentrate on nonlinear inverse problems and assume throughout that $f \in H^{-1}(\Omega)$ in known.

Inverse problems are frequently considered in their "output least squares" formulation, i.e. one considers

$$(P) \qquad \min_{q \in Q_{ad}} |u(q) - z|_Z^2,$$

where $|\cdot|_Z$ denotes the norm in some appropriate normed linear space, Q_{ad} is the set of admissible parameters, and $u(q)$ denotes the solution of (1.1) as a function of the unknown coefficient, which in our case is a, c or the pair (a, c). It will be convenient to also introduce the attainable set V, which is defined as $V = \{u(q) : q \in Q_{ad}\}$. In general, the set V is not convex; for an example see e.g., [BK, Chapter 4].

In this paper we address the question of parameter identifiability i.e. the injectivity of the parameter-to-solution mapping $\Phi : q \to u(q)$, as well as aspects of stability, which refers to the continuous invertibility of Φ, or, more generally, the continuous dependence of the solutions q^* of (P) on the data z. We also discuss numerical schemes that allow one to approximate the unknown coefficient from the data z in a stable manner. To ensure such a stability the original inverse problem frequently requires some type if modification. In this context we describe the use of Tikhonov regularization for nonlinear inverse problems.

Progress has been made recently in several other directions of inverse problems related to the estimation of coefficients in distributed systems such as (1.1). These include identifiability and stability results when only boundary measurements are available [FV], [KV], the development of statistical tests to evaluate numerical results for the approximation of the unknown parameters when appropriate assumptions on the statistical distribution of the data z are made [F], and the development of a theory of projections onto non-convex sets and its application to nonlinear least squares problems where the non-convex set is given by the attainable set V. For further references of related work we refer to an earlier survey article [K2] and to the monograph [BK], for example.

2. Identifiability

While identifiability of one coefficient such as a or c from knowledge of u was studied previously by several authors, see [G], [K1], [R], identifiability of more than one coefficient in an elliptic equation has only been considered recently. In this section we survey some results from [BaK].

Let us fix a pair of reference coefficients $\bar{q} = (\bar{a}, \bar{c})$. It can easily be seen that it is in general impossible to determine $\bar{q}$ uniquely from knowledge of just one solution $u(\bar{q})$ and we therefore search for conditions such that knowledge of the states corresponding to two inhomogeneities f^1 and f^2 or two boundary conditions allows to determine $\bar{q}$ uniquely. We shall in fact obtain certain stability estimates.

To describe the first class of results, let $u(a, c)$ denote the solution of (1.1) with f replaced by f^1 and analogously let $v(a, c)$

be the solution of (1.1) with f replaced by f^2. Further we put

$$D = u(\bar{q}) \bigtriangledown v(\bar{q}) - v(\bar{q}) \bigtriangledown u(\bar{q}).$$

We shall require the following hypotheses.

(H1) (a, c) and $(\bar{a}, \bar{c})$ are in Q_{ad}, where
$Q_{ad} = \{(\tilde{a}, \tilde{c}) \in W^{1,4} \times H^1 : \tilde{a}(x) \le m, |a|_{H^1} \le E, |c|_{H^1} \le E\}$,
with E and m given constants,

(H2) $u(q), v(q), u(\bar{q})$, and $v(\bar{q})$ are solutions of (1.1) in H^2,

(H3) $(a - \bar{a})^2 D \cdot \nu \ge 0$ a.e. on ∂r, with ν the outer normal to $\partial\Omega$,

(H4) there exists a constant $\rho > 0$ such that $|v(\bar{q})(x)| \ge \rho$ for all $x \in \Omega$.

Theorem 1. If (H1)-(H4) hold, then for any $\lambda \in R$ there exists a constant K depending continuously on $(\lambda, \rho, m, E, |u(\bar{q})|_{H^2}, |v(\bar{q})|_{H^2})$, such that

$$\begin{aligned} &< (a - \bar{a})^2, \exp(\lambda \tfrac{u(\bar{q})}{v(\bar{q})})(\bigtriangledown \cdot D + \lambda \tfrac{|D|^2}{v(\bar{q})^2}) >_{L^2} \\ &\le K(|u(q) - u(\bar{q})|_{H^1} + |v(q) - v(\bar{q})|_{H^1}) + \\ &\quad K\delta(|u(q) - u(\bar{q})|_{H^2} + |v(q) - v(\bar{q})|_{H^2}, \end{aligned}$$

where $\delta = \max\{|a(x) - \bar{a}(x)| : x \in \partial\Omega\}$.

The a-priori bound $|c|_{H^1} \le E$ in (H1) can be replaced by $|c|_{L^2} \le E$ for Theorem 1. It is, however, required for the following result, which provides an a-priori bound on a and c in terms of the solutions of (1.1), if additional hypotheses are satisfied.

(H5) There exist constants $\lambda_0 \in R$ and $\kappa > 0$, such that
$(\bigtriangledown \cdot D + \lambda_0 \frac{|D|^2}{v(\bar{q})^2}(x) \ge \kappa$ for almost all $x \in \Omega$,

(H6) $v(\bar{q}) \in C^1$,

(H7) $a \bigtriangledown v(q) \cdot \nu = \bar{a} \bigtriangledown v(\bar{q}) \cdot \nu$ on $\partial\Omega$.

Theorem 2. If (H1)-(H7) hold then there exists a constant $K = K(\lambda_0, \kappa, \rho, m, E, |u(\bar{q})|_{H^2}, |v(\bar{q})|_{W^{1,\infty}})$ such that

$$\begin{aligned} |a - \bar{a}|_{L^2} + |c - \bar{c}|_{L^2} &\le K(|u(q) - u(\bar{q})|_{H^1} + |v(q) - v(\bar{q})|_{H^1} \\ &+\delta(|u(q) - u(\bar{q})|_{H^2} + |v(q) - v(\bar{q})|_{H^2}))^{\frac{1}{4}}. \end{aligned}$$

Observe that this result provides conditions for the injectivity of the mapping $(a,c) \to (u(a,c), v(a,c))$ at $(\bar{a}, \bar{c})$. Condition (H5) is satisfied, for example, if $u(\bar{q}) \in C^2, v(\bar{q}) \in C^2$ (which in turn implies that $f^1 v(\bar{q}) - f^2 u(\bar{q}) \in C$), and

$$\inf_{x\in\Omega}(f^1 v(\bar{q}) - f^2 u(\bar{q}))(x) > 0.$$

The proofs of Theorem 1 and 2 are based on a variational technique, employing a variation of a transformation of the state variables u and v that was originally used in [KL].

In a different approach to derive results comparable to Theorem 1 and 2 one first eliminates c from (1.1) and then studies the resulting hyperbolic equation in a. We consider (1.1) with a Neumann boundary condition:

$$\text{(1.2)} \qquad \begin{aligned} &-\nabla\cdot(a\nabla u) + cu = f_i \quad \text{in}\,\Omega \\ &a\nabla u\cdot\nu = g_i \quad \text{on}\,\partial\Omega, \end{aligned}$$

for $i = 1,2$, where $(f_i, g_i) \in H^1(\Omega)^* \times H^{-\frac{1}{2}}(\partial\Omega)$. The set of admissible parameters is now restricted so as to guarantee existence of a solution $u(q)$ if q is in

$$\tilde{Q}_{ad} = \begin{cases} (a,c) \in (H^1 \cap L^\infty) \times H^1 : m_1 \le a(x) \le m_2 \,\text{a.e;} \\ k \le c(x)\,\text{a.e.}, |a|_{H^1} \le E, |c|_{H^1} \le E\}, \end{cases}$$

where $0 < m_1 \le m_2, 0 < k$ and E are given constants such that $\tilde{Q}_{ad}$ is nonempty.

Theorem 3. Assume that q and $\bar{q}$ are in $\tilde{Q}_{ad}$ and that $(u(\bar{q}), v(\bar{q})) \in C^2 \times C^2$. If there exists a constant $\mu > 0$ such that

$$|v(\bar{q})(x)| \ge \mu \,\text{and}\, |D(x)| \ge \mu \,\text{on}\, \bar{\Omega},$$

then we have

$$|a-\bar{a}|_{L^2} + |c-\bar{c}|_{L^2} \le \hat{m}(|u(a,c) - u(\bar{a},\bar{c})|_{H^1}^{\frac{1}{2}} + |v(a,c) - v(\bar{a},\bar{c})|_{H^1}^{\frac{1}{2}})$$

for some constant $\hat{m}$ independent of $(a,c) \in \tilde{Q}_{ad}$.

The condition $|D(x)| \geq \mu > 0$ on $\bar{\Omega}$ holds for example, if $\{\nabla u(\bar{q})(x), \nabla v(\bar{q})(x)\}$ are linearly independent for all $x \in \bar{\Omega}$.

3. Stability

In this section we focus on the question of the stability of the solutions of the least squares problem (P) with respect to perturbation of z. Our analysis if based on the fact that if a higher order (here second order) sufficient optimality condition holds at a local minimum of an optimization problem, and if this minimum is a regular point, then it is continuous with respect to perturbations in the problem data. To explain such a result more precisely consider the parameter dependent optimization problem

$$(P)_w \qquad \min f(x,w) \quad \text{with}\, g(x) \in -K,$$

where $w \in W$, and W is a metric space with metric $d(\cdot,\cdot)$. Here we use the mappings $f : D \times W \to R, g : Q \to Y$, with D an open subset of a Banach space Q. Further K is a closed convex cone with vertex in Y and Y is a Banach space. We are concerned with the continuous dependence of a local solution x_0 of $(\mathrm{P})_{w_0}$, $w_0 \in W$, with respect to perturbations of w_0. Throughout it is assumed that f and g are twice continuously Fréchet differentiable with respect to x at w_0 and that f is uniformly continuous with respect to w at w_0; i.e., for all $\epsilon > 0$ there exists $\delta > 0$ such that for all $x \in Q_{ad}$ and all $w \in W$

$$(3.1) \qquad d(w,w^0) < \delta\, \text{implies}\, |f(x,w) - f(x,w^0)| < \epsilon.$$

We denote by $\epsilon(\delta)$ the corresponding modulus of continuity. Thus, for some $\delta_0 > 0$ and all $\delta \in (0,\delta_0)$

$$\epsilon(\delta) := \inf\{\epsilon > 0 : \ (3.1)\, \text{holds for all}\, x \in Q_{ad}\}$$

is well defined and positive. Moreover, we have $\epsilon(\delta) \to 0^+$ for $\delta \to 0^+$. We also require x_0 to be a regular point with respect to the constraint $g(x) \in -K$, i.e.

$$(3.2) \qquad 0 \in \operatorname{int}(g(x_0) + g_x(x_0)Q + K),$$

where g_x denotes the Fréchet derivative with respect to x. These assumptions guarantee the existence of a Lagrange multiplier $y^* \in Y^*$, where Y^* is the topological dual of Y, such that $f_x(x_0, w_0) + y^* g_x(x_0) = 0$, $y^* \in K^*$, and $y^* g(x_0) = 0$, with $K^* = \{y^* \in Y^* :< y, y^* >\geq 0$, for all $y \in K\}$. The following second order optimality condition will be used:

(SC) there exists a seminorm M on Q with $M(x) \leq m_0|x|$ for some $m_0 > 0$ independent of $x \in Q$, and there are constants $\alpha > 0$ and $r > 0$ such that $[f_{xx}(x, w_0) + y^* g_{xx}(x)](v - x_0, v - x_0) \geq \alpha M(v - x_0)^2$, for all $v \in Q$ satisfying $g(v) \in -K$ and all x satisfying $g(x) \in -K$ and $M(x - x_0) \leq r$.

In the statement of the following theorem we employ the $M-$local extremal value function, which is defined by

$$\mu_r(w) = \inf\{f(x, w) : g(x) \in -K, \quad M(x - x_0) \leq r\}.$$

Theorem 4. Let x_0 be a local solution of $(\mathrm{P})_{w_0}$ satisfying (SC) and let $\delta_0 > 0$ be chosen sufficiently small. Suppose that for all w with $d(w, w_0) \leq \delta_0$ there exists x_w with $g(x_w) \in -K, M(x_w - x_0) \leq r$ and $\mu_r(w) = f(x_w, w)$. Then, for all w with $d(w, w_0) \leq \delta_0$, we have $M(x_w - x_0) < r$, i.e., x_w is a local (in the topology induced by the seminorm M) solution of $(\mathrm{P})_w$ and

$$M(x_w - x^0) \leq (\frac{2}{\alpha})^{\frac{1}{2}} \epsilon(d(w, w_0))^{\frac{1}{2}}.$$

This theorem, which is proved in [CK2] differs form related results on the stability of the solutions of abstract optimization problems, see e.g. [Aℓ], in that Q is endowed with two different topologies. One of them arises by choosing Q and its (strong) topology such that for $x \in D \subset Q$ the function $f(x, w)$ is well-defined and such that existence of a solution of $(\mathrm{P})_{w_0}$ is guaranteed. The second, weaker topology is chosen such that

(SC) can be shown. For ill-posed inverse problems these two topologies do not coincide.

We now describe the applicability of Theorem 4 to the problem of estimating the diffusion coefficient a in a two point boundary value problem. The equation under consideration is

$$\begin{aligned} -(au_x)_x + cu &= f \text{ in } (0,1), \\ u(0) = \beta_1, u(1) &= \beta_2, \end{aligned} \tag{3.3}$$

where $f \in L^2, c \in L^2, c \geq 0$, and $\beta_i \in R, i = 1,2$, are given. All function spaces in the remainder of this section are considered over the interval (0,1). We shall concentrate on a simple case here, referring to [CK2] for a more detailed investigation. Let us assume that the unperturbed observation z_0 is attainable by a coefficient $a^* \in H^1$ with $a^* \geq \nu > 0$, so that

$$z_0 = u(a^*). \tag{3.4}$$

Observe that the regularity properties of a, c and f imply that $u(a^*) \in H^2 \subset C^1$. We further assume that there are constants k_1 and k_2 such that

$$0 < k_1 < u_x(a^*)(x) < k_2 \text{ for all } x \in (0,1), \text{ and } 2k_1 > k_2. \tag{3.5}$$

Theorem 4 then allows to discuss the stability of the least squares method for the recovery of a^* under the assumption that the mean $\int_0^1 a^*(x)dx$ is known. We consider

$$\min |D(u(a) - z)|_{L^2}^2 \text{ over } a \in Q_{ad}, \tag{P_z}$$

where $Q_{ad} = \{a \in H^1 : a \geq \zeta, |a|_{H^1} \leq \gamma, \int_0^1 adx = \int_0^1 a^*dx\}$, and D stands for differentiation. It is assumed that $0 < \zeta < \int_0^1 a^*dx < \mu$ and $|a^*|_{H^1} \leq \gamma$. It is simple to argue existence of a solution a_z of (P_z) for any $z \in H^1$ and clearly a^* is a solution of (P_{z_0}). While we assumed that z_0 was attainable, it will not be necessary to require attainability of the perturbed observations z. To apply Theorem 4 to (P_z), we put

$$Q = H^1, W = H^1, Y = H^1 \times R \times R,$$

$$K = \{\phi \in H^1 : \phi \geq 0\} \times R^+ \times \{0\}, w_0 = z_0,$$

$$f(a,z) = |D(u(a)-z)|_{L^2}^2, \quad g(a) = (\zeta - a, |a|_{H^1}^2 - \gamma^2, \int_0^1 (a-a^*)dx)$$

and choose D as an open subset of Q_{ad}. Clearly f and g satisfy the smoothness requirements with respect to a and it can also be shown that the modulus of continuity of f with respect to z (see (3.1) and below) is $\epsilon(\delta) = K\delta$, with K independent of $\delta \in (0, \delta_0), \delta_0 > 0$ and fixed. The element a^* satisfies the regular point condition (3.2). This follows from $\zeta < \int_0^1 a^* dx < \mu$ and Lemma 2.10 in [CK2] (with $h_0 = 1$). Hence there exists a Lagrange multiplier $(\lambda^*, \mu^*, \eta^*) \in K^*$ with respect to the constraint $g(a) \in -K$ at a^*. To verify (SC) we calculate the second Fréchet derivative of the Lagrange functional at $a \in Q_{ad}$ in direction $(h, h) \in H^1 \times H^1$ with $\int_0^1 h dx = 0$:

$$\begin{aligned} &f_{aa}(a, z_0)(h,h) + < (\lambda^*, \mu^*, \eta^*), g_{aa}(a)(h,h) > = \\ &2|Du_a(a)(h)|^2 + 2 < D(u(a) - z_0), Du_{aa}(a)(h,h) > \\ &+ 2\mu^* |h|_{H^1}^2. \end{aligned} \tag{3.6}$$

We shall estimate the three terms on the right hand side of (3.6) separately. The last term is nonnegative, since $\mu^* \geq 0$. To estimate the first term , let us observe that

(3.7) there exists $r_1 > 0$ such that $|a - a^*|_{L^2} \leq r_1$ and $a \in Q_{ad}$ imply $0 < k_1 \leq u_x(a)(x) \leq k_2$ on (0,1).

In fact, if $\{a_n\}$ is any sequence in Q_{ad} converging strongly in L^2 to an element $\tilde{a} \in Q_{ad}$, then $\{a_n\}$ converges weakly in H^1 to $\tilde{a}$. Consequently $\{u(a_n)\}$ converges weakly in H^2 and hence strongly in C^1 to $u(\tilde{a})$. Together with (3.5) this implies (3.7). We shall use the operators $A(a) : H_0^1 \to H^{-1}, a \in H^1, a > 0$, given by $A(a)\phi = D(aD\phi)$. The operator $DA(1)^{-1}D$ from H^1 to L^2 has an extension as a bounded linear operator from L^2 to L^2, which coincides with the L^2-orthogonal projection P from L^2 onto $\{\phi \in L^2 : \int_0^1 \phi dx = 0\}$. In the following estimate κ

denotes a generic constant that is independent of $a \in Q_{ad}$ with $|a - a^*|_{L^2} \le r_1$ and $h \in H^1$ with $\int_0^1 h dx = 0$:

$$|Du_a(a;h)|_{L^2}^2 = |DA^{-1}(a)(hu_x(a))_x|_{L^2}^2 \ge \kappa|DA^{-1}(1)D(hu_x(a))|_{L^2}^2$$

$$= \kappa|P(hu_x(a))|_{L^2}^2 = \kappa(|hu_x(a)|_{L^2}^2 - (\int_0^1 hu_x(a)dx)^2).$$

Denoting $\Omega_+ = \{x \in (0,1) : h(x) \ge 0\}$ and $\Omega_- = \{x \in (0,1) : h(x) < 0\}$ we find

$$\int_0^1 hu_x(a)dx = \int_{\Omega_+} hu_x(a)dx + \int_{\Omega_-} hu_x(a)dx \le k_2 \int_{\Omega_+} hdx$$

$$+k_1 \int_{\Omega_-} hdx$$

$$= k_2 \int_0^1 hdx + (k_1 - k_2)\int_{\Omega_-} hdx \le (k_2 - k_1)\int_0^1 |h|dx.$$

Combining the last two estimates and the assumption $k_2 < 2k_1$ from (3.5) we obtain

$$|Du_a(a)(h)|_{L^2}^2 \ge \kappa|h|_{L^2}^2 \tag{3.8}$$

for all $a \in Q_{ad}$ with $|a - a^*|_{L^2} \le r_1$ and $h \in H^1$ with $\int_0^1 hdx = 0$.

Finally we have the estimate

$$|< D(u(a)-z_0), Du_{aa}(a)(h,h) >_{L^2}| = 2|< A^{-1}(a)A(1)(u(a)-z_0),$$

$$D(hDu_0(a)h) >|$$

$$\le \kappa_1|D(u(a) - z_0)|_{L^\infty}|h|_{L^2}|DA^{-1}(a)D(hDu_a(a))|,$$

from which we obtain

$$|< D(u(a) - z_0, Du_{aa}(a)(h,h) >_{L^2}|$$

$$\le \kappa_2|D(u(a) - z_0)|_{L^\infty}^2|h|_{L^2}^2, \tag{3.9}$$

for constants κ_i independent of $a \in Q_{ad}$ and $h \in H^1$. The factor $|D(u(a) - z_0)|_{L^\infty}$ in (3.9) can be made arbitrarily small uniformly for $a \in Q_{ad}$, with $|a - a^*| \leq r$, by choosing r sufficiently small. This fact, together with (3.6), (3.8) and (3.9) imply that there exist constants $r \in (0, r_1)$ and $\alpha > 0$ such that

$$f_{aa}(a, z_0)(h, h) + < (\lambda^*, \mu^*, \eta^*), g_{aa}(a)(h, h) > \geq \alpha |h|^2_{L^2}$$

for all $a \in Q_{ad}$ with $|a - a^*|_{L^2} \leq r$ and $h \in H^1$ with $\int_0^1 h dx = 0$. This implies the desired second order condition (SC). We have thus established that with (3.4) and (3.5) holding, Theorem 4 is applicable to (P$_z$). Thus there exist $\delta_0 > 0$ and a constant K such that $|D(z - z_0)|_{L^2} \leq \delta_0$ implies the existence of a local solution a_z of (P$_z$) with $|a_z - a^*|_{L^2} \leq r$ and any such local solutions satisfy $|a_z - a^*|_{L^2} \leq K|z - z_0|^{\frac{1}{2}}$.

4. A numerical estimation procedure

The most common approach to estimate a coefficient such as the diffusion coefficient a in

$$(4.1) \qquad \begin{aligned} -\mathrm{div}(a \,\mathrm{grad}\, u) + c\,u = f \quad \text{in} \quad \Omega, \\ u|\partial\Omega = 0, \end{aligned}$$

where $f \in L^2$ and $c \in L^2$ are known, centers around different versions of discretizing the least squares problem (P) and applying a nonlinear optimization routine to the resulting finite dimensional minimization problem. In spite of the fact that (4.1) is linear in a for given u and linear in u for known a, the resulting optimization problem (P) is nonconvex and might not have a solution unless some kind of compactness assumptions are made for Q_{ad}. The least squares formulation has the property of being versatile with respect to the availability of different types of data. An alternative approach to the output least squares formulation is given by the equation error technique. It is based on the observation that with u replaced by z in (4.1) one arrives at a hyperbolic equation for a given by

$$\tilde{e}(a) := -\mathrm{div}(a \, grad \, z) + cz - f = 0.$$

which can alternatively be formulated as an optimization problem

$$(P_{\tilde{e}}) \qquad \min_{a \in Q_{ad}} |\tilde{e}(a)|_Z^2,$$

where $|\cdot|_Z$ denotes some appropriately chosen norm. The choice $Z = (H^1)^*$, for example, was studied in [Ac]. We observe that $(\mathrm{P}_{\tilde{e}})$ has the advantage over (P) that it is quadratic in a; however, it requires differentiation of the data and in general $(\mathrm{P}_{\tilde{e}})$ is not radially unbounded, since ∇z may vanish at certain points of the domain Ω.

To combine the advantageous features of both, the output least squares and the equation error technique, a hybrid method was developed in [IK] and we briefly describe it here. In this method the unknown coefficient a and the state variable u are both considered as independent variables. We define the error function $e : H^2 \times H_0^1 \to H_0^1$ by

$$e(a,u) = (-\Delta)^{-1}(-\mathrm{div}(a\, grad\, u) + cu - f),$$

where Δ denotes the Laplacian from H_0^1 to H^{-1}. For an observation $z \in H_0^1$ consider the minimization problem

$$(4.2) \qquad \begin{array}{l} \min |u - z|_{H_0^1}^2 \\ e(a,u) = 0 \\ a \geq \zeta, |a|_{H^2} \leq \gamma. \end{array}$$

This problem, which involves equality as well as inequality constraints is now reformulated as an augmented Lagrangian problem. Let (a^*, u^*) denote a solution of (4.2), and let $(\lambda^*, \mu^*, \eta^*)$ stand for the Lagrange multipliers associated with the equality constraints $e(a,u) = 0$ and the inequality constraints $|a|_{H^2} \leq \gamma$ and $a \geq \zeta$. Under certain conditions it can be shown that (a^*, u^*) is also a solution of

$$(P_A) \qquad \min_{\substack{a \geq \zeta \\ (a,u) \in H^2 \times H_0^1}} \begin{array}{l} |u - z|_{H_0^1}^2 + < \lambda^*, e(a,u) >_{H_0^1} + k|e(a,u)|_{H_0^1}^2 \\ +\mu^* \hat{g}(a, \mu^*, k) + k\hat{g}(a, \mu^*, k)^2, \end{array}$$

where $\hat{g}(a,\mu,k) = \max(-\frac{\mu}{k}, g(a))$ and $g(a) = \frac{1}{2}(|a|_{H^2}^2 - \gamma^2)$. Thus the equality constraint and the inequality constraint with finite dimensional image space are eliminated from (4.2), and (P_A) contains only the simple inequality constraint $a \geq \zeta$ as explicit constraint. We stress the fact that it is not necessary to take the limit $k \to \infty$ in (P_A) to obtain (a^*, u^*) as a solution of (P_A). In (P_A) the Lagrange multipliers $\lambda^* \in H_0^1$ and $\mu^* \in R$ are unknown and in any practical computation they need to be determined as part of the algorithm. This is accomplished iteratively. Let $(\lambda_0, \mu_0) \in H_0^1 \times R$ be starting values. The iterative procedure

$$
\begin{aligned}
\lambda_{n+1} &= \lambda_n + k\, e(a_n, u_n) \\
\mu_{n+1} &= \mu_n + k\, \hat{g}(a_n, \mu_n, k)
\end{aligned}
\tag{4.3}
$$

was studied in [IK] for (4.1) and in [PK] for parabolic equations with a spatially and temporally dependent diffusion coefficient. In (4.3), (a_n, u_n) denote the solution of (P_A) with (λ^*, μ^*) replaced by (λ_n, μ_n). Extensive numerical testing has been carried out for the *augmented Lagrangian method* described above [IKK], [KK3]; the algorithm proved to be very robust with respect to noise added to the data, with respect to the choice for k and the initial choice for the coefficient a. As a start-up value for u one can take the data z (or an interpolation of pointwise data), and as a start-up value for λ^* one observes that the first order necessary optimality condition implies $\lambda^* = A^{-1}(a^*)\Delta(u^* - z)$, so that $\lambda_0 = 0$ is a reasonable choice. We observe that the iterative steps require the minimization of a cost functional that is quadratic in the variables a and u, so that the conjugate gradient algorithm can be employed efficiently.

The study of when a solution of (P) is also a solution of (P_A) as well as convergence of the iteration (4.3) and the sequence of solutions (a_n, u_n) is based on the following coercivity condition:

(4.4) there exists $\sigma > 0$ and $k_0 > 0$ such that the second Fréchet derivative L''_{k_0} of L_{k_0} satisfies $L''_{k_0}(a^*, u^*, -g(a^*))(h, v, y)(h, v, y) \geq \sigma(|h|^2_{H^2} + |v|^2_{H^1_0} +|y|^2)$, for all $(h, v, y) \in H^2 \times H^1_0 \times R$, where $L_k(a, u, w) = |u - z|^2_{H^1_0} + < \lambda^*, e(a, u) >_{H^1_0} + \mu^* g(a) + < \eta^*, \alpha - a >_{H^2} + \frac{k}{2}|e(a, u)|^2_{H^1_0} + \frac{k}{2}|g(a) + w|^2$.

Such a condition, which can be compared to the second order optimality condition of section 3, only holds in specific cases, as for instance, if $\mu^* > 0$ or if the set of admissible parameters is intersected with a finite dimensional space. In general it requires a change of the problem formulation. One possibility is given by the introduction of a regularization term; i.e. the cost functional in (4.2) and (P_A) is changed to

$$|u - z|^2_{H^1_0} + \beta N(a)^2$$

where $\beta > 0$ and $N(a)$ is an appropriately chosen norm or semi-norm on H^2. It is then necessary to investigate the convergence of the solutions (a^β, u^β) of the regularized problems as $\beta \to 0^+$. Under very general assumptions one can show that (a^β, u^β) converges to a solution of the unregularized problem as $\beta \to 0^+$ and that $|u(a^\beta) - z| \to |u(a^*) - z|$ like $o(\sqrt{\beta})$, [CK1]. These facts were used in the convergence analysis of the augmented Lagrangian method in [IK]. In the next section we obtain an improved rate of convergence under more stringent assumptions. Let us also mention that there is some arbitrariness in the choice of the norms for the output term $|u - z|$ and for imposing the equation errors term $e(a, u) = 0$ in (P_A). The choice that was made in this section has the advantage that it requires the same amount of differentiation of the data in all terms of the cost functional in (P_A). Other choices are discussed in [IKK].

5. Regularization for nonlinear inverse problems

Regularization techniques are frequently used to stabilize ill-posed inverse problems. If in the example of section 3 we desired to obtain stability of a in the H^1 norm, then a regularization term would be required. We have also pointed out the need for regularization in a different context in the previous section. While regularization is well studied for linear inverse problems much less is known for nonlinear problems. We describe a result on the rate of convergence of the solutions of the regularized problems, as the regularization parameter and the error level tend to zero.

We require some preliminaries. Let X and Z be Hilbert spaces and let $F : D \subset X \to Z$ be a nonlinear operator. Throughout it is assumed that F is continuous and weakly sequentially closed (i.e., for any sequence $\{x_n\}$ in D weak convergence of x_n to x in X and weak convergence of $F(x_n)$ to z in Z imply $x \in D$ and $F(x) = y$). We are concerned with solving

$$F(x) = z_0, \tag{5.1}$$

for some $z_0 \in Z$. Throughout this section the existence of a solution x^* of (5.1) is assumed (attainability of z_0). A regularized least squares formulation is used to solve (5.1) and it is assumed that only an approximation z_δ to z_0 is known. Hence we consider

$$\min |F(x) - z_\delta|_Z^2 + \beta|x - x_e|^2 \text{ over } x \in D, \tag{5.2}$$

where $\beta > 0$. In (5.2) the element x_e plays the role of an a-priori estimator to the solution of (5.2). We call x^* an x_e–minimum norm solution of (5.1) if

$$|x^* - x_e| = \min\{|x - x_e| : F(x) = z_0\}.$$

The assumptions on F guarantee the existence of solutions x_β^δ of (5.2) as well as of an x_e–minimum norm solution of (5.1). We have the following rate of convergence result for $x_\beta^\delta \to x^*$ as $\delta \to 0, \quad \beta \to 0$.

Theorem 5. In addition to the above assumptions on F, assume that D is convex, and that $|z_\delta - z_0|_Z \leq \delta$. Moreover, let the following conditions hold:

(i) F is Fréchet differentiable,

(ii) there exists $L > 0$ such that $\| F'(x_0) - F'(x) \| \leq L|x_0 - x|_X$ for all $x \in D$,

(iii) there exists $w \in Z$ such that $x^* - x_e = F'(x^*)^* w$, and

(iv) $L|w|_Z < 1$.

Then for the choice $\beta \sim \delta$ we obtain

$$|x_\beta^\delta - x^*| = O(\sqrt{\delta}).$$

Here $F'(x^*)^*$ denotes the adjoint of $F'(x^*)$ and $\beta \sim \delta$ stands for $\beta = O(\delta)$ and $\delta = O(\beta)$. A slightly more general version of the above theorem, which also allows for error in solving (5.2), is proved in [EKN].

In [EKN] it is shown that Theorem 5 is applicable to the mapping $F : D \subset H^1 \to H_0^1$, given by $F(a) = u(a)$, where $u(a)$ is the solution of

$$\begin{aligned} -(a\, u_x)_x + c\, u &= f \text{ in } (0,1), \\ u(0) = u(1) &= 0, \end{aligned}$$

with $c \geq 0, c \in L^2, f \in L^2$ and $D = \{a \in H^1 : a(x) \geq \zeta > 0\}$. We assume that z_0 is attainable, i.e. there exists $a^* \in D$, such that $u(a^*) = z_0$. The assumptions of Theorem 5 can be verified for this parameter estimation problem, provided that an estimator a_e is chosen such that

$$\bar{w} := \int_0^\bullet \frac{B(a_e - a^*)}{u_x(a^*)}(s)ds \in H^2 \cap H_0^1, \tag{5.3}$$

where B is the Neumann operator in L^2 given by $B\phi = -\phi_{xx} + \phi$ and $\operatorname{dom} B = \{\phi \in H^2 : \phi'(0) = \phi'(1) = 0\}$. The function w in Theorem 5 is then given by $w = (a^* \bar{w}_x)_x + c\bar{w}$, and $|w|_{L^2}$ has to be sufficiently small for (iv) to hold.

We have carried out numerous numerical tests to check the practical validity of applying Theorem 5 to parameter estimation problems. It turns out, that the convergence rate predicted by Theorem 5 can be observed numerically, and that the choice of an appropriate estimator is essential [Ge]. Let us consider as a specific numerical example the case

$$-(a\,u_x)_x = f \text{ in } (0,1)$$
$$u(0) = u(1) = 0,$$

$a(x) = 1 + \sin \pi x$ and $f(x) = 2\pi^2(2(1+\sin \pi x)\sin 2\pi x - \cos \pi x \cdot \cos 2\pi x)$. The discretization of the state variable u was carried out with cubic B-splines on a uniform grid of (0,1) with mesh size $\frac{1}{N}$ and the discretization of the parameter space was accomplished on the same grid with piecewise linear splines. The resulting finite dimensional least squares problem was solved with the Levenberg Marquardt algorithm for $N = 2, 4, \ldots, 36$, with the choice for $\beta = \frac{1}{500N^2}$ and $\delta = 100\beta$.

The abscissa in the figures below gives the value for $\ell n N$, while the ordinate shows the values for $\ell n(|a^* - a_\beta^{N,\delta}|_{H^1})$, where a^* is the "true parameter" $1 + x$ and $a_\beta^{N,\delta}$ is the solution of the discretized least squares problem with regularization parameter $\beta = \beta(N)$ and error level $\delta = \delta(N)$. Using the results of [EKN], [N] one expects a convergence rate of $O(\frac{1}{N})$, if the estimator a_e satisfies the requirements of Theorem 5. Figure 1 shows the results for an estimator which satisfies these assumptions and the rate is the desired one. The estimator for the calculations in Figure 2 is $a_e(x) = 2 - x$ which does not satisfy the requirements. Finally, for the calculations in Figure 3 we again started with the bad estimator $a_e(x) = 2 - x$, but iteratively we used the result $a_{\beta(N-2)}^{N-2,\delta(N-2)}$ of mesh size $\frac{1}{N-2}$ as an estimator for mesh size $\frac{1}{N}$. We see that this leads to a definite improvement over the results in Figure 2.

References

[Ac] R.C. Acar: Identification of the coefficient in elliptic equations, preprint.

[Aℓ] W. Alt: Lipschitzian perturbations of infinite dimensional problems, in "Mathematical Programming with Data Perturbations II,"(A.V. Fiacco, ed.), Lecture Notes in Pure and Applied Mathematics, Vol. 85, Marcel Dekker, Inc. New York, 1983, 7-21.

[BK] H.T. Banks and K. Kunisch: Estimation Techniques for Distributed Parameter Systems, Birkhäuser Boston, to appear 1989.

[BaK] J. Baumeister and K. Kunisch: Identifiability and stability of diffusion and restoring force in elliptic systems, submitted.

[CG] C. Chicone and J. Gerlach: A note on the identifiability of distributed parameters in elliptic equations, SIAM J. Math. Anal. 18 (1987), 1378-1384.

[CK1] F. Colonius and K. Kunisch: Output least squares stability in elliptic systems, Applied Mathematics and Optimization 19(1988), 33-63.

[CK2] F. Colonius and K. Kunisch: Stability of perturbed optimization problems with application to parameter estimation, submitted.

[EKN] H.W. Engl, K. Kunisch and A. Neubauer: Tikhonov regularization for the solution of nonlinear illposed problems I, to appear in Inverse Problems.

[F] B.G. Fitzpatrick: Statistical methods in parameter identification and model selection, Ph.D. Thesis, Division of Applied Mathematics, Brown University, Providence, Rhode Island 1988.

[FV] A. Friedman and M. Vogelius: Identification of small inhomogeneities of extreme conductivity by boundary measurements: a continuous dependence result, Arch. Rat. Mech. Anal., to appear.

[G] G. Geymayer: Regularisierungsverfahren und deren Anwendung auf inverse Randwertprobleme, Master Thesis, Technical University of Graz, Austria, 1988.

[IK] K. Ito and K. Kunisch: The augmented Lagrangian method for parameter estimation in elliptic systems, to appear in SIAM J. on Control and Optimization.

[IKK] K. Ito, M. Kroller and K. Kunisch: A numerical study of the augmented Lagrangian method for the estimation of parameters in elliptic systems, submitted to SIAM J. on Sci. and Stat. Computing.

[K1] K. Kunisch: Inherent identifiability of parameters in elliptic differential equations, J. Math.Anal. Appl. 132(1988), 453-472.

[K2] K. Kunisch: A survey of recent results on the output least squares formulation of parameter estimation problems, Automatica 24(1988), 531-593.

[KK] M. Kroller and K. Kunisch: A numerical study of an augmented Lagrangian method for the estimation of parameters in elliptic systems: Noisy data, outliers and discontinuous coefficients, Technical Report 123 (1988), Technical University of Graz, Austria.

[KL] R.V. Kohn and B.D. Lowe: A variational method for parameter estimation, RAIRO Math. Mod. and Num. Anal. 22(1988), 119-158.

[KP] K. Kunisch and G. Peichl: Estimation of a temporally and spatially varying diffusion coefficient in a parabolic system by an augmented Lagrangian technique, submitted.

[KV] R. Kohn and M. Vogelius: Determining conductivity by boundary measurements II, Interior Results, Comm. Pure Appl. Math. 38(1985), 643-667.

[M] H. Moritz: Advanced Physical Geodesy, Wichman, Karlsruhe, 1980.

[N] A. Neubauer: Tikhonov regularization for the solution of nonlinear ill-posed problems, II, to appear in Inverse Problems.

[R] G.R. Richter: An inverse problem for the steady-state diffusion equation, SIAM J. Appl. Math. 41(1981), 210-221.

[Y] W.W.-G. Yeh: Review of parameter identification procedures in ground water hydrology: The inverse problem, Water Res. Rev. 22(1986), 95-108.

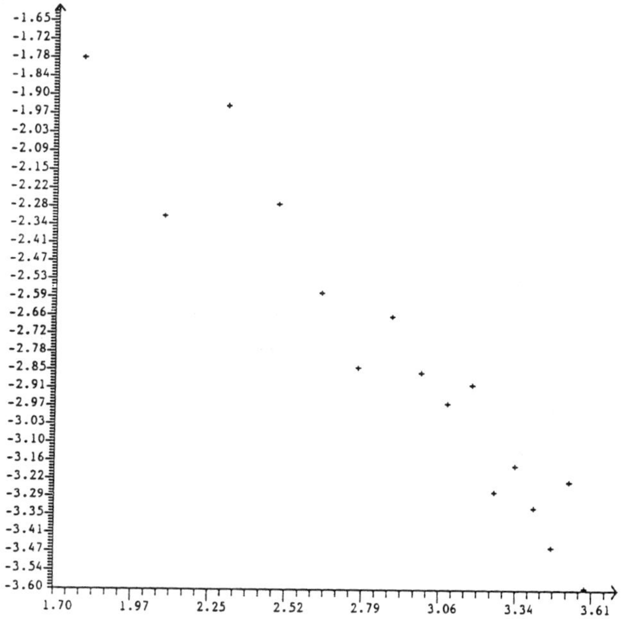

Figure 1

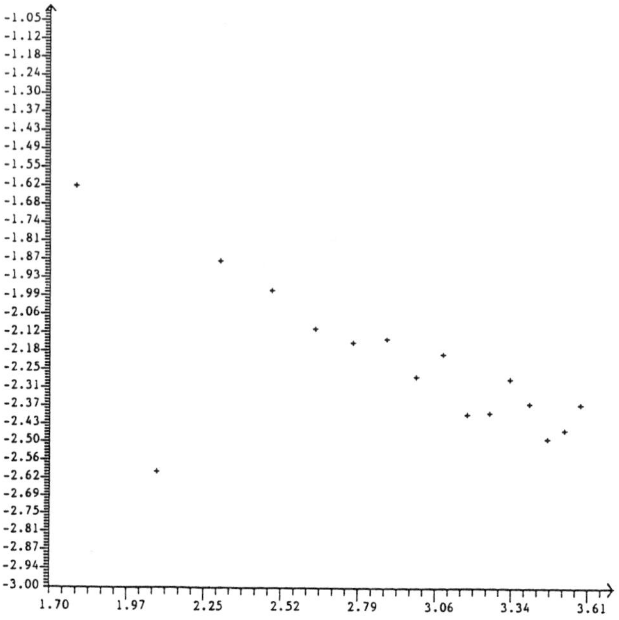

Figure 2

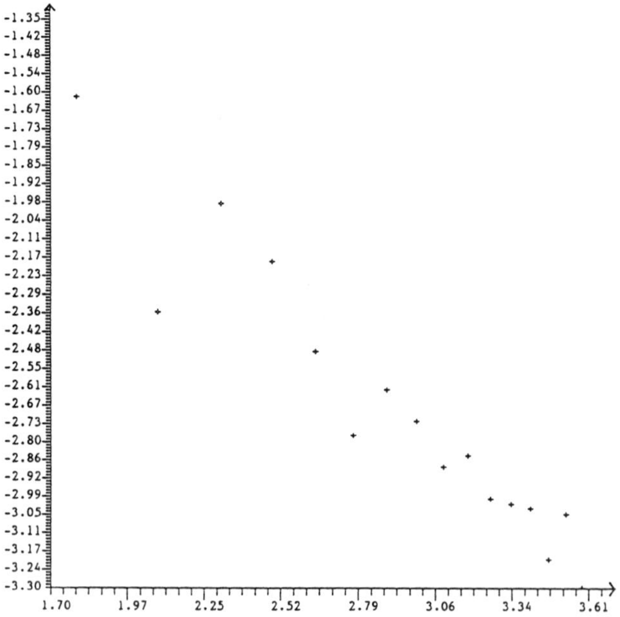

Figure 3

FULLER'S PHENOMENA

I.A.K. Kupka

0) Introduction: In his paper [F], Fuller studied the following optimal control problem. In the state space $\mathbb{R}^2$ with coordinates (x, y) he considers the control system: $\frac{dx}{dt} = u \quad \frac{dy}{dt} = x$ where the control u is restricted to the segment $[-1, +1]$. Given a point a in $\mathbb{R}^2$, he wants to determine the trajectories $(\hat{x}, \hat{y}, \hat{u}) : [0, \hat{T}] \longrightarrow \mathbb{R}^2 \times [-1, +1]$ of the system, starting at a, ending at 0 and minimizing the cost $\frac{1}{2} \int_0^{\hat{T}} \hat{y}(t)^2 dt$.

Fuller discovered that if the initial point a lied on either curve

$$\Gamma_+, \Gamma_- \ \textit{where}\ \Gamma_\varepsilon = \{(x, y) \big| \varepsilon x < 0; \quad y = \varepsilon \xi x^2\}$$

then there existed a unique solution $(x_a, y_a, u_a) : \ [0, T_a] \longrightarrow \mathbb{R}^2 \times [-1, +1]$ to the above problem and it had an infinite sequence of switching points tending to 0. ξ was the unique positive root of the polynomial $x^4 + \frac{x^2}{12} - \frac{1}{18}$.

This property, for an optimal trajectory to have an infinite set of switching points in finite time, property we shall call the Fuller's phenomenon, raises serious difficulties for the construction of optimal control synthesis as defined for example in [G] and [S] . Hence it is important to study the Fuller phenomenon thoroughly. In particular it is crucial to know how general this phenomenon is.

In the following paper we shall study a few aspects of the Fuller phenomenon, in particular its generality and stability. For a start, we shall consider some examples.

(1) Discussion of the Fuller example.

1. Let us take a closer look at Fuller's original example: in $\mathbb{R}^2$ we have the linear control system: $\frac{dx}{dt} = u \quad \frac{dy}{dt} = x$. The control space being the interval $[-1, +1]$. The cost function c is $\frac{1}{2}y^2$. The optimal curves are the projections of the extremals, that is the curves

$$(x, y, p, q, u) : [0, T] \longrightarrow \mathbb{R}^2 \times \mathbb{R}^2 \times [-1, +1]$$

satisfying the following conditions:

a) $\frac{dx}{dt} = u \quad \frac{dy}{dt} = x \quad \frac{dp}{dt} = -\frac{\partial H_\lambda}{\partial x} = -q, \ \frac{dq}{dt} = -\frac{\partial H_\lambda}{\partial y} = \lambda y$ where $H_\lambda : \mathbb{R}^2 \times \mathbb{R}^2 \times \mathbb{R} \longrightarrow \mathbb{R}$ is the function:

$$H_\lambda(x, y, p, q, u) = pu + qx - \frac{1}{2}\lambda y^2$$

b) for almost all $t \in [0, T]$

$$H_\lambda(x(t), y(t), p(t), q(t), u(t)) = \sup_{-1 \leq v \leq +1} H_\lambda(x(t), y(t), p(t), q(t), v)$$

c) λ is a constant which takes either the values 1 or the value 0.

This is just the statement of the maximum principle. Condition b) above implies that on an extremal, $u(t) = \text{sign}(p(t))$ for all t, such that $p(t) \neq 0$. I claim that if for no $t \in [0, T[$, do we have simultaneously $x(t) = y(t) = p(t) = q(t) = 0$, then the only possible accumulation point of the zeros of p is T. This is trivial in the case when $\lambda = 0$. If $\lambda = 1$, since $\frac{dp(t)}{dt} = -q(t)$, $\frac{d^2p(t)}{dt^2} = -y(t)$, $\frac{d^3p(t)}{dt^3} = -x(t)$ at any zero t of p, one of these derivates of p does not vanish and hence t is an isolated zero of p.

From now on, we shall consider only the ordinary extremals, that is those for which $\lambda = 1$. I claim that the projection $(\hat{x}, \hat{y}, \hat{u}) : [0, T] \longrightarrow \mathbb{R}^4 \times [-1, +1]$, not passing through 0, is an optimal trajectory. To see this let $(x, y, u) : [0, T] \longrightarrow \mathbb{R}^2 \times [-1, +1]$, be any other trajectory such that $x(0) = \hat{x}(0)$, $x(T) = \hat{x}(T)$, $y(0) = \hat{y}(0)$, $y(T) = \hat{y}(T)$. Set $\Delta x = x - \hat{x} \quad \Delta y = y - \hat{y} \quad \Delta u = u - \hat{u}$. If $c, \hat{c}$ denote the costs of (x, y, u) and $(\hat{x}, \hat{y}, \hat{u})$ respectively, we have:

$$c - \hat{c} = \int_0^T \hat{y}(t)\Delta y(t)dt + \frac{1}{2}\int_0^T (\Delta y(t))^2 dt.$$

Using the relations: $\frac{d\hat{q}}{dt} = \hat{y}, \quad \frac{d\hat{p}}{dt} = -\hat{q}$,

$$\Delta y(0) = \Delta y(T) = \Delta x(0) = \Delta y(T) = 0,$$

and integration by parts, we get:

$$\int_0^T \hat{y}(t)\Delta y(t)dt = -\int_0^T \hat{p}(t)\Delta u(t)dt.$$

At any t such that $\hat{p}(t) \neq 0$, $\hat{u}(t) = \text{sign } p(t)$ and hence sign $\Delta u(t) = -\text{sign } p(t)$. Since the set of zeros of $\hat{p}(t)$ on $[0, T[$ is discrete $\int_0^T \hat{y}(t)\Delta y(t)dt > 0$. Hence $c - \hat{c} > 0$.

Let us now proceed to study the structure of the extremals. In the space $\mathbb{R}^4 = \mathbb{R}^2 \times \mathbb{R}^2$ let Σ denote the hyperplane $p = 0$. Σ divides $\mathbb{R}^4$ into two half spaces E_+ E_- where $E_\varepsilon = \{x, y, p, q\} \Big| \varepsilon p > 0\}$, $\varepsilon = +$ or $-$. In E_ε, the extremals are the trajectories of the hamiltonian field $\vec{H}_\varepsilon$ of the function $H_\varepsilon = \varepsilon p + qx - \frac{1}{2}y^2$. An extremal $z = (\hat{x}, \hat{y}, \hat{p}, \hat{q}, \hat{u}) : [0, T] \longrightarrow \mathbb{R}^2 \times \mathbb{R}^2 \times [-1, +1]$ is entirely determined by the set $SW(z)$ of its switching times, that is the set of all t's such that $z(t) \in \Sigma$ i.e. $p(t) = 0$. By what has been proved above, such a set is either finite or a sequence $\{t(n)|n \geq 1\}$, $t(n) \leq t(n+1)$, tending to T.

Now, we are going to introduce a device to enable us, to see the structure of the set of the switching points, that is the points $z(t(n))$, $n \geq 1$. Let $\Sigma_\varepsilon, \varepsilon = +, -$, be the half plane in Σ of all $z = (x, y, 0, q)$ such that $\varepsilon q < 0$ and let Σ_0 be the plane $\{p = q = 0\}$, common boundary of Σ_+ on Σ_-. We can define two smooth mappings $\sigma_\varepsilon : \Sigma_\varepsilon \longrightarrow \Sigma$, $\varepsilon = +$ or $-$, as follows: let $z \in \Sigma_\varepsilon$. The extremal $\hat{z}(t)$ starting at z, is a trajectory of the hamiltonian field $\vec{H}_\varepsilon$ and lies in E_ε until it hits Σ again, and as we shall see below, it always does. The first point at which this happens is $\sigma_\varepsilon(z)$.

If $p(\hat{z}(t)) = \hat{p}(t)$, then:

$$\hat{p}(t) = -t\left[\frac{\varepsilon t^3}{24} + \frac{x_0 t^2}{6} + \frac{y_0 t}{2} + q_0\right]$$

where $z = (x_0, y_0, 0, q_0)$. The polynomial in the brackets has the sign ε at $+\infty$ and sign (q_0) at 0 . Since $\varepsilon q_0 < 0$, this polynomial has always at least one root in the half line $]0, +\infty[$.

Now we can exhibit the device defining the switching points of an extremal, at least if it does not meet Σ_0. Denoting by ζ the first point where it meets Σ, ζ belongs to Σ_ε. The switching set ordered by the time of their occurrence is the part of the orbit $\{\sigma_\varepsilon(\zeta),\ \sigma_{-\varepsilon}\sigma_\varepsilon(\zeta), \sigma_\varepsilon\sigma_{-\varepsilon}\sigma_\varepsilon(\zeta), \ldots\}$ contained in the extremal. (It can happen that the whole orbit is contained in the extremal as we shall see later.)

An important feature of the extremal system in the Fuller problem is that it is quasi-homogeneous under the action of the multiplicative group G_m of real positive numbers, defined as follows: if $g \in G_m$, $x \longrightarrow gx$, $y \longrightarrow g^2y$, $p \longrightarrow g^4p$, $q \longrightarrow g^3q$, $u \longrightarrow u$, $t \longrightarrow gt$. Under these transformations, the equations satisfied by x, y, p, q, u, are invariant, H is multiplied by g^4 and the symplectic form by g^5. As a consequence, both σ_+, σ_- commute with the action of G_m on $\mathbb{R}^4$. In particular σ_+, σ_- map orbits of the G_m-action on $\mathbb{R}^4$ onto other orbits of G_m.

Fuller discovered that there existed a pair of orbits $\Gamma_+, \Gamma_-, \Gamma_\varepsilon \subset \Sigma_\varepsilon$ $\varepsilon = +$ or $-$, such that $\sigma_+(\Gamma_+) = \Gamma_-$, $\sigma_-(\Gamma_-) = \Gamma_+$. Hence if z_0 is a point on Γ_ε and $z(t)$ is the extremal starting at z_0, its odd switching points z_{2n+1} will lie on $\Gamma_{-\varepsilon}$, its even switching points on Γ_ε. Moreover, there exists a $\rho \in G_m$ such that $z_2 = \rho \cdot z_0$ (if $z_2 = (x_2, y_2, 0, q_2)$, $z_0 = (x_0, y_0, 0, q_0)$, $x_2 = \rho x_0$, $y_2 = \rho^2 y_0$, $q_2 = \rho^3 q_0$) since z_2, z_0, lie on the same orbit Γ_ε. It turns out that $\rho = \frac{1-2\xi}{1+2\xi}$, hence $\rho < 1, (\frac{2}{11} < \xi < \frac{1}{5})$. The invariance of the extremal system under G_m implies that the extremal originating at z_0 is invariant under the action of the semigroup $\{\rho^n | n \geq 1\}$ of G_m. In particular, the switching set of that extremal is the sequence (z_n) such that $z_{2n} = \rho^n \cdot z_0$, $z_{2n+1} = \rho^n \cdot z_1$. The corresponding switching times are:

$$t(n) = \frac{1+\sqrt{\rho}}{1-\sqrt{\rho}}\left(1 - \frac{1}{\rho^{n/2}}\right) |x_0|$$

where $x_0 = x(z_0)$.

Now let us indicate how one can prove these results. First, it is clear that the field

$\vec{H}_\varepsilon$, $\varepsilon = +, -$, has the following three independent first integrals:

$$I_{1\varepsilon} = y - \frac{\varepsilon x^2}{2}, \quad I_{2\varepsilon} = q - \varepsilon yx + \frac{x^3}{3}, \quad I_{3\varepsilon} = H_\varepsilon = \varepsilon p + qx - \frac{1}{2}y^2.$$

These first integrals determine the mapping σ_ε as follows: let $z \in \Sigma_\varepsilon.\sigma_\varepsilon(z)$ is the unique point $z' \in \Sigma$ such that:

a) $I_{k,\varepsilon}(z') = I_{k,\varepsilon}(z) \quad k = 1,2,3$

b) $\varepsilon(x(z') - x(z)) > 0.$

c) If $z'' \in \Sigma$ and $I_{k,\varepsilon}(z'') = I_{k,\varepsilon}(z)$ for $k = 1,2,3$, then $\varepsilon(x(z'') - x(z')) > 0$.

We should remember that the maximum principle, for the two point boundary problem states that the function $\sup(H_+, H_-)$ is zero along any extremal corresponding to an optimal trajectory. Hence we shall look for Γ_ε in the subset $\mathcal{H} = \{z \in \Sigma | H_+(z) = H_-(z) = 0\}$ of $\Sigma.\mathcal{H}$ is G_m invariant.

Since σ_ε commutes with the action of G_m on Σ, it induces a mapping $\hat{\sigma}_\varepsilon : \hat{\Sigma}_\varepsilon \to \hat{\Sigma}$ where $\hat{\Sigma}_\varepsilon, \hat{\Sigma}$ are respectively the quotients of Σ_ε and $\Sigma - \{0\}$ by the G_m-action. Let $\hat{\mathcal{H}}$ be the quotient of $\mathcal{H}$. To Γ_ε corresponds a point $\hat{\Gamma}_\varepsilon$ on $\hat{\mathcal{H}}$. We have that $\hat{\sigma}_\varepsilon(\hat{\Gamma}_\varepsilon) = \hat{\Gamma}_{-\varepsilon}$, $\hat{\sigma}_{-\varepsilon}(\hat{\Gamma}_{-\varepsilon}) = \hat{\Gamma}_\varepsilon$. A natural system of coordinates on the subset $\{x \neq 0\}$ of $\hat{\Sigma}$ is the system η, χ, where $\eta = \frac{y}{x^2}$, $\chi = \frac{q}{x^3}$. If we introduce the function J_ε:

$$J_\varepsilon = (\chi - \varepsilon\eta + \frac{1}{3})^2 / (\varepsilon\eta - \frac{1}{2})^3$$

then if $\zeta \in \hat{\Sigma}_\varepsilon$ and $\zeta' = \hat{\sigma}_\varepsilon(\zeta)$, we have $J_3(\zeta') = J_\varepsilon(\zeta)$. This is a consequence of a) above. Hence if the pair $(\hat{\Gamma}_+, \hat{\Gamma}_-)$ exists, we have $J_+(\hat{\Gamma}_+) = J + (\hat{\Gamma}_-), J_-(\hat{\Gamma}_-) = J_-(\hat{\Gamma}_+)$. We can simplify the search for $\hat{\Gamma}_+ < \hat{\Gamma}_-$ using the following final remark. If r is the reflexion $(\eta, \chi) \to (-\eta, \chi)$, then $J_\varepsilon \circ r = -J_{-\varepsilon}$. If we can find a $\zeta \in \hat{\mathcal{H}}$ such that $J_+(\zeta) = J_+(r(\zeta))$, then we could take $\hat{\Gamma}_+ = \zeta$ and $\hat{\Gamma}_- = r(\zeta)$. Any point ζ of the form $(\eta, \frac{1}{2}\eta^2)$ is on $\hat{\mathcal{H}}$ and the condition $J_+(\zeta) = J_+(r(\zeta))$ boils down to the equation $\eta^4 + \frac{1}{12}\eta^2 - \frac{1}{18} = 0$. If ξ is the unique positive solution of this equation, then it can easily be checked that the curves $\Gamma_+ = \{(x, y, 0, q) | y = \xi x^2, \quad q = \frac{1}{2}\xi^2 x^3, \ x \leq 0\}$ and

$\Gamma_- = \{(x,y,0,q)|y = -\xi x^2, \quad q = \frac{1}{2}\xi^2 x^3,\ x \geq 0\}$ are the curves we are looking for. This concludes our study of Fuller's example. Before going to the next example, let us formalize the Fuller phenomenon in a definition. Let us recall that a switching point of an absolutely continuous curve is any point in any neighborhood of which the derivate of the curve is discontinuous.

Definition 0: Given an optimal control problem defined on a state space M which is a smooth manifold. A Fuller pair at a point m_0 in the cotangent space T^*M of M is a pair Γ_+, Γ_- of smooth regular curves having both m_0 as starting point, with the following property: for any point $z_0 \in \Gamma_+ \cup \Gamma_-$, say $z_0 \in \Gamma_\varepsilon$, there exists an extremal $z : [0, T(z_0)] \longrightarrow T^*M$ of the problem, starting at z_0, ending at m_0, such that its switching times form an increasing sequence $\{t(n)|n \geq 1\}$ converging to $T(z_0)$ and:

(i) For all integers n, $z\,(t(2n)) \in \Gamma_\varepsilon, \quad z(t(2n+1)) \in \Gamma_{-\varepsilon}$

(ii) There exists a constant $k > 1$, depending only on the pair (Γ_+, Γ_-) such that the limits of both sequences $\{k^{2n}[t(2n+1) - t(2n)]|n \in \mathbb{N}\}$ and $\{k^{2n}[t(2n) - t(2n-1)]|n \in \mathbb{N}\}$ exist.

The point m_0 will be called a Fuller point. If x_0 is a point in M such that there exists a Fuller point m_0 in the cotangent space $T^*_{x_0}M$ at x_0 and the pair (Γ_+, Γ_-) associated to m_0 projects onto regular smooth curves γ_+, γ_- in M, then x_0 will also be called a Fuller point.

2) More general examples.

In our next example, the state space is $\mathbb{R}^3$ with coordinates (x,y,z): The control space is the interval $[-1,+1]$ and the dynamics is: $\frac{dx}{dt} = u$, $\frac{dy}{dt} = x$, $\frac{dz}{dt} = 1$ and the cost function $c(x,y,z)$:

$$c(x,y,z) = \frac{1}{2}ay^2 - \frac{1}{2}bx^2y - q(x,z)$$

where $q(x,z) = c_0xz^3 + c_1x^2z^2 + c_2xz + c_3x^4$, a, b, c_i $i = 0,1,2,3$, are constants. The special case where $a = 1$, $b = c_0 = c_1 = c_2 = 0$ and $c_3 = \mu$, μ parameter, was analyzed by

Ryan in [R].

The corresponding hamiltonian is:

$$H_\lambda = up + qc + r - c(x, y, z).$$

As in the Fuller's example, this problem is quasi homogeneous. Under the following action of G_m, $c \to gx$, $y \to g^2 y$, $z \to gz$, $p \to g^4 p$, $q \to g^3 q$, $r \to g^4 r$, $u \to u$, $t \to gt$, the extremal system is invariant,H is multiplied by g^4 and the symplectic form by g^5.

Let m_0 denote the point $(0,0,0,0,0,1)$ in $\mathbb{R}^6 = \mathbb{R}^3 \times \mathbb{R}^3$ (at m_0, $x = y = z = p = q = 0$, $r = 1$). Now it can be proved that there exists a semi-algebraic subset F in $\mathbb{R}^6$ with non-empty interior, such that if the parameter sequence $(a, b, c_0, c_1, c_2, c_3)$ belongs to F, the problem stated above has a Fuller pair at m_0. In particular the parameter sextuple $(1,0,0,0,0,0)$ corresponds to Fuller's example. The proof of this result is contained in [K2]. It follows the lines of our discussion of Fuller's example. The Fuller pair is composed of two orbits of the G_m action on $\mathbb{R}^6$. To find them we consider the quotient $\hat{\mathcal{H}}$ of the manifold $\mathcal{H} = \{p = H_+ = 0\}$ by G_m and determine a couple of points $\hat{\Gamma}_+, \hat{\Gamma}_-$ in $\hat{\mathcal{H}}$ such that $\hat{\sigma}_+(\hat{\Gamma}_+) = \hat{\Gamma}_-$ and $\hat{\sigma}_-(\hat{\Gamma}_-) = \hat{\Gamma}_+$.

3) Setting for the general case.

In the next sections we want to discuss how general the Fuller phenomenon is. First we have to understand what is the proper setting for it.

We start with a control system $\frac{dx}{dt} = F(x, u)$ having as state space , an open subset M in an euclidean space $\mathbb{R}^d$ and as control set, a subset U of another euclidean space $\mathbb{R}^e$. Finally ew have a cost function $c : M \times U \to \mathbb{R}$. In order for the parametrized vector field $F : M \times U \to \mathbb{R}^d$ and the function c to have asufficiently smooth behaviour in u, x, we require that F and c have extensions to a set $M \times U'$ where U' is an open neighborhood of U, that are smooth. (Smooth will mean either infinitely differentiable or real analytic.)

To the problem of minimizing the cost c, among all trajectories of the system having given boundary points A, B in M, the maximum principle associates the extremal system:

let $H_\lambda : M \times \mathbb{R}^d \times U \longrightarrow \mathbb{R}$ be the function $H_\lambda(x,p,u) = \langle p, F(x,u) - \lambda c(x,u)$. λ is a parameter. H_λ defines the extremal system as follows:

$$EX \left\{ \begin{array}{l} \frac{dx}{dt} = \frac{\partial H_\lambda}{\partial p}(x,p,u), \frac{dp}{dt} = -\frac{\partial H_\lambda}{\partial x}(x,p,u) : \\ H_\lambda(x,p,u) = \sup\{H_\lambda(x,p,v) | v \in U\} \end{array} \right.$$

λ is a constant equal to 0 or 1.

In the following discussion, for the sake of simplicity, we shall consider only the ordinary extremal system, that is the one corresponding to $\lambda = 1$ and we denote H_1 simply by H.

In the generic situation there exists an open dense set $\mathcal{O}$ of $M \times \mathbb{R}^d$ such that the second condition of EX determines a unique smooth function $\check{u} : \mathcal{O} \longrightarrow U$ with the following properties:

(i) $H(x,p,\check{u}(x,p)) = \sup\{H,(x,p,v) | v \in U\}$ for all $(x,p) \in \mathcal{O}$

(ii) if $(x,p,w) \in \mathcal{O} \times U$ and $H(x,p,w) = \sup\{H(x,p,v) | v \in U\}$ then $w = \check{u}(x,p)$.

On $\mathcal{O}$ the system EX is equivalent to the following hamiltonian system

$$\left\{ \begin{array}{l} \frac{du}{dt} = \frac{\partial \mathcal{H}}{\partial p}(x,p), \quad -\frac{dp}{dt} = \frac{\partial \mathcal{H}}{\partial x}(x,p) \\ u = \check{u} \end{array} \right.$$

where $\mathcal{H}(x,p) = H(x,p,\check{u}(x,p))$.

Hence in $\mathcal{O}$ the extremals do not have switching points.

Now let us consider the complement of $\mathcal{O}$. Take a point $m_0 \in \Sigma = M \times \mathbb{R}^d - \mathcal{O}$. Then, either of the following two situations occurs: (i) for any open neighborhood ω of m_0, denoting by $\{\omega_j | j \in J\}$ the connected components of $\omega \cap \mathcal{O}$, one of the restrictions $\check{u}|w_j$ does not have a smooth extension to ω.

(ii) All the restrictions $\check{u}|\omega_j$, $j \in J$ have smooth extensions to ω but these extensions do not coincide.

Here we want to discuss the simplest instance of the second occurrence: the case when there are two connected components. Take $J = \{+,-\}$. We are in the following situation: there exist two smooth functions $\check{u}_+, \check{u}_- : \omega \rightarrow U$ such that, if $H_+, H_- : \omega \rightarrow \mathbb{R}$

denote the functions, $H_\varepsilon(x,p) = H(x,p,\check{u}_\varepsilon(x,p))$, $\varepsilon = +,-$, then, for all $(x,p) \in \omega_\varepsilon$, all $v \in U$, $v \neq \check{u}_\varepsilon(x,p)$, we have: $H_\varepsilon(x,p) > H(x,p,v)$, $\varepsilon = +$ or $-$.

It is clear that the common boundary $\Sigma \cap \omega$ of ω_+ and ω_- is just the set of all $z \in \omega$ such that $H_+(z) = H_-(z)$. We shall assume moreover that 0 is a regular value of the function $H_+ - H_-$. As a consequence, $\Sigma \cap \omega$ is a smooth hypersurface.

Denote by $\vec{H}_+, \vec{H}_-$ the hamiltonian fields associated to H_+, H_-. Let the bracket $\{f,g\}$ denote the Poisson bracket of two functions f,g. Since $\vec{H}_+(H_+ - H_-) = -\{H_-, H_+\}$ $= \{H_+, H_-\} = \vec{H}_-(H_+ - H_-)$, $\vec{H}_+$ and $\vec{H}_-$ are tangent to $\Sigma \cap \omega$ at the same points. If $\vec{H}_+, \vec{H}_-$ are not tangent to $\Sigma \cap \omega$ at m_0, then, in a possibly smaller neighborhood ω' of m_0, the extremals in ω' have at most one switching point, located on $\Sigma \cap \omega$. Hence if we want Fuller type phenomena to happen at m_0, we have to assume that $\vec{H}_+, \vec{H}_-$ are both tangent to $\Sigma \cap \omega$ at m_0.

The preceding discussion can be well illustrated by Fuller's example: we can take $M = \mathbb{R}^2$, $\omega = \mathbb{R}^4 = \mathbb{R}^2 \times \mathbb{R}^2$. u_+, u_- are respectively the constant functions 1 and -1,

$$\begin{aligned} H_+(x,y,p,q) &= p + qx - \frac{1}{2}y^2 \\ H_-(x,y,p,q) &= -p + qx - \frac{1}{2}y^2 \end{aligned}$$

Σ is the hyperplane $\{p = 0\}$, ω_+, ω_- the half spaces $\{p > 0\}$, $\{p < 0\}$ respectively. For m_0 we take the point 0. Since $\{H_+, H_-\} = 2q$, it is clear that $\vec{H}_+, H_-$ are tangent to Σ at m_0. In fact the order of contact of $\vec{H}_+$ and $\vec{H}_-$ with Σ at m_0 is 3.

It turns out that in order to have a Fuller phenomenon at m_0, the order of the contacts of $\vec{H}_+, \vec{H}_-$ with $\Sigma \cap \omega$ has to be high enough. For example, if these *orders* are both 1, or equivalently, if the double brackets $\{H_+, \{H_+, H_-\}\}(m_0)$ and $\{H_-, \{H_+, H_-\}\}(m_0)$ are both non zero, our paper [K1] shows that there are no Fuller phenomena at m_0.

In the following sections we are going to state sufficient conditions on a pair H_+, H_-, for the existence of a Fuller pair at m_0. One of the main tools to do this, is a normal form we are going to discuss in the next section.

4) Normal forms

Consider the euclidean space $\mathbb{R}^{2d}$ with coordinate system $x^1, \ldots, x^d, p_1, \ldots, p_d$ and with the symplectic structure defined by the form $\Sigma_{i=1}^d dx^i \wedge dp_i$. Let $H_+, H_- : M \to \mathbb{R}$ be two smooth functions defined an open neighborhood M of 0. Then, we can state the following proposition.

Proposition 0: Assume that:

(i) all Poisson brackets of H_+, H_- of length up to 4 are zero at 0, i.e.

$$0 = H_+(0) = H_-(0) = \{H_+, H_-\}(0) = \{H_+, H_-\}(0) = \{H_-, \{H_+, H_-\}\}(0)$$
$$= ad^3 H_+(H_-)(0) = ad^3 H_-(H_+)(0) = adlt\_ad^2 H_+(H_-)(0)(adf(f) = \{f, g\})$$

(ii) The differentials $dH_+(0), dH_-(0), d\{H_+, H_-\}(0)$ are linearly independent.

Then in an open neighborhood N of zero, there exists a symplectic system of coordinates $X^1, \ldots, X^d,\ P_1, \ldots, P_d$ such that:

$$H_\varepsilon = \varepsilon P_1 + X^1 P_2 + P_3 - \frac{1}{2} a(X^2) + \frac{1}{2} b(X^1)^2 X^2 + \sum_{i=1}^{4} c_i (X^1)^i (X^3)^{4-i} + R$$

where the coefficients $a, b, c_i,\ 1 \le i \le 4$ have the values given below and the remainder term R is of the form: $R = (X^1)^2 X^1 X^2 R_2, R_1, R_2$ are smooth functions defined on N and their orders at 0 are at least 3 in the gradation defined by the weights:

$$w(X^1) \;=\; w(X^3) \;=\; 1, \quad w(X^2) \;=\; 2, \quad w(X^k) = 3 \; if \; k \ge 4$$

$$w(P_1) \;=\; w(P_3) \;=\; 4, \quad w(P_2) \;=\; 3, \quad w(P_k) = 3 \; if \; k \ge 4$$

The values of $a, b, c_i,\ 1 \le i \le 4$ are

$$a = \{ad^2 F(G), adF(G)\}(0), \qquad b = \{ad^2 G(F), adG(F)\}(0),$$

$$c_1 = \tfrac{1}{6} ad^4 F(G)(0), \qquad c_2 = \tfrac{1}{4} ad^2 F\, ad^2 G(F)(0)$$

$$c_3 = \tfrac{1}{6} adF ad^3 G(F)(0), \qquad c_4 = \tfrac{1}{24} ad^4 G(F)(0)$$

Here $F = \frac{1}{2}(H_+ + H_-)$ $G = \frac{1}{2}(H_+ - H_-)$, and $adf(g) = \{g, f\}$. If we forget about the remainder term R, we see that our hamiltonians are exactly those occuring in the example of section 2. Our method for finding Fuller's pairs in the general case will be to use the above normal form find conditions on the coefficients a, b, c_i, $1 \leq i \leq 4$, for the system without the remainder term to have a Fuller's pair. This will be done using the quasi-homogeneity of the normal form without remainder term to reduce the dimensions. Then a perturbation argument proves the existence of a Fuller's pair of curves for the original system.

In the next section, we state our main result on the Fuller's phenomenon.

5) Main theorem

Let M be a symplectic manifold and m_0 a point on M. To each pair of smooth functions k, h defined in a neighborhood of m_0, we associate the following vector $\alpha(k, h, m_0)$ in $\mathbb{R}^6$: set $f = \frac{1}{2}(k + h)$ $g = \frac{1}{2}(k - h)$. The components of $\alpha(k, h, m_0)$:

$$\{ad^2 f(g), adf(g)\}(m_0),\ \{ad^2 g(f), adg(f)\}(m_0),\ \frac{1}{6}ad^4 f(g)(m_0)\},\ \frac{1}{4}ad^2 f ad^2 g(f)(m_0),$$
$$\frac{1}{6}adf ad^3 g(f)(m_0),\ \frac{1}{24}ad^4 g(f)(m_0).$$

Theorem 0: There exists a semi-algebraic set $\mathcal{F}$ in $\mathbb{R}^6$ having a non empty interior with the following property:

Let H_+, H_- be any pair of smooth functions defined in a neighborhood of m_0 in M and satisfying the following assumptions:

a) The differentials $dH_+(m_0)$, $dH_-(m_0)$, $d\{H_+, H_-\}(m_0)$ are independent.

b) All Poisson brackets of H_+, H_- of length up to 4 are zero at m_0.

c) $\alpha(H_+, H_-, m_0) \in \mathcal{F}$.

Then the couple H_+, H_- has a Fuller pair of curves at m_0.

Remark 0: b) means that, $H_+(m_0), H_-(m_0), \{H_+, H_-\}(m_0), \{H_+, \{H_+, H_-\}\}(m_0)$, $\{H_-, \{H_+, H_-\}\}(m_0)$, $ad^3 H_+(H_-)(m_0), ad^3 H_-(H_+)(m_0), adH_+ ad^2 H_-(H_+)(m_0)$ are all zero.

The proof of the theorem just stated is too long and too involved to be given here. We refer the reader to our paper [K2] for it.

The preceding theorem gives us to answer the question of how general the Fuller's phenomenon is. It is the object of the next corollary. To state it, we need to recall some well know concepts.

Let M be a smooth manifold and U is a closed subset of a euclidean space $\mathbb{R}^N$. Denote by OS the set of all pairs (V,c) of a smooth vector field V on M parametrized by U, $V : M \times U \rightarrow TM$, and a smooth function $c : M \times U \rightarrow \mathbb{R}$. (Smooth is given the meaning defined at the beginning of section 3). We endow the set OS with the C^∞ topology. Now we can state our corollary.

Corollary 0: Assume that the dimension of M is at least 4. Then there exists an open non empty subset Φ of OS with the property: for any system (V,c) in Φ there exist a non empty submanifold $S(V,c)$ of codimension 4 in M such that all points in $S(V,c)$ are Fuller points for (V,c).

Proof: It is not hard to give a proof. To show how one goes about it, we treat the case when M is an open neighborhood of 0 in $\mathbb{R}^4$ and U is the interval [-1,+1].

Let us start with the following control system defined on $\mathbb{R}^4$, u being the control

$$\frac{dx^1}{dt} = u, \quad \frac{dx^2}{dt} = 1, \quad \frac{dx^3}{dt} = x^1 + B_1(x^1)^2, \quad \frac{dx^4}{dt} = D_1(x^1)^2 + D_2x^1x^2$$

B_1, D_1, D_2 are constants.

As cost function we choose the function

$$c_0 = \frac{1}{2}a(x^3)^2 - \frac{1}{2}b(x^1)^2x^3 - c_1x^1(x^2)^3 - c_2(x^1)^2(x^2)^2 - c_3(x^1)^3x^2 - c_4(x^1)^4$$

$$-A_1(x^1)^2x^4 - E_1(x^1)^3x^3 - A_2x^1x^2x^4 - E_2x^1(x^2)x^3$$

where a, b, c_i, $1 \leq i \leq 4$, A_1, A_2, E_1, E_2 are constants chosen as follows:

(i) $(a, b, c_1, c_2, c_3, c_4)$ ia an interior point of the set F.

(ii) Both determinants Δ_1 and Δ_2 are non-zero

$$\Delta_1 = \det \begin{vmatrix} A_1 & D_1 \\ A_2 & D_2 \end{vmatrix}$$

$$\Delta_2 = \det \begin{vmatrix} 24c_4 & 6c_3 & E_1 \\ b+4c_3 & 4c_2 & aB_1 \\ a+2c_2 & 6c_1 & 6E_2 \end{vmatrix}$$

Let us call S_0 the preceding system and let us denote by H_+, H_-, the hamiltonians $H_\varepsilon = up_1 + p_2 + (x^1 + B_1(x_1)^2)p_3 + [D_1(x^1)^2 + D_2 x^1 x^2]p_4 - \hat{c}(x^1, x^2, x^3, x^4)$. The subset of $\mathsf{R}^4 \times \mathsf{R}^4$ of all points where the poisson brackets of length less or equal to 4 of H_+, H_- are zero is reduced to the point 0 in a neighborhood of 0: this set is the zero set of the functions, $F, G, \{F,G\}$, $ad^2F(G)$, $ad^2G(F)$, $ad^3F(G)$, $adFad^2G(F)$, $ad^3G(F)$, ($F = \frac{1}{2}(H_+ + H_-)$, $G = \frac{1}{2}H_+ - H_-)$. An easy computation, using condition (ii), shows that the differential at 0 of these eight functions are independent.

Condition (i) implies that S_0 has a Fuller pair at $m_0 = 0$. Since the point $\alpha = (a, b, c_1, c_2, c_3, c_4)$ lies in the interior of F and since the differentials of the eight functions above are independent, the theorem 0 shows that any system S, formed by a controlled vector-field F on R^4 and a cost function $c : \mathsf{R}^4 \to \mathsf{R}$ which are sufficiently near to the system S_0 in the C^4-topology, will have a Fuller pair at a point in a neighborhood of zero.

References

[F] A.T. FULLER: "Study of an optimum non linear system" J. Electronic Control **15** (1963) pp 63-71.

[G] R. GAMKRELIDZE: "Principles of optimal control theory", Plenum (1978).

[K1] I.A.K. KUPKA: "Geometric theory of extremals in optimal control problems: I. The fold and maxwell case." T.A.M.S. vol. **299** no 1 (Jan 1987) pp 225-243.

[K2] I.A.K. KUPKA: "The ubiquity of Fuller's phenomena" to appear in the proceedings of the workshop on Optimal Control at Rutgers University. H. Sussmann editor, M.

Dekker publisher.

[R] E.P. RYAN: "Optimal relay and saturating system synthesis" I.E.E. Control Engineering Series no. 14, Peter Peregrinus (1982).

[S] H.J. SUSSMANN: in "Differential geometry control theory" R.W. Brockett, R.S. Millman, H.J. Sussmann ed., Birkhäuser PM no. 27 (1983).

FUNNEL EQUATIONS AND MULTIVALUED INTEGRATION PROBLEMS FOR CONTROL SYNTHESIS

A.B. Kurzhanski, O.I. Nikonov

I.I.A.S.A., Laxenburg, Austria and

Institute of Mathematics & Mechanics, Sverdlovsk, USSR

This paper deals with the problem of synthesizing a feedback control strategy for a linear controlled system subjected to unknown but bounded input disturbances and convex state constraints (see [1-7]). While seeking for the solution in the form of an "extremal strategy" as introduced by N.N. Krasovski, it is shown that the respective sets of solubility states that are crucial for the solution of the control problem could also be treated as cross-sections of trajectory tubes for some specially designed "funnel equations". The set-valued solutions to these could be then presented in the form of specially derived multivalued integrals.

1. The Problem of Synthesizing "Guaranteed" Control Strategies.

Consider a controlled system

$$\dot{x} \in u - Q(t) , \tag{1.1}$$

$$t \in T = [t_0, t_1] , \quad x \in \mathbf{R}^n$$

under restrictions

$$u \in P(t) , \tag{1.2}$$

$$x(t) \in Y(t) , \tag{1.3}$$

$$x(t_1) \in \mathcal{M} . \tag{1.4}$$

Here (1.1) is the equation for the controlled process, u is the *control parameter* restricted by the set-valued function $P(t)$ as in (1.2), (1.3) is the *state constraint*, (1.4) is the *terminal condition*. The functions $P(t)$, $Q(t)$ are set-valued, with values in conv $(\mathbf{R}^n)$ and measurable in t, $Y(t)$ is set-valued with values in cl$(\mathbf{R}^n)$, continuous in t. The notions of continuity and measurability of multivalued maps are taken in the sense of [12]. The set $\mathcal{M} \in \text{conv}\,(\mathbf{R}^n)$.

Here and further we assume the notations:

- $\text{conv}(\mathbf{R}^n)$ for the set of convex compact subsets of $\mathbf{R}^n$,
- $\text{cl}(\mathbf{R}^n)$ for the set of closed convex subsets of $\mathbf{R}^n$,
- $\text{comp}(\mathbf{R}^n)$ for the set of compact subsets of $\mathbf{R}^n$, I - for the unit matrix.

The set-valued function $Q(t)$ describes the range of uncertainty in the process assuming that the system is affected by unknown but bounded input disturbances $v(t)$, so that

$$\dot{x} = u - v(t)$$

with

$$v(t) \in Q(t) \ . \tag{1.5}$$

The aim of this paper is to describe a certain unified scheme for solving the following problem of *"guaranteed" control synthesis*:

Specify a *synthesizing strategy*

$$u = u(t,x)$$

so that for *every* solution to the system

$$\dot{x}[t] \in u(t,x[t]) - Q(t)$$

the inclusions

$$u(t,x[t]) \subseteq P(t) \ ,$$

$$x[t] \in Y(t) \ ,$$

$$x[t_1] \in \mathcal{M} \ ,$$

would be satisfied for any initial condition $x^0 = x[t_0]$, from a given set $X^0 : x^0 \in X^0$.

The class $\mathcal{U}$ of strategies $u(t,x)$ within which the problem is to be solved is taken to consist of set-valued functions $u(t,x)$ with values in $\text{conv}\,(\mathbf{R}^n)$, measurable in t and upper semicontinuous in x. The inclusion $u(t,x) \in \mathcal{U}$ here ensures the existence of a solution to the equation

$$\dot{x} \in u(t,x) - Q(t) \ , \ \ x^0 \in D$$

for any $D \in \mathrm{conv}(\mathbf{R}^n)$. The solution control strategy $u(t,x)$ should thus guarantee the inclusions (1.1), (1.3), (1.4) whatever are the disturbance $v(t)$, and the vector $x^0 \in X^0$.

A crucial point in the solution of the control synthesis problem is to find the set $W[t_0] = \{x_0\}$ of all initial states x^0 that assure the solvability of the problem (so that $W[t_0]$ would be the largest of all sets X^0 that ensure the solution). A similar question may be posed for any instant of time $\tau \in (t_0, t_1)$. This leads to a multivalued function $W[\tau]$, $\tau \in T$.

As demonstrated in [1] the knowledge of function $W[t]$, in particular for the "linear-convex" problems (1.1), (1.2), (1.4), allows to devise a solution $u(t,x)$ in the form of an "extremal strategy" of control. However, the basic scheme also remains true for the problem (1.1)-(1.4) with a state constraint. The main accents of this paper are not on the discussion of the relevance of the extremal strategy (which will nevertheless be specified below) but rather on the unified formal scheme for describing $W(t)$. It will be shown in the sequel that $W[t]$ may be defined as the solution to an evolution "funnel equation" which allows a representation in the form of a special multivalued integral that generalizes some of the conventional multivalued integrals (Auman's integral, the convolution integral, Pontriagin's alternated integral [7]).

The treatment of equation (1.1) rather than

$$\dot{x} \in A(t)x + u - Q(t)$$

causes no loss of generality, provided $P(t)$ does not reduce to a constant transformation.

2. The Basic "Funnel" Equation

In this paragraph we introduce a formal evolution equation whose solutions are set-valued functions which will later be shown to describe the required sets $W[t]$.

Assume the multivalued functions $P(t)$, $Q(t)$, $Y(t)$ are to be given as in § 1. With P', $P'' \in \mathrm{conv}(\mathbf{R}^n)$ we introduce the standard "Hausdorff semidistance" as

$$h_+(P', P'') = \min_{r \geq 0}\{r : P' \subseteq P'' + rS\}$$

where $S = \{x : (x,x) \leq 1,\ x \in \mathbf{R}^n\}$ is a unit ball in $\mathbf{R}^n$ ((x,x) is the inner product in $\mathbf{R}^n$).

Definition 2.1. A multivalued map $Z(t)$ with compact values will be said to be h_+ - *absolutely continuous* on an interval $[t_0, t_1]$ if $\forall \varepsilon > 0$, $\exists \delta > 0$:

$$\sum_i (t_i'' - t_i') < \delta \Longrightarrow \sum_i h_+(Z(t_i'), Z(t_2'')) < \varepsilon$$

where $\{(t_i',t_i'')\}$ - is a finite or countable number of nonintersecting subintervals of $[t_0,t_1]$.

Consider the *"funnel equation"*

$$\lim_{\sigma\to 0} \sigma^{-1} h_+(Z(t-\sigma) - \sigma Q(t), (Z(t)\cap Y(t)) - \sigma P(t)) = 0 \tag{2.1}$$

with boundary condition

$$z(t_1) \subseteq \mathcal{M} \ , \mathcal{M} \in \text{comp}(\mathbf{R}^n) \ . \tag{2.2}$$

Definition 2.2. An *h_+-solution* to equation (2.1) will be defined as an h_+-absolutely continuous set-valued function $Z(t)$ with values in comp$(\mathbf{R}^n)$ that satisfies (2.1) almost everywhere on $[t_0,t_1]$.

In general the h_+-solution to (2.1), (2.2) is nonunique. The unicity may be achieved by selecting a "maximal" solution to (2.1), (2.2) in the sense of the partial order $\leq$ for the set of all h_+-solutions $\{Z(\cdot)\}$ to (2.1), (2.2), $t\in[\tau,t_1]$ introduced by assuming that $Z_1(\cdot) \leq Z_2(\cdot)$ iff $Z_1(t) \subseteq Z_2(t)$ for all $t\in[\tau,t_1]$.

Lemma 2.1. If $W[\tau]\neq\phi$ for some $\tau\in(t_0,t_1)$, then the variety of all solutions to (2.1), (2.2), $t\in[\tau,t_1]$ is nonvoid and has a unique maximal element with respect to the partial order $\leq$.

3. The Formal Solution

The solution to the problem of synthesizing controls that guarantee the restrictions (1.1), (1.2), (1.4) is given by the following theorem:

Theorem 3.1 (i) *The solution to the problem of control synthesis for the system (1.1)-(1.4) in the class $u\in\mathcal{U}$ from the initial position $x_\tau^0 = x[\tau]$ does exist if and only if $W[\tau]\neq\phi$ and*

$$x_\tau^0 \in W[\tau] \ , \tag{3.1}$$

(ii) *The condition $W[\tau]\neq\phi$ is fulfilled if and only if on the interval $[\tau,t_1]$ there exists an h_+-solution $Z(t)$ to equation (2.1) with boundary condition (2.2); then $W[t]$ is the unique maximal solution to (2.1) with respect to the partial order $\leq$,*

(iii) *the guaranteed synthesizing strategy that resolves (1.1)-(1.4) is given in the form*

$$u(t,x) = \begin{cases} P(t), \text{if } x\in W[t] \\ \partial_\ell \, \rho(-\ell^0|P(t)), \text{ if } x\overline{\in} W[t] \end{cases} \tag{3.2}$$

Here $\rho(\ell|Z) = \max\ \{(\ell,z)|z\in Z\}$ *is the support function* for set Z, $\partial_\ell f(\ell,t)$ stands for the *subdifferential* of $f(\ell,t)$ in ℓ, $\ell^0 = \ell^0(t,x)$ is a unit vector that solves the problem

$$d(x,Z(t)) = (\ell^0,x) - \rho(\ell^0|Z(t)) = \max_{\|\ell\|=1}\{(\ell,x) - \rho(\ell|Z(t))\}\ , \tag{3.3}$$

and the symbol $d(x,Z(t))$ stands for the Euclid distance from point x to set $Z(t)$.

Extremal strategies of type (3.2), (3.3) were introduced by N.N. Krasovski (see [1] , [6]).

The proof of Theorem 3.1 is based on the following assertions.

Lemma 3.1. Suppose $Z(t)$ is an h_+-solution to equation (2.1), (2.2) for the interval $[\tau,t_1]$. Then

$$Z(t) \subseteq Y(t), \quad t\in[\tau,t_1]\ .$$

For any h_+-solution $Z(t)$ to (2.1), (2.2) we may define an extremal strategy $u_z(t,x)$ according to (3.2), (3.3), (substituting W for Z).

Lemma 3.2. The multivalued map $u = u_z(t,x)$ is such that $u\in U$, i.e. $u_z(t,x)$ is measurable in t and upper semicontinuous in x.

This ensures the existence of solutions to equation

$$\dot{x}\in u_z(t,x) - Q(t) \tag{3.4}$$

for any $x^0\in W[t_0]$.

Lemma 3.3. Assume the inclusion (3.4) is generated by a strategy $u_z(t,x)$ for a given h_+-solution $Z(\cdot)$ of the evolution equation (2.1), (2.2). Then, for almost all values of $t\in T$, in the domain $d(x,Z(t)) > 0$ the following estimate is true along the solutions to (3.4)

$$\frac{d}{dt}(d[t]) = (l^0,u[t]-v[t]) - \frac{\partial}{\partial t}\rho(l^0|Z(t)) \leq (l^0,u[t]-v[t])$$

$$+\ \rho(-l^0|P(t)) - \rho(-l^0|Q(t))$$

where $u[t]\in P(t)$, $v[t]\in Q(t)$.

Lemma 3.4. For a solution $x^0[t]$ of (3.4) the initial condition $x^0[\tau]\in Z(\tau)$ yields $x^0[t]\in Z(t)$ for any $t\in[\tau,t_1]$.

4. Multivalued Integration

Once the evolution equation (2.1), (2.2) is given it is possible to define the crossection $W[t]$ of the solution tube $W(\cdot)$, $W(t_1)\subset\mathcal{M}$ as a certain *multivalued integral*. As we shall see in the sequel this integral generalizes a whole range of "simpler" multivalued integrals. Let us proceed with constructing the corresponding integral sums. Suppose $\mathbf{M}[t',t'']$ stands for the set of $(n\times n)$-matrix valued functions continuous on $[t',t'']$, and $\mathbf{M}$ for the set of square matrices of dimension n. Introduce a subdivision P_m of the interval $[\tau,t_1]$ as

$$P_m{:}\tau = \tau_0 < \tau_1 < \cdots < \tau_m = t_1$$

$$\Delta_m = \max\{|\tau_i - \tau_{i-1}|,\ i=0,\ldots,m\}$$

and define integral sums of the following three types:

(1)

$$X_m^{(1)}(P_m,\mathcal{M}) = \mathcal{M}$$

$$X_{i-1}^{(1)} = \cap\{[\int_{\tau_{i-1}}^{\tau_i} (I - \int_{\tau_{i-1}}^{t} M(\xi)d\xi)P(t)\,dt \tag{4.1}$$

$$+ \int_{\tau_{i-1}}^{\tau_i} M(t)\,Y(t)\,dt + (I - \int_{\tau_{i-1}}^{\tau_i} M(\xi)\,d\xi)X_i^{(1)}]$$

$$\dot{-}\ [-\int_{\tau_{i-1}}^{\tau_i} (I - \int_{\tau_{i-1}}^{t} M(\xi)\,d\xi)\,Q(t)\,dt]\ |\ M(\cdot)\in\mathbf{M}[\tau_{i-1},\tau_i]\}$$

$$i = m, m-1,\ldots,1$$

(2)

$$X_m^{(2)}(P_m,\mathcal{M}) = \mathcal{M}$$

$$X_{i-1}^{(2)} = \cap\{-\int_{\tau_{i-1}}^{\tau_i} P(t)\,dt + MY(\tau_i) + (I-M)X_i^{(2)}] \tag{4.2}$$

$$\dot{-}\ [-\int_{\tau_{i-1}}^{\tau_i} Q(t)\,dt]\ |M\in\mathbf{M}|\ \}$$

$$i = m, m-1,\ldots,1$$

(3)

$$X_m^{(3)}(P_m,\mathcal{M}) = \mathcal{M}$$

$$X_{i-1}^{(3)} = [-\int_{\tau_{i-1}}^{\tau_i} P(t)\,dt + X_i^{(3)} \cap Y(\tau_i)]$$

$$\dot{-}\ [-\int_{\tau_{i-1}}^{\tau_i} Q(t)\,dt]$$

Symbol $\dot{-}$ stands for the "geometrical" ("Minkowski") difference, i.e. for sets **A**,**B** given

$$\mathbf{C} = \mathbf{A} \dot{-} \mathbf{B} = \{\mathbf{c}{:}\mathbf{c} + \mathbf{B} \subseteq \mathbf{A}\}\ .$$

The convergence of the integral sums with $m \rightarrow \infty$ to a value that does not depend on the subdivision P_m is ensured by the following assumption:

Assumption 4.1. There exists a function $\beta(t)$, continuous in t, $t \in T$ and such that $\beta(t) > 0$ for $t \in [t_0,t_1)$ and that for any subdivision P_m of the interval T the following inclusion is true

$$X_i^{(3)}(P_m,\mathcal{M}) \cap Y(\tau_i) \dot{-} \beta(\tau_i)S \neq \phi$$

$$(i = 0,1,\cdots,m)$$

Theorem 4.1. Under Assumption 4.1 the limit

$$J(\tau,t_1,\mathcal{M}) = \lim_{\substack{m \rightarrow \infty \\ \Delta_m \rightarrow 0}} X_0^{(i)}(P_m,\mathcal{M})$$

depends neither on the sequence of subdivisions P_m *nor on the index* $i = 1,2,3$. *The following equality is true*

$$W[\tau] = J(\tau,t_1,\mathcal{M})\ .$$

The definition of the integral $J(\tau,t_1,\mathcal{M})$ is therefore correct.

We will now follow several particular cases starting from the simplest one.

5. Attainability Domains for Control Systems

Assume $Q(t) = \{0\}$, $Y(t) \equiv \mathbf{R}^n$. Then $W[\tau]$ is the attainability domain for system

$$\dot{x} = u, \quad u \in P(\tau), \quad x(t_1) = \mathcal{M}$$

written in backward time from t_1 to t_0 and evolving from set $\mathcal{M}$. A funnel equation for differential inclusions in the absence of state constraints was studied in [8,9] in terms of the *Hausdorff distance* $h(Z_1, Z_2)$, where

$$Z_1, Z_2 \in \text{comp } \mathbf{R}^n \,, \quad h(Z_1,Z_2) = \max \, \{h_+(Z_1,Z_2),\, h_-(Z_1,Z_2)\}$$

and

$$h_-(P',P'') = \min \, \{r : P'' \subseteq P' + rS \,|\, r \geq 0\} \; .$$

The funnel equation for $W[\tau]$ is as follows

$$\lim_{\sigma \to +0} \sigma^{-1} h(Z(t-\sigma),\, Z(t) - \sigma P(t)) = 0 \tag{5.1}$$

$$Z(t_1) = \mathcal{M} \; .$$

Lemma 5.1. Under conditions $Q(t) \equiv \{0\}$, $Y(t) \equiv \mathbf{R}^n$, the set $W[t]$, $W[t_1] = \mathcal{M}$ is the unique solution to equation(5.1)and also the unique maximal solution to equation

$$\lim_{\sigma \to 0} \sigma^{-1} h_+(Z(t-\sigma),\, Z(t) - \sigma P(t)) = 0$$

$$Z(t_1) = \mathcal{M} \; .$$

It may be represented as a "multivalued Lebesque integral" ("Aumann's integral")

$$W[t] = \; J(t,t_1,\mathcal{M}) = \mathcal{M} + \int_{t_1}^{t} P(\xi)\, d\xi$$

6. "Viability" Tubes.

Consider the particular case $Q(t) \equiv \{0\}$ (a system with state constraints in the absence of uncertainty). Then $W[\tau]$ is the set of states from each of which there exists a "viable" trajectory (relative to constraint (1.3)) that ends in $\mathcal{M}$. In other words, for each $x^0 \in W[\tau]$ there exists a control $u[t]$ restricted by (1.2) that generates a trajectory $x[t] \in Y(t)$ for all $t \in T$ and such that $x[t_1] \in \mathcal{M}$. ($W[\tau]$ is also the attainability domain for system (1.1)-(1.4), $Q(t) \equiv \{0\}$ in backward time.)

The evolution equation (2.1) here forms to be

$$\lim_{\sigma \to +0} \sigma^{-1} h_+(Z(t-\sigma), (Z(t) \cap Y(t)) - \sigma P(t)) = 0 \ . \tag{6.1}$$

Theorem 6.1. Under condition $Q = \{0\}$ the multivalued function $Z = W[t]$ is the only maximal solution to equation (6.1), (2.2) with respect to the partial order $\leq$. This solution may be presented in the form of a multivalued convolution integral

$$W[\tau] = J(\tau,t_1,\mathcal{M}) \tag{6.2}$$

$$= \cap \{ \int_\tau^{t_1} [S(t)P(t) - \dot{S}(t)Y(t)]dt + S(t_1)\mathcal{M} \,|\, M(\cdot) \in \mathbf{M}[\tau,t_1]\}$$

where $S(t)$ and $M(t)$ are connected through the equation

$$\frac{dS(t)}{dt} = - M(t), \ S(\tau) = I, \ M(\cdot) \in \mathbf{M}[\tau,t_1] \ .$$

The intersection (6.2) is taken over all matrix functions $M(\cdot) \in \mathbf{M}[\tau,t_1]$. The integral (6.2), introduced in [4] is a "multivalued" version of the convolution integral described in [12].

Another version of the funnel equation for $W[t]$, as given in [4], is the following:

$$\lim_{\sigma \to 0} \sigma^{-1} h(Z(t-\sigma), (Z(t) - \sigma P(t)) \cap Y(t)) = 0 \ . \tag{6.3}$$

Here we use the Hausdorff distance and the equation is true if either the function $Y(t)$ has a convex graph, where

$$\text{graph } Y(t) = \{t,x{:}x \in Y(t);\ t \in T\}$$

or if the support function $\rho(\ell | Y(t)) = f(t,\ell)$ is continuously differentiable in t,ℓ. The transition to the Hausdorff semidistance in the context of equation (6.1) allows to drop the additional requirements on $Y(t)$ but yields no unicity of solution. The latter is regained, however, if we consider the maximal ($\leq$) solution to (6.1).

7. Solution Sets for Game-theoretic Control Synthesis.

Assume $Y(t) \equiv \mathbf{R}^n$ so that there is no state constraint. Then the initial problem transforms to one of synthesizing a control strategy in a differential game with fixed time with terminal cost being the distance to set $\mathcal{M}$. A guaranteed solution strategy $u(t,x)$ should ensure $d(x[t_1],\mathcal{M}) = 0$ so that $x[t_1] \in \mathcal{M}$.

The funnel equation for the set $W[t]$ that would generate a solution strategy of type (3.2) is now as follows

$$\lim_{\sigma \to +0} \sigma^{-1} h_+(Z(t-\sigma) - \sigma Q(t),\, Z(t) - \sigma P(t)) = 0\ . \tag{7.1}$$

The formulae (4.1)-(4.3) for the integral sums (with $Y(t) \equiv \mathbf{R}^n$) now coincide with the "alternated sums" introduced in [7].

Theorem 7.1. With $Y(t) \equiv \mathbf{R}^n$ the multivalued map $W[\tau]$ is the unique maximal ($\leq$) solution to equation (7.1), (2.2). Under Assumption 4.1 the function $W[\tau]$ is continuous and satifies (7.1) for all $t \in T$. It may be presented through the "alternated integral" of L.S. Pontriagin [7], as

$$W[\tau] = J(\tau,t_1,\mathcal{M}) = \int_{\mathcal{M},t_1}^{\tau} (P(t)dt \dot{-} Q(t)dt)\ . \tag{7.2}$$

Assuming that $P(t)$, $Q(t)$ are of "similar type" (i.e.,$0 \in P(t)$, $0 \in Q(t)$ and $P(t) = \alpha Q(t)$, $\alpha > 0$), the Hausdorff semidistance h_+ in (7.1) may be substituted for the distance h and the integral (7.2) transforms into

$$J(\tau,t_1,\mathcal{M}) = \int_{\mathcal{M},t_1}^{\tau} (P(t) \dot{-} Q(t))dt$$

We then arrive at the "regular case" for the respective differential game [1].

References

[1] Krasovski, N.N., Subbotin, A.I. Positional Differential Games. Nauka, Moscow, 1974 (English translation in Springer-Verlag, 1988).

[2] Kurzhanski, A.B. Control and Observation Under Conditions of Uncertainty. Nauka, Moscow, 1977.

[3] Kurzhanski, A.B., Filippova, T.F. On the description of the set of viable trajectories of a differential inclusion. Soviet Math. Doklady Vol. 289, Nr. 1, 1986.

[4] Kurzhanski, A.B., Filippova, T.F. On the Set-Valued Calculus in Problems of Viability and Control of Dynamic Processes: the Evolution Equation. IIASA Working Paper WP-88-91, 1988.

[5] Aubin, J.P., Cellina, A. Differential Inclusions, Springer-Verlag, 1984.

[6] Krasovski, N.N. Game Problems on the Encounter of Notions. Nauka, Moscow, 1970.

[7] Pontriagin, L.S. Linear Differential Games of Pursuit, Mat. Sbornik, Vol. 112 (154), 3 (7), 1980.

[8] Panasiuk, A.I., Panasiuk, V.I. Mat. Zametki, Vol. 27, No. 3, 1980.

[9] Tolstonogov, A.A. Differential Inclusions in Banach Space. Nauka, Novosibirsk, 1986.

[10] Akilov, G.P., Kutateladze, S.S. Ordered Vector Spaces. Nauka, Novosibirsk, 1978.

[11] Filippov, A.F. Differential Equations with Discontinuous Right-hand Side. Nauka, Moscow, 1985.

[12] Ioffe, A.D., Tikhomirov, V.M. The Theory of Extremal Problems. Nauka, Moscow, 1974.

Output Tracking Through Singularities

F. Lamnabhi-Lagarrigue , P.E. Crouch and I. Ighneiwa

ABSTRACT. In this paper we consider the tracting problem for single-input, single-output nonlinear systems, affine in the control and where f, g and h are analytic. In the tracking problem one is given a function y and considers conditions under which there exists a control u_d, and corresponding state trajectory x_d, so that the output of the initialized system coincides with y_d on some time interval [0, T]. If T is alowed to be arbitrarily small then the solution is well studied. We are concerned here with one problem which occurs when trying to extend these results to the case where T is specified a priori. There are many applications of singular tracking, namely in robotics. This study pursues the analysis begun by Hirschorn and Davis. It is introduced further structure which is important in quantifying smoothness of solutions to the problem. Specific results are given in the particular case of time-varying linear systems.

1. INTRODUCTION

In this paper we consider the tracking problem for S.I.S.O. nonlinear systems, affine in the control and having the form

$$\dot{x} = f(x) + u\, g(x), \quad x \in \mathbb{R}^n, \qquad x(0) = x_o$$
$$y = h(x) \tag{1}$$

We shall say that a function, or vector field on an open set $U \subset \mathbb{R}^n$ is *analytic* if it is the restriction of a complex analytic function, or vector field on $\mathbb{C}^n$. In particular we allow analytic functions, or vector fields on $U \subset \mathbb{R}^n$, to have singularities. We say that an analytic function or vector field is *regular (analytic)* in a given region if it has no singularities in that region and that a function is *regular (analytic)* at a point if it has a convergent power series at that point. Our discussion could be set on a general manifold rather than $\mathbb{R}^n$ without any extra work, but no added insight is obtained in such a generalization.

In the tracking problem one is given a function y_d and considers conditions under which there exists a control u_d, and corresponding state trajectory x_d, so that the output of the initialized system (1) coincides with y_d on some interval [0, T], with the added constraint that both x_d and u_d do not "blow up" on the given interval. If T is allowed to be arbitrarily small, then the solution is well studied (see Hirschorn [5,6], Singh [11,12], Nijmeijer [9] and Grasse [3] as well as the references therein for details). We are concerned here with one problem which occurs when trying to extend these results to the case where T is specified a priori.

In the present work we shall assume that y_d is a regular analytic function on $(-\infty, \infty)$. If α denotes the relative degree of system (1), i.e., the least non-negative integer k such that

$$y^{(k)}(t) = \frac{d^k y}{dt^k}(t)$$

depends explicitly upon the control u, we have

$$y^{(\alpha)}(t) = f^{\alpha} h(x(t)) + u(t)\, g\, f^{\alpha-1} h(x(t)) \tag{2}$$

Here $f^{\alpha} h$, $g\, f^{\alpha-1} h$ etc. indicate repeated Lie derivatives of functions by vector fields. If $g\, f^{\alpha-1} h(x_o) \neq 0$ and

$$y_d^{(k)}(0) = f^k h(x_o), \qquad 0 \leq k \leq \alpha - 1 \tag{3}$$

then y_d can be tracked on a maximal time interval $[0, t_s)$, $t_s > 0$, for which $g\, f^{\alpha-1}h(x_d(t)) \neq 0$, $t \in [0, t_s)$ and x_d, the solution of system (1) subject to the feedback control

$$u_D(x, y_d^{(\alpha)}, t) = \frac{y_d^{(\alpha)}(t) - f^{\alpha} h(x(t))}{g\, f^{\alpha-1} h(x(t))} \tag{4}$$

exists on $[0, t_s)$.

We shall make the following definition.

DEFINITION 1 (Hirschorn and Davis [5])

x_s is a singular point for output tracking in case $g f^{\alpha-1} h(x_s) = 0$. The set of all such singular points x_s is denoted by the subset S of $\mathbb{R}^n$. □

If we assume that the trajectory x_d above is uniformly bounded on $[0, t_s)$, then since u_d and x_d are necessarily regular analytic on $[0, t_s)$, $x_s = \lim_{t \to t_s^-} (x_d(t))$ exists. It follows that $x_s \in S$ and $\lim_{t \to t_s^-} (g f^{\alpha-1} h(x_d(t))) = 0$.

In discussing extensions of the curves x_d and u_d past t_s it is clear that we may expect solutions with singularity at t_s. We therefore make the following definition

DEFINITION 2

A function u is said to be *piecewise regular analytic* on a bounded interval (a, b) $\subset \mathbb{R}$ if it consists of a finite concatenation of regular analytic functions u_i over intervals $(a_i, b_i) \subset \mathbb{R}$, $1 \leq i \leq N$, $a_1 = a$, $b_N = b$, $b_i = a_{i+1}$, $1 \leq i \leq N-1$, such that the following limits exist:

$$u_i^- = \lim_{t \to b_i^-} u_i(t), \quad u_{i+1}^+ = \lim_{t \to a_{i+1}^+} u_{i+1}(t), \qquad 1 \leq i \leq N-1.$$

We do not insist that $u_i^- = u_{i+1}^+$, $1 \leq i \leq N-1$. A function u on an infinite interval is piecewise regular analytic if it is piecewise regular analytic on any bounded subinterval.

An *admissible control* function for system (1) on a bounded interval (a, b) $\subset \mathbb{R}$ is any piecewise regular analytic function. We make a similar definition for an admissible control function on an unbounded interval.

An *admissible state trajectory* x, for system (1), corresponding to an admissible control function u, is a continuous function such that if (a, b) is an interval on which u is regular analytic, then x is regular analytic on (a, b) and satisfies the differential equation there. □

DEFINITION 3

We say that y_d can be *tracked through the singular point $x_s \in S$ at time t_s*, if there exists an admissible pair $(\bar{u}_d, \bar{x}_d)$ defined on some interval $[0, t_s + \varepsilon)$ such that $u_d(t) = \bar{u}_d(t)$, $x_d(t) = \bar{x}_d(t)$, $t \in [0, t_s)$, and $\bar{x}_d(t_s) = x_s$.

We say that y_d can be *tracked through the singular set S at $x_s \in S$ at time t_s* if, in addition, $\bar{x}_d(t) \notin S$, for $t \neq t_s$. □

We refer to an admissible pair $(\bar{u}_d, \bar{x}_d)$ in definition (3) as a solution to the (singular) tracking problem posed by y_d at time t_s. In general solutions will not be unique.

To simplify the problem of tracking through singularities, we now assume that x_o in system (1) is a member of S and $t_s = 0$. We modify definition (3) accordingly: y_d can be tracked through the singular point $x_o \in S$ if there exists an admissible pair (u_d, x_d) defined on some interval $(-\varepsilon, \varepsilon)$ such that $x_d(0) = x_o$. y_d can be tracked through the singular set S at $x_o \in S$ if in addition $x_d(t) \notin S$ for $t \neq 0$. Again any such admissible pair $(\bar{u}_d, \bar{x}_d)$ will be referred to as a solution to the (singular) tracking problem. This latter problem is indeed simpler since the former more general case is a non local problem involving constraints on u_d and x_d at two points $x_o \in S$ and $x_s \in S$. Indeed, even more general problems of tracking through higher numbers of singular points can be posed.

It is this simpler problem of tracking through the singular point $x_o \in S$, that Hirschorn and Davis [7, 8] consider and that we also pursue in this paper. Indeed, we demonstrate that the complexity of the problem is far greater than one is probably led to understand from the paper by Hirschorn and Davis [7, 8]. The main insight for our study comes from the realization that when the control u_D in (4) is substituted in the system equation (1) we obtain a singular differential equation which may be written in the form

$$g\, f^{\alpha-1} h(x)\dot{x} = g\, f^{\alpha-1} h(x)\, f(x) + (y^{(\alpha)} - f^{\alpha}h(x))\, g(x) \tag{6}$$

This equation may be considered as a special case of a general singular differential equation

$$e(x, t)\, \dot{x} = F(x, t), \quad x \in \mathbb{R}^n \tag{7}$$

where F and e are mappings, $F : \mathbb{R}^n \times \mathbb{R} \to \mathbb{R}^n$, $e : \mathbb{R}^n \times \mathbb{R} \to \mathbb{R}$, whose component functions are regular analytic.

There are many standard texts treating various aspects of singular equations, but the following are useful: Ince [9], Hille [4], Erdelyi [2] and Wasow [13]. Most analysis concerns the complex version of equation (7), but this ultimately yields information about the real version. In case $e(x, t) = e(t)$ and $F(x, t) = F(t)x$, where $F(t)$ is an nxn matrix valued function, then equation (7) is linear. Even in this case the theory is not complete in certain aspects, so that any study of the tracking problem must necessarily be incomplete in some areas.

In section 2 we review and develop some ideas in Hirschorn and Davis [7]. In section 3 we give our definitions and results for the singular tracking problem. In particular we introduce an integer r called the *rank of a singular point $x_o \in S$ with respect to the function y_d*. This integer depends on y_d in general, unlike the degree β of the singularity $x_o \in S$ introduced in Hirschorn and Davis [7]. In Section 4 linear time-varying systems are considered. In Section 5 we present an example of the theory. In a future paper [1] the multiplicity of solutions will be also determined in a rather general context.

2. SINGULAR TRACKING: THE DEGREE OF SINGULARITY

In this section we review aspects of the paper by Hirschorn and Davis [7] while introducing notation and concepts for the singular tracking problem. In particular we wish to find conditions under which y_d can be tracked through $x_o \in S$ at $t = 0$. Let $x_d(t) = x_d(t,s,z)$ be any solution of equation (1), if one exists, where $u_d(t) = u_D(x_d, y_d^{(\alpha)}, t)$ and $x_d(s,s,z) = z$. Clearly if $z \notin S$ then $x_d(t)$ exists on some interval $t \in (s-\varepsilon, s+\varepsilon)$ and is regular analytic there. However, if $s = 0$ and $z = x_o$, x_d may or may not exist and if it does exist it will not be unique in general. We assume, however, that u_d and x_d are admissible in the sense of definition 2.

We introduce some further notation. Write $\bar{u}^k(t)$ for the k+1 vector $(u(t), u^{(1)}(t), \ldots, u^{(k)}(t))'$, and similarly $\bar{y}^k(t)$ for the k+1 vector $(y(t), y^{(1)}(t), \ldots, y^{(k)}(t))'$. Differentiating the output y of system (1) along a sufficiently smooth solution x and correspondingly smooth control u we obtain

$$\begin{aligned} y^{(j)} &= a_j(x) = f^j h(x) \quad , \quad 0 \le j \le \alpha - 1 \\ y^{(\alpha)} &= a_\alpha(x) + u\, b(x) = f^\alpha h(x) + u\, g\, f^{\alpha-1} h(x) \\ y^{(\alpha+j)} &= c_j(x, \bar{u}^{j-1}) + u\, b_j(x, \bar{u}^{j-1}) + d_j(x, \bar{u}^j) \end{aligned} \tag{8}$$

where

$$\frac{d^j}{dt^j} a_\alpha(x(t)) = c_j(x(t), \bar{u}^{j-1}(t)), \qquad c_o = a_\alpha$$

$$\frac{d^j}{dt^j}\, b(x(t) = b_j(x(t), \bar{u}^{j-1}(t)), \qquad\qquad b_o = b$$

and
$$\frac{d^j}{dt^j}\, u(t)\, b(x(t) = d_j(x(t), \bar{u}^j(t)) + u(t) b_j(x(t), \;\; \bar{u}^{j-1}(t))$$

for $j \geq 1$. The definition of singular point $x_o \in S$ implies that $b(x_o) = 0$.

DEFINITION 4 (Hirschorn and Davis [7]).

The degree $\beta(x_o)$ of the singularity $x_o \in S$ is the minimum integer such that the map

$$\bar{u}^{\beta-1} \rightarrow b_\beta(x_o, \bar{u}^{\beta-1})$$

is not the zero map. Set $\beta(x_o) = 0$ if $x_o \notin S$. □

Since

$$\frac{d^j}{dt^j}\, (u(t)\, b(x(t)) = \sum_{k=1}^{j} \binom{j}{k} \frac{d^k}{dt^k}\, u(t)\, \frac{d^{j-k}}{dt^{j-k}}\, b(x(t))$$

we see that the maps

$$\bar{u}^k \rightarrow d_k(x_o, \bar{u}^k)\,, \quad 1 \leq k \leq \beta$$

are all the zero maps. If we now assume that u_d and hence x_d satisfy the additional regularity requirements that $u_d \in C^{\beta-1}$ and

$$\lim_{t \to 0^\pm} \left(\frac{d^\beta}{dt^\beta}\, u_d(t)\right)$$

exists, while y_d is tracked through $x_o \in S$, then from (8) we obtain the following necessary conditions:

$$y_d^{(j)}(0) = a_j(x_o), \qquad\qquad 0 \leq j \leq \alpha - 1 \tag{9}$$

$$y_d^{(\alpha+j)}(0) = c_j(x_o, \bar{u}_d^{j-1}(0)), \qquad 0 \leq j \leq \beta - 1 \tag{10}$$

$$y_d^{(\alpha+\beta)}(0) = c_\beta(x_o, \bar{u}_d^{\beta-1}(0)) + u_d(0)\, b_\beta(x_o, \bar{u}_d^{\beta-1}(0)). \tag{11}$$

Note that equations (10) and (11) represent polynomial equations in the components of $\bar{u}_d^{\beta}(0)$, which may or may not have a solution for a given y_d. This already indicates the non-uniqueness properties of solutions of the singular tracking problem. Hirschorn and Davis [7] approach the problem of determining $\lim_{t\to 0^\pm} u_d(t)$ assuming $x_d(t) \notin S$, for $t \neq 0$, by analyzing the identities

$$u_d(t) = u_D(x_d(t), y_d^{(\alpha)}, t) = \frac{y_d^{(\alpha)}(t) - a_\alpha(x_d(t))}{b(x_d(t))}, \qquad t \neq 0. \tag{12}$$

Assume that conditions (9)-(11) are satisfied for some $\bar{v}^{\beta-1} = (v_o, v_1, \ldots, v_{\beta-1})' \in \mathbb{R}^\beta$, which also satisfies $b_\beta(x_o, \bar{v}^{\beta-1}) \neq 0$ and that there exists an admissible solution (u_d, x_d) to the tracking problem, with $x_d(t) \notin S$, for $t \neq 0$. Let $u_r \in C^\infty$ be a control satisfying

$$\frac{d^j}{dt^j} u_r(0) = v_j, \qquad 0 \le j \le \beta - 1.$$

when system (1) is solved with this control while initialized at $x_o \in S$, it yields a solution $x(t) = r(t)$ which satisfies $r(t) \notin S$, for $t \neq 0$ and t sufficiently small since

$$\frac{d^\beta}{dt^\beta} b(r(t))\Big|_{t=0} = b_\beta(x_o, \bar{v}^{\beta-1}) \neq 0. \tag{13}$$

Consider the following possible method for determining $\lim_{t\to 0^+} u_d(t)$

$$\begin{array}{ccc}
\dfrac{y_d^{(\alpha)}(t) - a_\alpha(x_d(t, \sigma, r(\sigma)))}{b(x_d(t, \sigma, r(\sigma)))} & \xrightarrow[?]{\lim_{\sigma\to 0+}} & \dfrac{y_d^{(\alpha)}(t) - a_\alpha(x_d(t, 0, x_o))}{b(x_d(t, 0, x_o))} \\
\downarrow \lim_{t\to\sigma^+} & & ?\downarrow \lim_{t\to 0^+} \\
\dfrac{y_d^{(\alpha)}(\sigma) - a_\alpha(r(\sigma))}{b(r(\sigma))} & \xrightarrow{\lim_{\sigma\to 0+}} & v_o
\end{array}$$

L'Hopitals rule applied β times to

$$\lim_{\sigma \to 0^+} \frac{y_d^{(\alpha)}(\sigma) - a_\alpha(r(\sigma))}{b(r(\sigma))}$$

yields the limit v_o. The left limit $t \to \sigma^+$ also follows from the continuity of the map $t \to x_d(t, \sigma, r(\sigma))$. However, the upper limit $\sigma \to 0^+$ depends on the validity of the statement

$$\lim_{\sigma \to 0^+} x_d(t, \sigma, r(\sigma)) = x_d(t, 0, x_o) \tag{14}$$

where $x_o \in S$. Whereas the curve $t \to x_d(t, \sigma, r(\sigma))$, $\sigma \neq 0$, is uniquely determined by u_D, as we noted above, the curves $t \to x_d(t, 0, x_o)$ are not necessarily unique. Moreover, in general the map

$$(\sigma, z) \to x_d(t, \sigma, z)$$

is not continuous at $(0, x_o)$, as would be the case if $x_o \notin S$. We give a simple example of this behaviour.

The equation $\dot{x} - \frac{x}{t} = -2$ has many solutions $x(t, \sigma, r(\sigma))$ satisfying $x(0, 0, 0) = 0$,

$$x(t, 0, 0) = \begin{cases} at - 2t \ln t, & t \geq 0 \\ bt - 2t \ln(-t), & t \leq 0 \end{cases}$$

where a and b are arbitrary real numbers. If $a \neq b$ then x is continuous, and the unique analytic solution $x(t, \sigma, r(\sigma))$ satisfying $x(\sigma, \sigma, r(\sigma)) = r(\sigma)$, $\sigma > 0$, is given by

$$x(t, \sigma, r(\sigma)) = r(\sigma)\frac{t}{\sigma} - 2\,t(\ln t - \ln \sigma).$$

Only if $a = \lim_{\sigma \to 0^+} \left(\frac{r(\sigma) + 2\sigma \ln \sigma}{\sigma}\right)$ exists does $\lim_{\sigma \to 0^+} (x(t, \sigma, r(\sigma)))$ exist.

Because of these difficulties we consider another approach to the singular tracking problem in the next section.

3. SINGULAR TRACKING: THE RANK OF SINGULARITY

In this section we again assume that we wish to find conditions under which y_d can be tracked through $x_o \in S$ at $t = 0$. Differentiating the output y of system (1) along a sufficiently smooth solution x and correspondingly smooth control u we obtain the equations (8), which this time we write in the form

$$
\begin{aligned}
y^{(k)} &= a_k(x) \qquad 0 \le k \le \alpha - 1 \\
y^{(\alpha+k)} &= a_{\alpha+k}(x, \bar{u}^{k-1}) + u^{(k)}\, b(x),\ k \ge 0.
\end{aligned}
$$

Again $b(x_o) = 0$ and we denote by $\bar{a}_j(x,\ \bar{u}^{j-\alpha-1})$ the vector valued function

$$(a_o(x),\ a_1(x),\ \ldots,\ a_j(x,\ \bar{u}^{j-\alpha-1}))'.$$

Clearly $\bar{a}_j(x,\ \bar{u}^{j-\alpha-1}) = \bar{a}_j(x)$ for $j \le \alpha$. We now make our principal definition.

DEFINITION 5

Given a singular point $x_o \in S$ and function y_d, whenever possible, let $r = r(x_o, y_d) \ge 0$ be the largest integer for which there exists $\bar{v}^{r-1} = (v_o, v_1, \ldots, v_{r-1})' \in \mathbb{R}^r$ such that the equations

$$y_d^{(k)}(0) = a_k(x_o,\ \bar{v}^{k-\alpha-1}), \qquad 0 \le k \le \alpha+r \tag{15}$$

are satisfied. If equations (15) are satisfied for each $k \ge 0$ and a corresponding $\bar{v}^{k-\alpha-1}$ we set $r = \infty$. If equations (15) are not satisfied for any $r \ge 0$ then we say that r is *undefined.* We refer to *$r(x_o, y_d)$ as the rank of the singularity x_o determined by y_d.* □

Note that in general for a given x_o, $r(x_o, y_d)$ is dependent on y_d, although in particular cases r may be independent of y_d. Moreover, we distinguish between the cases $r = 0$ and r undefined. If $r = 0$ then we have

$$y_d^{(k)}(0) = a_k(x_o), \qquad 0 \le k \le \alpha. \tag{16}$$

If r is undefined the equations (16) are not all satisfied. Given a singular point $x_o \in S$ and function y_d such that $r(x_o, y_d) \ge 1$ we set

$$E_k(x_o, y_d) = \{\bar{v}^{k-1} \in \mathbb{R}^k;\ \ \bar{y}_d^{k+\alpha}(0) = \bar{a}_k(x_o, \bar{v}^{k-1})\}$$

for $1 \le k \le r$. Necessarily $E_k(x_o, y_d) \neq \emptyset$.

DEFINITION 6

The singular systems of order k, $0 \le k \le r(x_o, y_d)$, determined by the singular point x_o and function y_d, are given by the equations

$$\begin{aligned} \dot{x} &= f(x) + u_o g(x) & x(0) &= x_o \\ \dot{u}_o &= u_1 & u_o(0) &= v_o \\ &\vdots & &\vdots \\ \dot{u}_{k-2} &= u_{k-1} & u_{k-2}(0) &= v_{k-2} \end{aligned} \tag{17}$$

$$\dot{u}_{k-1} = \frac{y_d^{(\alpha+k)} - a_{\alpha+k}(x, u_o, \ldots, u_{k-1})}{b(x)} \qquad u_{k-1}(0) = v_{k-1}$$

$$y = h(x)$$

where $\bar{v}^{k-1} \in E_k(x_o, y_d)$.

The singular system of order $k = 0$ is simply the closed loop system

$$\begin{aligned} \dot{x} &= f(x) + \left(\frac{y_d^{(\alpha)} - a_\alpha(x)}{b(x)}\right) g(x), \qquad x(0) = x_o \\ y &= h(x) \end{aligned}$$

□

PROPOSITION 1

Consider a singular point $x_o \in S$ for system (1). A necessary condition for tracking y_d through x_o is that the rank $r(x_o, y_d)$ is defined. If $r(x_o, y_d) \ge 1$ and $r \ge k \ge 1$ then there exists a C^1 solution $(x, u_o, \ldots, u_{k-1})$ of a singular system of order k on $(-\varepsilon, \varepsilon)$ with $x(0) = x_o$, $(u_o(0), \ldots, u_{k-1}(0))' \in E_k(x_o, y_d)$ if and only if y_d can be tracked through x_o on $(-\varepsilon, \varepsilon)$ with solution (u_d, x_d) satisfying $u_d \in C^{k-1}$ and $x_d \in C^k$.

If $r < \infty$ then the maximum possible smoothness of solutions to the singular tracking problem is $x_d \in C^{r+1}$ and $u_d \in C^r$.

PROOF The necessity of equations (16) in order to track y_d is clear, and so $r(x_o, y_d)$ must be defined. If y_d can be tracked through x_o with $x_d \in C^{k+1}$, $u_d \in C^k$, then a C^1 solution to a singular system of order k exists. Conversely if there exists such a solution, we show that the output y of the singular system coincides with y_d. Differentiating $y(t) = h(x(t))$ along a solution of system (17) we obtain from equations (15)

$$y^{(j)}(0) = a_j(x_o, u_o(0), ..., u_{j-\alpha-1}(0)) = y_d^{(j)}(0) \tag{18}$$

for $0 \le j \le \alpha + k - 1$. Differentiating one further time we obtain

$$\begin{aligned} y^{(\alpha+k)}(t) &= \frac{d}{dt}\,(a_{\alpha+k-1}(x, u_o, ..., u_{k-2}) + u_{k-1}b(x)) \\ &= a_{\alpha+k}(x, u_o, ..., u_{k-1}) + b(x)\left(\frac{y_d^{(\alpha+k)} - a_{\alpha+k}(x, u_o, ..., u_{k-1})}{b(x)}\right) \\ &= y_d^{(\alpha+k)}(t). \end{aligned}$$

Thus from (18) we obtain $y^{(j)}(t) = y_d^{(j)}(t)$ for $t \in (-\varepsilon, \varepsilon)$, $0 \le j \le \alpha + k - 1$ and hence $y = y_d$ as claimed. Clearly $u_o = u_d \in C^k$ provides the desired control component for a solution to the singular tracking problem.

If $r < \infty$ and we assume that there exists a solution to the tracking problem with $u_d \in C^{r+1}$, $x_d \in C^{r+2}$ then on some interval containing $t = 0$

$$y_d^{(\alpha+r+1)} = a_{\alpha+r+1}(x_d, \bar{u}_d^r) + u_d^{(r+1)}b(x_d).$$

However, by definition of r, $y_d^{(\alpha+r+1)}(0) \neq a_{\alpha+r+1}(x_o, \bar{u}_d^r(0))$ and since $u_d^{r+1}(0)$ exists and $b(x_o) = 0$ we obtain a contradiction. We conclude that the maximum possible smoothness is $x_d \in C^{r+1}$, $u_d \in C^r$. □

From the above result we observe that in case $r(x_o, y_d) = \infty$ it is feasible to obtain C^∞ solutions to the tracking problem and even solutions, which are regular analytic at $t = 0$.

4 LINEAR TIME VARYING CASE

In this section we give a precise analysis of the smoothness of the solutions in the time-varying linear case,

$$\begin{cases} \dot{t} = 1 & t(0) = 0 \\ \dot{z} = A(t)z + d(t)u & z(0) = z_o \\ y = c(t)'z \end{cases} \tag{19}$$

where $(t, z) \in \mathbb{R}^n$. Some very rudimentary facts about linear singular equations which arise in our examples are recalled.

a) Singular differential equations

Singular differential equations constitute a vast subject area, so we give here only a few pertinent remarks for our paper. The results can be found in Wasow [13]. Most of the theory is developed in the complex domain so we shall work in this domain in this section. Although in general we must consider arbitrary nonlinear singular equations our examples mostly reduce to linear singular equations so we direct our attention to these.

Consider a differential equation

$$\dot{v} = A(t)v + f(t) \qquad v \in \mathbb{C}^n, \quad t \in \mathbb{C} \tag{20}$$

where $A(t)$ is an analytic matrix valued function, and $f(t)$ in an analytic vector valued function. If $t = 0$ is a point about which $A(t)$ and $f(t)$ are regular then there exists a fundamental matrix solution $V(t)$, $\det V \neq 0$, which is regular at the origin and satisfies

$$\dot{V} = A(t)V \qquad V \in \mathbb{C}^{nxn}, \quad t \in \mathbb{C} \tag{21}$$

and for any $v_o \in \mathbb{C}^n$ equation (20) has a unique solution

$$v(t) = V(t)V(0)^{-1} v_o + V(t) \int_0^t V(s)^{-1} f(s) ds \tag{22}$$

where the integral is taken along any path from 0 to t in a simply connected region in which V is regular.

If now A(t) has an isolated singularity at t = 0, a basic question concerns the existence of a fundamental matrix solution V(t) satisfying equation (21) in a neighbourhood of t = 0. If A(t) has an isolated simple pole at t = 0 then we may rewrite equation (21) as

$$t \dot{V} = \bar{A}(t)V \tag{23}$$

where $\bar{A}(t)$ is regular in a neighbourhood of the origin. t = 0 is said to be a regular singular point. Equation (23) possesses a fundamental matrix solution

$$V(t) = P(t)\, t^B \tag{24}$$

where P(t) is regular in a neighbourhood of t = 0, B is a constant matrix whose eigenvalues do not differ by a positive integer and

$$t^B = e^{(\ln t)B}.$$

Solutions of equation (20), if they exist, can be computed as before using equation (22), as long as the integral can be suitably interpreted. (f may also have a singularity at t = 0). Elements of the matrix $P(t)\, t^B$ may in general be written in the form

$$t^\lambda \sum_{j=0}^{r} (\log t)^j \phi_j(t)$$

where $\phi_j(t)$ are analytic functions which are regular in a neighbourhood of 0.

If on the other hand A(t) has an isolated pole of order k > 1 at t = 0 then we may rewrite equation (20) as

$$t^k \dot{V} = \bar{A}(t)\, V \tag{25}$$

where $\bar{A}(t)$ is regular in a neighbourhood of the origin. $t = 0$ is said to be an *irregular singular point* of rank $k - 1$.

A fundamental solution of the equation (24) does exist but cannot in general be expressed in the form of equation (23); in particular solutions are no longer expressed in terms of convergent power series expansions but the more general "asymptotic expansions", see Wasow [13] or Erdelyi [2].

A scalar differential equation

$$p_n(t)\, u^{(n)}(t) + p_{n-1}(t)\, u^{(n-1)}(t) + \dots + p_o(t)\, u(t) = 0$$

where $p_k(t)$ are analytic functions which are regular in a neighbourhood of $t = 0$, may be transformed into a first order vector equation whose fundamental matrix solution satisfies equation (21), and the structure of solutions examined as above. If $p_n(0) \neq 0$ then $A(t)$ is regular in a neighbourhood of 0 and solutions of the scalar equation are also. A theorem of Fuchs gives conditions for a regular singularity at $t = 0$ and may be expressed in the form

$$p_j(t)/p_n(t) = t^{-(n-j)}\, q_j(t)\,, \qquad 0 \leq j \leq n-1$$

where $q_j(t)$ are regular in a neighbourhood of $t = 0$. If these conditions are not satisfied we obtain an irregular singularity at $t = 0$.

For example $t = 0$ is a regular singularity of the equation

$$t\,\dot{u} - a\,u = 0$$

with solution $u = b\, t^a$. Another example of a regular singularity, taken from Ince [9] is given by the equation

$$\ddot{u} - \frac{\dot{u}}{t+1} + \frac{u(3t+1)}{4t^2(t+1)} = 0$$

This equation has a regular singularity at $t = 0$ with fundamental solutions

$$u_1 = t^{1/2} \quad \text{and} \quad u_2 = t^{1/2} \ln t + t^{3/2}$$

On the other hand the equation

$$t^2 \dot{u} + a\, u = 0$$

has an irregular singularity at $t = 0$ and solution $u = c\, e^{-a/t}$.

b) Smoothness of solutions

Let us now analyse the time-varying linear systems (19). Letting

$$s_k'(t) = s_{k-1}(t)'A(t) + \frac{d}{dt}\, s_{k-1}(t)'$$
$$s_o(t) = c(t)$$

we may write

$$y^{(k)}(t) = s_k(t)'z, \qquad 0 \leq k \leq \alpha\text{-}1$$
$$y^{(\alpha)}(t) = s_\alpha(t)'z + u(t)\, s_{\alpha-1}(t)'\, d(t).$$

Thus $\quad (b(x) =)\ b(t) = s_{\alpha-1}(t)'\, d(t) \quad$ and

$$(u_D(x, y_d^{(\alpha)}, t) =)\ \frac{y_d^{(\alpha)} - s_\alpha(t)'z}{b(t)} = u_D(z,\ y_d^{(\alpha)}, t) \tag{26}$$

Assuming $(0, z_o) \in S \subset \mathbb{R} \times \mathbb{R}^{n-1}$ in system (19), we see that $b(0) = 0$. Since we assume analytic data, without singularities, $b(t) \neq 0$ for $t \neq 0$ and small enough. In particular if y_d can be tracked through the singularity $(0, z_o)$, it will also be tracked through the singular set at $(0, z_o)$.

Because of the linearity of system (19), we may often decompose any solution to the singular tracking problem into two components. The first component consists of a particular solution to the tracking problem, and the second component consists of any solution to the problem of tracking the zero function through $(0, 0) \in S \subset \mathbb{R} \times \mathbb{R}^{n-1}$. Denote a fixed particular solution by the pair (z^P, u^P), $z^P(0) = z_o$, and any solution, tracking the zero function, by (z^o, u^o), $z^o(0) = 0$. Clearly $z_d = z^P + z^o$, $u_d = u^P + u^o$, will also provide a solution to the tracking problem.

As long as $r((0, z_o), y_d)$ is defined we may explicitly write out a particular solution to the tracking problem as follows. Substituting the control (26) into system (19) we obtain the equations

$$\dot{z} = (A(t) - \frac{d(t)\, s_d'(t)}{b(t)})\, z + d(t)\frac{y_d^{(\alpha)}(t)}{b(t)}, \qquad z(0) = z_o \tag{27}$$

It is tempting to consider classifying the singularity of this equation by classifying the singularity of the matrix

$$A^*(t) = A(t) - \frac{d(t)\, s_\alpha(t)'}{b(t)}$$

at t=0. However, as we show the example, this does not necessarily give an answer consistent with the solutions of the tracking problem, and in particular the input-output method of obtaining solutions as indicated in the introduction. Let $P(t)$ be an analytic matrix valued function which is regular at t=0 and $\det P(t) \neq 0$ for $t \neq 0$. Define next for $t \neq 0$

$$A_p(t) = \frac{d}{dt}(P(t)^{-1})\, P(t) + P(t)^{-1} A^*(t)\, P(t).$$

DEFINITION 7

Define $p(z_o) \geq 0$ to be the minimum integer for which there exists a matrix valued function $P(t)$ as above such that the following limit exists

$$\lim_{t \to 0} (t^p A_p(t)), \tag{28}$$

and z_o is in the range of $P(0)$. □

Note, by taking $P(t) \equiv$ identity matrix, $A_p \equiv A^*$ and so there are integers which satisfy the requirements of the definition. If w_o satisfies $z_o = P(0)\, w_o$ and $\tilde{A}_p(t) = t^p A_p(t)$, consider the equation

$$t^p \dot{w} = \tilde{A}_p(t)\, w + t^p P(t)^{-1} d(t)\, \frac{y_d^{(\alpha)}(t)}{b(t)}, \qquad w(0) = w_o. \tag{29}$$

Clearly solutions of equation (27) and (29) are related by $z(t) = P(t)\, w(t)$. Since $\bar{A}_p(t)$ is analytic and regular at t=0, the differential equation

$$t^p\dot{V} = \bar{A}_p(t)\, V, \qquad V \in \mathbb{R}^{n\times n} \tag{30}$$

is in the general form of the linear differential equation (25) discussed above. There therefore exists a fundamental solution V(t), which may be expressed in terms of power series at t=0, in the regular case p=1, or a more general asymptotic expansion in the case $p > 1$. We may now write out a solution to (27) using equation (29) giving an expression for z^p in the form

$$z^p(t) = P(t)\, V(t)\, V(0)^{-1}\, w_o + \int_0^t P(t)\, V(t)\, V(\sigma)^{-1} P(\sigma)^{-1}\, d(\sigma)\, \frac{y_d^{\alpha}(\sigma)}{b(\sigma)}\, d\sigma$$

The corresponding expression for u^p may now be obtained by substituting this expression into equation (26). The smoothness of the particular solution (z^p, u^p) can be studied using the resulting expressions in particular cases.

Note that for time varying linear systems (19), the degree of singularity $\beta(x)$ as in definition 4 depends only on t and for a singular point $(0, z_o)$, β is defined as the smallest integer β such that the following limit exists

$$\lim_{t\to 0} \left(\frac{t^{\beta}}{b(t)}\right)$$

It seems, however, that for linear systems it would be better to define the degree of the singularity $(0, z_o)$ to be the integer p-1, where $p(z_o)$ is the integer given in definition 7. The the degree would then agree with the rank of the associated homogeneous singular equation (30).

Solutions (z^o, u^o) to the problem of tracking the zero function through the singular point (0, 0) for system (19) may be analysed as follows: In this case $v_d \equiv 0$ certainly constitutes an analytic control which is regular at t=0, and also constitutes a component of a solution to the problem of tracking the zero function through (0, 0). Smoothness of the composite solutions $(z_d, u_d) = (z^P + z^o, u^P + u^o)$ now depends on that of the solution (z^P, u^P) discussed earlier. This then demonstrates the existence of solutions to the singular tracking problem, which are not regular at t=0, at least in the linear case. In fact it can be shown [1] that a continuous multiplicity of solutions to the singular tracking problem are not regular at t = 0.

It is interesting to note that the matrix $[s_o(t) | \ldots | s_{n-1}(t)]$ is nonsingular at t=0 if and only if we may represent the input-output map of system (19) in the form

$$\begin{aligned} y^{(n)} + p_{n-1}(t)\, y^{(n-1)} + \ldots p_o(t) y \\ = q_o(t)\, u + \ldots q_m(t)\, u^{(m)} \end{aligned} \tag{31}$$

where $m = n-\alpha$, $q_m(t) = b(t)$, see [14]. Clearly the solutions (z^o, u^o) are obtained via solutions u^o to the homogeneous equation, or "zero dynamics",

$$u^{(m)} q_m(t) + \ldots + u^{(1)} q_1(t) + u\, q_o(t) = 0 \tag{32}$$

for which certain derivatives $u^{(k)}(0)$ may not be specified in the tracking problem. If the matrix $[s_o(t) | \ldots | s_{n-1}(t)]$ is singular at t=0 then we must consider higher order input-output equations.

5 EXAMPLE

$$\begin{cases} \dot{x}_1 = u\, t \\ \dot{x}_2 = 2x_1 - t^2 u \\ \dot{x}_3 = x_2 \\ y = t x_2 - 5x_3 \end{cases}$$

For this time varying system

$$\begin{aligned} y^{(1)} &= 2tx_1 - 4x_2 - t^3 u \\ y^{(2)} &= -6x_1 - 3t^2 u - t^3 u^{(1)} \\ y^{(3)} &= -t^3 u^{(2)} \\ y^{(4)} &= -3t^2 u^{(2)} - t^3 u^{(3)} \\ y^{(5)} &= -6tu^{(2)} - 6t^2 u^{(3)} - t^3 u^{(4)} \\ y^{(5)} &= -6u^{(2)} - 18\, tu^{(3)} - 9t^2 u^{(4)} - t^3 u^{(5)} \end{aligned}$$

Again we have a tracking singularity at t=0. Let $x_o = (x_1(0), x_2(0), x_3(0))' = 0$, then for

$r(x_o, y_d)$ to be defined we require $y_d(0) = y_d^{(1)}(0) = 0$. If also $y_d^{(k)}(0) = 0$ for $2 \le k \le 5$ then $r(x_o, y_d) = \infty$ and if $y_d^{(k)}(0) \neq 0$ but $y_d^{(j)}(0) = 0$ for $2 \le j < k$ then $r(x_o, y_d) = k\text{-}2$ for $k \le 5$. Substituting the control

$$u_D = \frac{2x_1}{t^2} - \frac{4x_2}{t^3} - \frac{y_d^{(1)}}{t^3}$$

into the system equations we obtain equation (27) for our specific example

$$\frac{d}{dt}\begin{pmatrix} x_1 \\ x_2 \\ x_3 \end{pmatrix} = \begin{pmatrix} 2/t & -4/t^2 & 0 \\ 0 & 4/t & 0 \\ 0 & 1 & 0 \end{pmatrix}\begin{pmatrix} x_1 \\ x_2 \\ x_3 \end{pmatrix} + \begin{pmatrix} -1/t^2 \\ 1/t \\ 0 \end{pmatrix} y_d^{(1)}, \qquad x_o = 0$$

Clearly

$$A^*(t) = \begin{pmatrix} 2/t & -4/t^2 & 0 \\ 0 & 4/t & 0 \\ 0 & 1 & 0 \end{pmatrix}$$

defines a system of linear equations with an irregular singularity at t=0. However, the integer p in definition7 turns out to be equal to 1, which can be seen by selecting the matrix

$$P(t) = \begin{pmatrix} 1 & 0 & 0 \\ 0 & t & 0 \\ 0 & 0 & 1 \end{pmatrix}$$

from which we calculate

$$A_p(t) = \begin{pmatrix} 2/t & -4/t & 0 \\ 0 & 3/t & 0 \\ 0 & t & 0 \end{pmatrix}$$

Clearly $x_o = 0$ is in the range of P(0). $\bar{A}_p(t) = t\,A_p(t)$ now defines the system of linear equations (30) with a regular singularity at t=0, and so we expect solutions of the tracking problem which reflect this fact. Indeed, the input-output equations with $y = y_d$ do have regular singularities at t=0, having the form

$$u^{(2)} = - y_d^{(3)} / t^3 \qquad (33)$$

or

$$u^{(3)} + 3\, u^{(2)} / t = - y_d^{(4)} / t^3$$

Using equation (33) it is easy to see that for $\mathbf{y_d = t^6}$, $r(x_o, y_d) = \infty$, and the open loop controls are given by

$$u_d = -60t^2 + a\,t + b$$

where a and b are arbitrary reals.

On the other hand if $\mathbf{y_d = t^5}$, $r(x_o, y_d) = 3$ and the open loop controls are given by

$$u_d = -60t \ln t + a\,t + b, \qquad t > 0$$

where a and b are again arbitrary reals, which in turn shows that the corresponding state trajectories $x_d \in C^1$. Even though $r(x_o, y_d) = 3$ this expression for u_d shows that the corresponding singular system of order 3 in definition6 does not have a solution. Indeed, only the singular system of order 1 has a solution.

REFERENCES

[1] P.E. Crouch, I. Ighneiwa and F. Lamnabhi-Lagarrigue, On the singular tracking problem, Preprint 89.

[2] A. Erdelyi, *Asymptotic Expansions*, Dover, 1956.

[3] K.A. Grasse, Sufficient conditions for the functional reproducibility of time varying input-output systems, *SIAM J. Control Optimiz.*, 26, pp.230-249, 1988.

[4] E. Hille, *Ordinary Differential Equations in the Complex Domain*, Wiley, 1976.

[5] R.M. Hirschorn, Invertibility of nonlinear control systems, *SIAM J. Control Optimiz.*, 17, pp.289-297, 1979.

[6] R.M. Hirschorn, Output tracking in multivariable nonlinear systems, *IEEE Trans. Aut. Contr.,* AC-26, pp.593-595, 1981.

[7] R.M. Hirschorn and J. Davis, Output tracking for nonlinear systems with singular points, *SIAM J. Control Optimiz.*, 26, pp.547-557, 1987.

[8] R.M. Hirschorn and J. Davis, Global output tracking for nonlinear systems, *SIAM J. Control Optimiz.*, 26, pp.132

[9] E.L. Ince, *Ordinary Differential Equations*, Dover, New York, 1956.

[10] H. Neimeijer, Invertibility of affine nonlinear control systems: a geometric approach, *Systems and Control Letters*, 2, pp.163-168, 1982.

[11] S.N. Singh, Reproductibility in nonlinear systems using dynamic compensation and output feedback, *IEEE Trans. Aut. Contr.*, AC-27, pp.955-958, 1982.

[12] S.N. Singh, Generalized functional reproductibility condition for nonlinear systems, *IEEE Trans. Aut. Contr.*, AC-27, pp.958-960, 1982.

[13] W. Wasow, *Asymptotic expansions for ordinary differential equations*, Interscience Publishers, J. Wiley, Pure and Applied Mathematics Vol.XIV, New York, 1965.

[14] L.M. Silverman, Properties and applications of inverse systems, *IEEE Trans. Aut. Contr.*, pp. 436-437, 1968.

Algebraic Riccati equations arising in boundary/point control: A review of theoretical and numerical results
Part I: Continuous case

Irena Lasiecka and Roberto Triggiani
Department of Applied Mathematics
Thornton Hall
University of Virginia
Charlottesville, Virginia 22903

1. Introduction

Consider the following optimal control problem: Given the dynamical system

$$y_t = Ay+Bu; \; y(0) = y_0 \in Y \tag{1.1}$$

minimize the quadratic functional

$$J(u,y) = \int_0^\infty [\|Ry(t)\|_Z^2+\|u(t)\|_U^2]dt \tag{1.2}$$

over all $u \in L_2(0,\infty;U)$, with y solution of (1.1) due to u. Throughout the paper, we shall make the following standing assumptions on (1.1), (1.2):

(i) Y, U, Z are Hilbert spaces;

(ii) $A: Y \supset \mathcal{D}(A) \to Y$ is the generator of a s.c. semigroup e^{At} on Y, $t > 0$, generally unstable on Y, i.e., with $\omega_0 = \lim[(\ln\|\exp(At)\|)/t] > 0$ as $t \to +\infty$ in the uniform norm $\mathcal{L}(Y)$, so that $\|e^{At}\| \le Me^{(\omega_0+\varepsilon)t}$, $\forall$ $\varepsilon > 0$, $t \ge 0$, and M depending on $\omega_0 + \varepsilon$; we then consider throughout the translation $\hat{A} = -A+\omega I$, $\omega =$ fixed $> \omega_0$, so that $\hat{A}$ has well-defined fractional powers on Y and $-\hat{A}$ is the generator of an s.c. semigroup $e^{-\hat{A}t}$ on Y satisfying $\|e^{-\hat{A}t}\| \le \hat{M}e^{-\hat{\omega}t}$, $t \ge 0$; $\hat{\omega} = \omega-\omega_0-\varepsilon > 0$;

(iii) $B: U \supset \mathcal{D}(B) \to [\mathcal{D}(A^*)]'$, the dual of $\mathcal{D}(A^*)$ with respect to the Y-topology, A^* being the Y-adjoint of A; moreover, it is assumed that

$$(\hat{A})^{-\gamma} B \in \mathcal{L}(U;Y) \quad \text{for some constant } 0 \leq \gamma \leq 1; \tag{1.3}$$

(iv) the operator R is bounded,

$$R \in \mathcal{L}(Y,Z). \tag{1.4}$$

Our motivation comes from, and is ultimately directed to, partial differential equations with boundary or point control (in any space dimensions), such as they arise in 'flexible structures' problems; see the examples of section 3 below. Accordingly, we shall distinguish two general classes of not necessarily mutually exclusive dynamics, for which different treatments must be applied in order to capture optimal properties thereof. Each of the two classes will be singled out and modelled by one of the two ('regularity') assumptions below.

First class. The first class satisfies the assumption:

$$\text{(H.1)} \quad \begin{cases} \text{the s.c. semigroup } e^{At} \text{ is analytic on } Y, \\ t > 0, \text{ and the constant } \gamma \text{ appearing in} \\ \text{(1.3) is strictly less than 1: } \gamma < 1. \end{cases} \tag{1.5}$$

Second class. The second class satisfies the assumption: for any $0 < T < \infty$, there exists $c_T > 0$ such that

$$\text{(H.2)} \quad \begin{cases} \int_0^T \|B^* e^{A^* t} y\|_U^2 dt \leq c_T \|y\|^2, \; y \in \mathcal{D}(A^*), \text{ so that the} \\ \text{operator } B^* e^{A^* t} \text{ admits a continuous extension--} \\ \text{denoted henceforth by the same symbol--} \\ \text{satisfying } B^* e^{A^* t}\text{: continuous } Y \to L_2(0,T;U), \end{cases} \tag{1.6}$$

where

$$(B^* v, u)_U = (v, Bu)_Y; \quad v \in \mathcal{D}(B^*), \; u \in U. \tag{1.7}$$

To fix our ideas at the outset, the first class covers parabolic-like problems; not only the usual heat equations/diffusion equations, but also wave-like or plate-like problems with high degree of damping ('structural damping'), see section 3.1 below. Instead, the second class covers undamped, or conservative, or mildly damped wave-like or plate-like partial differential equations (e.g., with viscous damping) with boundary or point control. We shall refer to (H.2) = (1.6) as to an 'abstract' trace theory property, for this is what it amounts to in partial differential problems.

For either class we shall need an assumption that guarantees the existence of a unique optimal pair $\{u^0, y^0\}$ of the optimal control problem (1.2).

(F.C.C) $\left\{\begin{array}{l}\underline{\text{Finite Cost Condition}}\text{: For every } y_0 \in Y, \\ \text{there exists } u \in L_2(0,\infty;U) \text{ such that the} \\ \text{corresponding functional in (1.2)} \\ \text{satisfies } J(u,y(u)) < \infty.\end{array}\right.$ (1.8)

The main problems of interest in this paper are:

(p_1) pointwise (in time) feedback representation, via a Riccati operator, of the optimal control u^0 in terms of the optimal solution y^0, such as given by

$$u^0(t) = -B^*Py^0(t), \quad \text{a.e. in } t > 0, \tag{1.9}$$

where the operator P is a solution of an appropriate Algebraic Riccati Equation (ARE) formally written as

$$PA+A^*P+R^*R-PBB^*P = 0, \tag{1.10}$$

and to be properly interpreted in a technical sense, describe below;

(p_2) numerical approximations of the ARE for the computation of the operator P.

The present paper focuses only on the case where the input operator B is unbounded and such as it arises in both point control problems and, above all, the more challenging boundary control problems for p.d.e. Thus, the paper is based only on the natural "regularity" assumptions (H.1), or else (H.2) (not mutually exclusive) for the dynamics and the non-smoothing assumption (1.3) for the observation R. Such unboundedness of the operator B contributes to a number of mathematical difficulties in the study of the Riccati feedback synthesis--as expressed by (1.9)--of the optimal control problem (1.1), (1.2). These difficulties are present at two general levels. (i) One is the abstract level, which is aimed at a general theory of existence and uniqueness of the solution P to the ARE and the consequent Riccati synthesis, and which must overcome the difficulty inherent to the gain operator B^*P. We shall see that for the class of dynamics modelled by assumption (H.2), the gain operator B^*P is inherently unbounded in the most interesting cases of conservative waves and plates problems. Thus, the classical arguments with B bounded are no longer available, and new approaches must be devised. (ii) A second level of difficulty arises in the verification of the abstract assumptions of regularity and Finite Cost Condition, particularly for the class modelled by assumption (H.2), in specific, "concrete" p.d.e. problems. Here, p.d.e. techniques are required, which were brought to bearing on these problems only very recently. These techniques succeeded in showing the required optimal regularity results (assumption (H.2)) of several waves/plates boundary control problems, as well as their exact controlability/uniform stabilization properties, which are needed to verify the Finite Cost Condition for these systems. These regularity/exact controllability/uniform stabilization results will be reviewed in section 3 in the context of each specific dynamics.

In this paper, we review some of the most significant recent developments and provide an up to date status in the field of the Algebraic Riccati Equation (subject to space limitations). In section 2, we treat the two classes of dynamics--that subject to the regularity hypothesis (H.1), and that subject to the regularity hypothesis (H.2)--separately. This distinction is necessary in order to extract best results from each class. Indeed, these two dynamical classes require very different techniques, as they display peculiarly different properties which escape any meaningful and non-artificial or shallow "unification." Only statements of the main results will be given, with reference to the literature for a more complete treatment. Section 3 provides several p.d.e. examples of point and boundary control problems where we verify <u>all</u> the required assumptions. Some of these examples in section 3.1 are new and appear here for the first time (again with references provided for a more detailed treatment).

Corollaries 2.4 and 2.5--although almost immediate consequences of the results of [FLT]--are explicitly states here for the first time. They point out that for the most interesting and typical dynamics modelled by assumption (H.2) (conservative waves and plates) the gain operator B^*P is unbounded. This conclusion rules out, as inapplicable to these systems, other treatments dealing with unbounded B, which however required additional restrictions (on the observation R, on the finite cost conditions, etc.), which are incompatible for these systems; e.g., the results of [PS], see Remark 2.3 at the end of Section 2.

Part I of this paper deals with the continuous case. It is followed by Part II which deals with an approximation theory thereof.

2. Abstract Algebraic Riccati Equations: Existence and uniqueness

In this section we shall discuss the solvability of the following abstract Algebraic Riccati Equation (ARE)

$$(Px, Ay)_Y + (PAx, y)_Y + (Rx, Ry)_Z - (B^*Px, B^*Py)_U = 0;$$
$$\forall\ x, y \in \mathcal{D}(A). \qquad (2.1)$$

As is well known, a solution P of this equation (if it exists and it possesses certain regularity properties) provides the sought-after feedback operator which occurs in the representation (1.9) of the optimal control law. In the case where B is an unbounded operator, the obvious difficulty is related to the interpretation of the 'gain operator' B^*P, which *a priori* need not be even densely defined on Y (even when the existence of a "Riccati operator" $P \in \mathcal{L}(Y)$ is asserted). The crux of the matter is therefore this: To prove that the Riccati operator P possesses certain 'regularity' properties which will guarantee a proper definition of the gain operator B^*P, at least as an unbounded operator with dense domain in Y for the representation (1.9) to be meaningful. We begin with the first class ('analytic') which has received more attention in the literature.

2.1 Algebraic Riccati Equation for the first class subject to the analyticity assumption (H.1) = (1.5)

Theorem 2.1 ([D-I], [F.2], [L-T.7]).
I. **Existence**. For the first class covered by hypothesis (H.1) = (1.5) and subject to the Finite Cost Condition (1.8), there exists a self-adjoint, non-negative definite solution $0 \le P = P^* \in \mathcal{L}(Y)$ of the ARE (2.1) such that:

(i) $(\hat{A}^*)^{1-\varepsilon}P \in \mathcal{L}(Y), \quad \forall\ \varepsilon > 0;$ (2.2)

and indeed ε may be taken ε = 0 if the original A is self-adjoint or normal or has a Riesz basis of eigenvectors on Y; thus, P is compact if A has compact resolvent;

(ii) $B^*P \in \mathcal{L}(Y,U);$ (2.3)

(iii) $J(u^0,y^0) = (Py_0,y_0)_Y;$ (2.4)

(iv) $u^0(t) = -B^*Py^0(t)$ for all $0 < t < \infty;$ (2.5)
where $y^0(t) = y^0(t;y_0)$ and $u^0(t) = u^0(t;y_0)$.

II. **Regularity of the optimal pair**. For each fixed $y_0 \in Y$, the functions $y^0(t;y_0)$ and $u^0(t;y_0)$ are analytic in t as Y-valued or U-valued functions, a consequence of the analyticity of the feedback semigroup $e^{A_P t}$ below in (2.11) and of (2.3).

Remark 2.0. In the case where (1.1) models a second order parabolic equation on a bounded domain $\Omega \subset R^n$ with Dirichlet boundary control, the following additional regularity properties of the optimal pair hold true [L-T.7]:

$$\text{if } y_0 \in L_2(\Omega) \Rightarrow \begin{cases} e^{-\omega t}y^0(\cdot;y_0) \in H^{1-2\varepsilon,\frac{1}{2}-\varepsilon}(Q_\infty), \\ \qquad Q_\infty = (0,\infty)\times\Omega, & (2.6) \\ e^{-\omega t}u^0(\cdot;y_0) \in H^{\frac{3}{2}-2\varepsilon',\frac{3}{4}-\varepsilon'}(\Sigma_\infty), \\ \qquad \forall\ \varepsilon' > \varepsilon > 0,\ \Sigma_\infty = (0,\infty)\times\Gamma; & (2.7) \end{cases}$$

$$\text{if } y_0 \in H^{\frac{1}{2}-\rho}(\Omega) \Rightarrow \begin{cases} e^{-\omega t}y^0(\cdot;y_0) \in H^{\frac{3}{2}-2\rho,\frac{3}{4}-\rho}(Q_\infty), \\ \qquad \rho > 0, & (2.8) \\ e^{-\omega t}u^0(\cdot;y_0) \in H^{2-2\rho',1-\rho'}(\Sigma_\infty), \\ \qquad \rho' > \rho > 0. & (2.9) \end{cases}$$

III. __Uniqueness__. In addition to the assumption of part I, we assume that the following so-called 'detectability condition' (D.C.) holds:

$$\text{(D.C):} \begin{cases} \text{There exists } K \in \mathcal{L}(Z,Y) \text{ such that the} \\ \text{s.c. semigroup } e^{(A+KR)t} \text{ generated by } A+KR \\ \text{is exponentially (uniformly) stable on } Y. \end{cases} \qquad (2.10)$$

Then

(a) the solution P to the ARE (2.1) is unique within the class of non-negative self-adjoint operators in $\mathcal{L}(Y)$, which satisfy the regularity requirement (2.3);

(b) the s.c., analytic semigroup $e^{A_P t}$ generated by $A_P = A-BB^*P$ is exponentially (uniformly) stable on Y:

$$\|e^{A_P t}\|_{\mathcal{L}(Y)} \le M_P e^{-\omega_P t}, \quad t > 0 \qquad (2.11)$$

for some constants M_P, $\omega_P > 0$. ∎

Two distinct, yet complementary, approaches are available to prove Theorem 2.1 (existence and uniqueness): (i) a variational approach [L-T.7] and (ii) a so-called 'direct' approach [D-I], [F.2]. The variational argument in [L-T.7] starts from the control problem as the primary issue and constructs an explicit candidate for the Riccati operator (in terms of the data of the problem with the help of the optimal solution), which is then shown to satisfy the ARE (2.1). In contrast, the direct approach as in [D-I], [F.2] takes the direct study of well-posedness (existence and uniqueness) of the ARE as the primary object and only subsequently recovers the control problem (via dynamic programming) which generates the original ARE. In carrying out its task, the direct method begins actually with a direct study of the corresponding Differential (or Integral) Riccati Equation of the optimal problem over a finite interval $[0,T]$, $T < \infty$, and operates a limit process as $T \to \infty$ <u>on the</u> Differential Riccati Equation (in line with a classical approach, which now, however, has to overcome new technical difficulties, particularly the strong convergence of $B^*P_T(0)$ to B^*P). In both approaches, a key point consists in establishing that the gain operator B^*P (*a priori* not necessarily well defined) is, in fact, a bounded operator; see (2.3). In the variational approach, this latter property is accomplished by using analyticity of the free dynamics, together with a certain 'bootstrap' argument based on the Young inequality to show that the optimal pair is more regular, indeed $e^{-\omega t}u^0(t;y_0) \in C([0,\infty];U)$ and $e^{-\omega t}y^0(t;y_0) \in C([0,\infty];Y)$ for $y_0 \in Y$. (*A priori*, we only know that $u^0 \in L_2(0,\infty;U)$, while a general control $u \in L_2(0,T;U)$ need <u>not</u> produce in general a corresponding solution $y \in C([0,T];Y)$; a counterexample being obtained by a parabolic equation on Ω, with Dirichlet-boundary control where $U = L_2(\Gamma)$, and $Y = L_2(\Omega)$.) All this leads to the regularity property (2.2) via the explicit representation of P in terms of the optimal solution, which in turn leads to property (2.3). Instead, in case of the direct approach [D-I], [F.2], the boundedness (2.3) of the gain operator B^*P is established by proving first that the solution of the corresponding Differential Riccati Equation for the problem on $[0,T]$ possesses the desired regularity properties, and then by passing to the limit as $T \to \infty$. This, in turn, is accomplished in [F.2] by repeated applications of the Young's inequality to prove that the optimal trajectory is in $C([0,T];Y)$ for any $T > 0$; or in [D-I] by a direct study of the evolution equation via a fixed point argument.

<u>Remark 2.1.</u> It should be noted that the proof of Theorem 2.1 greatly simplifies and becomes rather straightforward, in fact, in case the constant γ appearing in assumption (1.3) can be taken to be $\gamma < \frac{1}{2}$, or even $\gamma = \frac{1}{2}$, if the operator A is self-adjoint or normal, or has a Riesz basis of eigenvectors. Indeed, in this case, standard analytic estimates give at the outset that <u>any</u> solution y to an $L_2(0,\infty;U)$-control function satisfies in fact the regularity property $y \in C([0,T];Y)$, $\forall\, T > 0$; $e^{-\omega t}y(t;y_0) \in C([0,\infty];Y)$ for $y_0 \in Y$. Thus, such property holds automatically true for the optimal y^0 in this case (while it is a distinctive property of y^0 to be proved when $\frac{1}{2} < \gamma < 1$, not shared by general solutions y to $L_2(0,\infty;U)$-controls u, as discussed above). As a consequence, one obtains immediately then that B^*P is bounded, by using the explicit representation of P in terms of the optimal solution.

Remark 2.2. The 'detectability' assumption (D.C.) = (2.10) guarantees not only uniqueness of the Riccati solution P to the ARE, but also the property that the resulting feedback semigroup $e^{A_P t}$ is exponentially stable as in (2.11); and, indeed, it is the latter property that is used to prove the former. The exponential decay of $e^{A_P t}$ is a particularly attractive feature in applications, for then the Riccati operator provides, constructively, a stabilizing feedback operator of the free dynamics $\dot{y} = Ay$ which may be, possibly, unstable to begin with.

2.2 Algebraic Riccati Equation for the second class subject to the 'trace' regularity assumption (H.2) = (1.6)

The study of the ARE is more complicated for the class of dynamics subject to assumption (H.2) = (1.6), rather than to assumption (H.1) = (1.5). Indeed, in this case, there is no smoothing effect of the free dynamics which will make up for the unboundedness of the operator B. And, in fact, in most of the interesting situations, the gain operator B^*P is intrinsically unbounded, see Corollaries 2.4, 2.5 below. This feature is in sharp contrast with the 'analytic' situation described in section 2.1, when assumption (H.1) instead is in force. Thus, boundedness of B^*P for the class (H.1) and unboundedness of B^*P for the class (H.2) in the most interesting situations is a distinguishing feature that tells the two cases apart. On the other hand, it should be noted that, in contrast with the situation of section 2.1, hypothesis (H.2) = (1.6) implies by duality the desired regularity $u \in L_2(0,T;U) \to y \in C([0,T];Y)$, which under the hypothesis (H.1) = (1.5) case $\frac{1}{2} < \gamma < 1$, is generally false, but can be proved to be true, however, for the optimal pair (u^0,y^0), as remarked in section 2.1. A rather complete theory for the Algebraic Riccati Equation under present assumptions was first given in [L-T.6] in the canonical case of the wave equation, or more generally second order hyperbolic equations, with Dirichlet control in $L_2(0,T;L_2(\Gamma))$, which was treated, however, by abstract operator-theoretic methods. This treatment was later put fully on an abstract space framework, and complemented by further results, in [FLT].

Theorem 2.2 ([L-T.6], [L-T.9], [FLT]).
I. **Existence**. For the second class covered by hypothesis (H.2) = (1.6) and subject to the Finite Cost Condition (1.8), there exists a self-adjoint, non-negative solution $0 \le P = P^* \in \mathcal{L}(Y)$ of the ARE (2.1) such that:

(i) $P \in \mathcal{L}(\mathcal{D}(A),\mathcal{D}(A_P^*)) \cap \mathcal{L}(\mathcal{D}(A_P),\mathcal{D}(A^*))$, (2.12)

where the operator

$$A_P = A-BB^*P \tag{2.13}$$

generates a s.c. semigroup on Y; thus, the ARE (2.1) holds true also for all $x,y \in \mathcal{D}(A_P)$;

(ii) $B^*P \in \mathcal{L}(\mathcal{D}(A),U) \cap \mathcal{L}(\mathcal{D}(A_P);U)$; (2.14)

(iii) $J(u^0,y^0) = (Py_0,y_0)_Y$; (2.15)

(iv) $u^0(t) = -B^*Py^0(t)$; (2.16)

where we write $y^0(t) = y^0(t;y_0)$, $u^0(t) = u^0(t;y_0)$, and (2.16) is understood a.e. in t if $y_0 \in Y$; while instead, if $y_0 \in \mathcal{D}(A_P)$, then (2.14) implies $y^0(t;y_0) \in C([0,T];\mathcal{D}(A_P))$, and by (2.16), $u^0(t;y_0) \in C([0,T];U)$ for any $T > 0$.

II. **Uniqueness**. In addition to the assumption of part I, we assume that the following 'detectability' condition (D.C.) holds true:

(D.C): There exists $K: Z \supset \mathcal{D}(K) \to Y$ densely defined such that

$$\|K^*x\|_Z \le C[|B^*x|_U+\|x\|_Y], \quad \forall\, x \in \mathcal{D}(B^*) \subset Y, \tag{2.17}$$

so that the operator

$$A_K = A+KR \quad \text{(interpreted as closed)} \tag{2.18}$$

is the generator of a s.c. semigroup e^{A_Kt} on Y, which is then assumed to be exponentially stable on Y:

$$\|e^{A_Kt}\|_{\mathcal{L}(Y)} \le M_K e^{-\omega_K t}, \quad t > 0, \tag{2.19}$$

for some M_K, $\omega_K > 0$. (For $R > 0$, we choose $K = -c^2R^{-1}$ with constant c sufficiently large, and the detectability assumption (2.17)-(2.19) is automatically satisfied.) Then

(a) the solution P to the ARE (2.1) is unique within the class of non-negative self-adjoint operators in $\mathcal{L}(Y)$ which satisfy the regularity properties (2.14);

(b) the s.c. semigroup $e^{A_P t}$ generated by A_P in (2.13) is exponentially (uniformly) stable on Y. ∎

The proof of Theorem 2.2 is given in [FLT] and follows the abstract treatment of the canonical case of second order hyperbolic equations with Dirichlet control [L-T.6]. It is based on a variational approach. The following comments, which constrast the technical methodology available in the case of Theorem 2.2 with that available in the case of Theorem 2.1, apply. A main difference between the two cases is that at present no Differential Riccati Equation on [0,T] is available under the assumption (H.2) = (1.6) with no smoothing of the operator R; i.e., with R subject only to assumption (1.4) of boundedness, in contrast with the situation available under assumption (H.1) = (1.5). (However, for a Differential Riccati Equation on a finite interval [0,T], under the assumption of minimal ("ε") smoothness of R, see the two companion papers [L-T.6] and [C-L] for scalar second-order hyperbolic equations with Dirichlet control, and first-order hyperbolic systems, respectively, each case being treated by abstract operator methods; and [DaP-L-T] under further smoothness of R, which yields also uniqueness of the Riccati solution.) Thus, an approach to the issue of existence of a solution of the ARE which is based on the classical idea of a limiting process as $T \to \infty$ on the Differential Riccati Equation is out of question (unlike the analytic case of assumption (H.1) = (1.5), in the direct approach [D-I], [F.2], as described in section 2.1). Therefore, a different strategy is now applied [L-T.6], [FLT]. As in the analytic case treatment of [L-T.7] under assumption (H.1), the existence of a solution to the ARE is now obtained under assumption (H.2) through the following steps: (i) First, one constructs an explicit candidate of a solution, in terms of the data of the problem (the optimal solution); (ii) next, one establishes the necessary regularity properties of such a candidate, as described in (2.12), (2.14), using its explicit represention; (iii) finally, one verifies that such candidate operator does satisfy the ARE (2.1).

As mentioned in the introduction of section 2.2, it is important to notice that, in contrast with the situation described by Theorem 2.1 under the analyticity assumption (H.1) = (1.5), the gain operator B^*P is now generally unbounded (in the most interesting situations). Indeed, this property follows from the next results.

Theorem 2.3. Let the hypotheses (H.2) = (1.6) for the dynamics and the Finite Cost Condition (1.8) hold true, as in Theorem 2.1, part I. In addition, assume the following exact controllability condition:

$$(E.C.) \begin{cases} \text{the equation } \dot{y} = A^*y+R^*v \text{ is exactly controllable} \\ \text{in } Y \text{ from the origin over some } [0,T],\ T < \infty, \\ \text{within the class of } L_2(0,T;Z)\text{-controls } v. \end{cases} \tag{2.20}$$

(We shall say, in short, that the pair $\{A^*,R^*\}$ is exactly controllable.) Then, the solution operator P to the ARE (2.1) guaranteed by Theorem 2.1, part I, is an isomorphism on Y. ■

Corollary 2.4. Under the assumptions (H.2) = (1.6), (1.8), and (2.20) of Theorem 2.3, we have: The operator $B: U \supset \mathcal{D}(B) \to Y$ is bounded if and only if the operator B^*P: from its domain in $Y \to U$ is bounded. ■

From Corollary 2.4, we see that for the second class subject to assumption (H.2) = (1.6), where moreover B is an unbounded operator (the interesting case), the requirement that the gain operator B^*P be bounded runs into conflict with the assumption (2.20) of exact controllability of the pair $\{A^*,R^*\}$. On the other hand, if (i) the original free dynamics e^{At} is an s.c. group uniformly bound for negative times (the case of all the interesting conservative wave and plate problems, which yield in fact unitary groups) and (ii) the desirable detectability condition D.C. = (2.17)-(2.19) of the pair $\{A,R\}$ holds true, then the pair $\{A^*,R^*\}$ is uniformly (exponentially) stabilizable (by the operator K^* in the notation of (2.17)). Then, a well-known result of D. Russell (1973) for time-reversible systems implies that the pair $\{A^*,R^*\}$ is exactly controllable (to the origin, or equivalently, from the origin). Thus:

Corollary 2.5. Assume hypothesis (H.2) = (1.6) for the dynamics, as well as the Finite Cost Condition (2.10) and the Detectability Condition (2.17)-(2.19). Assume, further, that the free dynamics e^{At} is an s.c. group uniformly bounded for negative times. Then the conclusion of Corollary 2.4 applies: B is bounded if and only if B^*P is bounded. ■

Remark 2.3. The following reference [P-S] also deals with the Algebraic Riccati Equation under the abstract hypothesis (H.2) = (1.6). In addition, however, [P-S] makes the following two further assumptions:

(i) an assumption of the smoothness on the observation operator $R \in \mathcal{L}(Y,Z)$ as expressed by the requirement

$$\int_0^T \|Re^{At}x\|_Y^2 dt \leq c_T \|x\|_V^2, \tag{2.21}$$

where V is a space strictly larger than Y and with weaker topology than Y, such that $B \in \mathcal{L}(U,V)$ and e^{At} generates a s.c. semigroup on both Y and V.

(ii) The assumption that the Finite Cost Condition holds true for all initial data in the strictly larger space V, not only on Y. (This assumption is <u>not</u> true in the cases of conservative waves and plates problems such as those of sections 3.2, 3.3, and 3.4). Under the above assumptions, [P-S] claims existence and, under the additional Detectability Condition on V, uniqueness of the solution P of the ARE, where P enjoys the following regularity property:

$$P \in \mathcal{L}(V,V'), \quad V' = \text{dual of } V \text{ with respect to Y-topology}, \tag{2.22}$$

which in turn implies boundedness of the gain operator

$$B^*P \in \mathcal{L}(Y,U). \tag{2.23}$$

In view of the above Corollaries 2.4, 2.5, we can say that: the results of [P-S] on the ARE are subsumed by the earlier Theorem 2.2, first given in the canonical case of the wave equation by abstract operator methods [L-T.6], and then fully cast in abstract setting in [FLT]; moreover, and in contrast with Theorem 2.2, the results of [P-S] cannot cover the important class of conservative waves and plates problems with B unbounded (point or boundary control) and R > 0, which is precisely the class which offers a main justification for introducing the abstract assumption (H.2) = (1.6) in the first place.

3. <u>Examples of partial differential equation problems satisfying (H.1) or (H.2)</u>

In this section, we illustrate the applicability of both Theorem 2.1 for the 'analytic' class subject to hypothesis (H.1) = (1.5), and of Theorem 2.2 for waves and plates problems which satisfy the ('abstract trace') hypothesis (H.2) = (1.6). In passing, some p.d.e. problems will be exhibited which satisfy both (H.1) and (H.2). Obvious candidates for the analytic class are heat or diffusion problems, for which we refer to [L.1], [L-T.7]. However, we prefer here to choose examples of plates with a strong degree of damping (structural damping), such as may arise in the study of flexible structures.

3.1 <u>Class (H.1): Structurally damped plates with point control or boundary control</u>

Theorem 2.1 fully applies to the case of general, say second order, parabolic equations on a bounded domain $\Omega \subset R^n$ with either Neumann or, the more challenging Dirichlet boundary control: see [L-T.7] for a detailed treatment of the latter case. Here, for lack of space, we shall concentrate only on newer examples of analytic classes, of interest in the study of flexible structures.

Example 3.1. The case $\alpha = \frac{1}{2}$ in [C-R], [C-T.1-2]. Consider the following model of a plate equation in the deflection $w(t,x)$.

$$\begin{cases} w_{tt}+\Delta^2 w-\Delta w_t = \delta(x-x^0)u(t) & \text{in } (0,T]\times\Omega = Q, \qquad (3.1a) \\ w(0,\cdot) = w_0; \quad w_t(0,\cdot) = w_1 & \text{in } \Omega, \qquad (3.1b) \\ w|_\Sigma \equiv \Delta w|_\Sigma \equiv 0 & \text{in } (0,T]\times\Gamma = \Sigma, \qquad (3.1c) \end{cases}$$

with load concentrated at the interior point x^0 of an open bounded (smooth) domain Ω of R^n, $n \leq 3$. Regularity results for problem (3.1), and other problems of this type, are given in [T.4]. Consistently with these results, the cost functional we wish to minimize is

$$J(u,w) = \int_0^\infty \{\|w(t)\|^2_{H^2(\Omega)} + \|w_t(t)\|^2_{L_2(\Omega)} + |u(t)|^2_{L_2(\Gamma)}\}dt, \tag{3.2}$$

where $\{w_0,w_1\} \in [H^2(\Omega) \cap H_0^1(\Omega)]\times L_2(\Omega)$.

Abstract setting. To put problems (3.1), (3.2) into the abstract setting of the preceding sections, we introduce the strictly positive definite operator

$$\mathcal{A}h = \Delta^2 h; \ \mathcal{D}(\mathcal{A}) = \{h \in H^4(\Omega)\colon h|_\Gamma = \Delta h|_\Gamma = 0\} \tag{3.3}$$

and select the spaces

$$Y \equiv \mathcal{D}(\mathcal{A}^{\frac{1}{2}})\times L_2(\Omega) = [H^2(\Omega) \cap H_0^1(\Omega)]\times L_2(\Omega); \ U = \mathbb{R}^1, \tag{3.4}$$

and finally define the operators

$$A = \begin{vmatrix} 0 & I \\ -\mathcal{A} & -\mathcal{A}^{\frac{1}{2}} \end{vmatrix}; \quad Bu = \begin{vmatrix} 0 \\ \delta(x-x^0)u \end{vmatrix}; \quad R = I \tag{3.5}$$

to obtain the abstract model (1.1), (1.2). We need to verify a few assumptions.

Assumption (1.3): $(-A)^{-\gamma}B \in \mathcal{L}(U,Y)$. It is easy to verify that assumption (1.3) is satisfied with $\gamma = 1$. Indeed, from (3.5), we require that

$$(-A)^{-1}Bu = \begin{vmatrix} \mathcal{A}^{-\frac{1}{2}} & \mathcal{A}^{-1} \\ -I & 0 \end{vmatrix} \begin{vmatrix} 0 \\ \delta(x-x^0)u \end{vmatrix} = \begin{vmatrix} \mathcal{A}^{-1}\delta(x-x^0)u \\ 0 \end{vmatrix} \in Y, \tag{3.6}$$

i.e., from (3.4), we require that $\mathcal{A}^{-1/2}\delta(x-x^0) \in L_2(\Omega)$, or that (#): $\delta(x-x^0) \in [\mathcal{D}(\mathcal{A}^{1/2})]'$, the dual of $\mathcal{D}(\mathcal{A}^{1/2})$ with respect to $L_2(\Omega)$. Since it is true that $\mathcal{D}(\mathcal{A}^{1/2}) \subset H^2(\Omega)$ for the fourth order operator $\mathcal{A}$ in (3.3) (in fact, regardless of the particular boundary conditions), and thus $[H^2(\Omega)]'$, $\subset [\mathcal{D}(\mathcal{A}^{1/2})]'$, then condition (#) is satisfied provided $\delta(x-x^0) \in [H^2(\Omega)]'$, i.e., provided $H^2(\Omega) \subset C(\bar{\Omega})$, which is indeed the case by Sobolev embedding provided $2 > \frac{n}{2}$, or $n < 4$, as required.

However, the above result is not sufficient for our purposes as--according to assumption (H.1) = (1.5)--we need to show that we can take $\gamma < 1$ in (1.3). As a matter of fact, we now show that assumption (1.3) holds true for any $\gamma > \frac{n}{4}$, which then for $n \leq 3$ yields $\gamma < 1$ as desired. To this end, we note that

$$(-A)^{-\gamma}B \in \mathcal{L}(U,Y) \text{ if and only if } B \in \mathcal{L}(U,[\mathcal{D}((-A^*)^{\gamma}]' \quad (3.7)$$

with duality with respect to Y. But $\mathcal{D}((-A^*)^{\gamma}) = \mathcal{D}((-A)^{\gamma})$: this follows since A is the direct sum of two normal operators on Y, with possibly an additional finite-dimensional component (if 1 is an eigenvalue of $\mathcal{A}$) [C-T.1], [C-T.2, Lemma A.1, case v(a) with $\alpha = 1/2$]. Moreover, [C-T.4, with $\alpha = 1/2$], we have

$$\mathcal{D}((-A^*)^{\gamma}) = \mathcal{D}((-A)^{\gamma}) = \mathcal{D}(\mathcal{A}^{1/2+\gamma/2}) \times \mathcal{D}(\mathcal{A}^{\gamma/2}),$$
$$0 < \gamma < 1 \quad (3.8)$$

(the first component does not really matter in the argument below). Thus, from (3.8) and B as in (3.5), it follows that (3.7) holds true, provided $\delta(x-x^0) \in [\mathcal{D}(\mathcal{A}^{\gamma/2})]'$ (duality with respect to $L_2(\Omega)$), where $\mathcal{D}(\mathcal{A}^{\gamma/2}) \subset H^{2\gamma}(\Omega)$, and hence, provided $\delta(x-x^0) \in [H^{2\gamma}(\Omega)]' \subset [\mathcal{D}(\mathcal{A}^{\gamma/2})]'$. But this in turn is the case, provided $H^{2\gamma}(\Omega) \subset C(\bar{\Omega})$; i.e., by Sobolev embedding provided $2\gamma > \frac{n}{2}$, as desired. We conclude:

<u>assumption</u> (1.13) $(-A)^{-\gamma}B \in \mathcal{L}(U,Y)$ <u>holds true for problem</u> (3.1) <u>with</u> $\frac{n}{4} < \gamma < 1$, $n \leq 3$.

<u>Assumption (H.1) = (1.5)</u>. The operator A in (3.5) generates an s.c. contraction semigroup e^{At} on Y, which moreover is analytic here for $t > 0$. (This is a special case of a much more general result [C-T.1-2]). This, along with the requirement $\gamma < 1$ proved above guarantees that problem (3.1) satisfies assumption (H.1) = (1.5).

Remark 3.1. Since the semigroup e^{At} is analytic on Y and also uniformly stable [C-T.2], we have by the just-verified property (1.3), in the norm of $\mathcal{L}(Y,U)$:

$$\|B^* e^{A^*t}\| = \|B^*(-A^*)^{-\gamma}(-A^*)^{\gamma} e^{A^*t}\| = O\left(\frac{1}{t^{\gamma}}\right), \quad 0 < t, \qquad (3.9)$$

with $\frac{n}{4} < \gamma < 1$, $n \leq 3$. This is a sharp estimate, which for n = 2,3 (the interesting cases) does not allow to conclude that assumption (H.2) = (1.6) holds true. Instead, (H.2) holds true only for n = 1.

Finite Cost Condition (1.8). With A as in (3.5), the semigroup e^{At} is uniformly (exponentially) stable in Y [C-T.2], and thus the Finite Cost Condition (1.8) holds true with $u \equiv 0$.

Remark 3.2. Suppose that instead of Eq. (3.1a), one has

$$w_{tt}+(\Delta^2+k_1)w-(\Delta+k_2)w_t = \delta(x-x^0)u(t) \quad \text{in } Q, \qquad (3.10)$$

along with (3.1b-c). Then, if $0 < k_1+k_2$ is sufficiently large, the generator A has finitely many unstable eigenvalues in $\{\text{Re } \lambda > 0\}$. Since e^{At} is analytic on Y, the usual theory [T.1] applies: The problem is stabilizable on Y if [T.1] and only if [M-T, Appendix] its projection onto the finite-dimensional unstable subspace is controllable.

For instance, if $\lambda_1,\ldots,\lambda_K$ are the unstable eigenvalues of A, assumed for simplicity to be simple, and $\Phi_1,\ldots,\Phi_K$ are the corresponding eigenfunctions in Y, then the necessary and sufficient condition for stabilization is that $\Phi_k(x^0) \neq 0$, $k = 1,\ldots,K$.

If $\lambda_1,\ldots,\lambda_K$ are not simple, then their largest multiplicity M determines the smallest number of scalar controls needed for the stabilization of (3.10), where now the right hand side is replaced by $\sum_{i=1}^{M} \delta(x-x^i)u_i(t)$, along with (3.1b-c). The necessary and sufficient condition for stabilization is now a well-known full rank condition [T.1].

Detectability Condition (2.10). This is satisfied since in our case R = I, see (3.5).

Conclusion. Theorem 2.1 applies to problem (3.1)-(3.2), $n \leq 3$, and provides existence and uniqueness of the solution to the ARE (2.1), with Riccati operator $P \in \mathcal{L}(Y,\mathcal{D}(A))$ (since A, as remarked above (3.8), is the direct sum of two normal operators on Y plus possibly a finite-dimensional component, in particular, A has a Riesz basis of eigenvectors on Y), where $\mathcal{D}(A) = \mathcal{D}(\mathcal{A})\times\mathcal{D}(\mathcal{A}^{1/2})$, see (3.3), (3.4) for the characterizations of these spaces. Thus, in particular, we have $B^*P \in \mathcal{L}(Y;U)$, where $B^*\begin{bmatrix} v_1 \\ v_2 \end{bmatrix} = v_2(x^0)$. (Note that Theorem 2.2 would apply as well for $n = 1$.)

Remark 3.3. Essentially the same analysis with minimal changes applies also to problem (3.1a-b), with the B.C. (3.1c) replaced now by $\frac{\partial w}{\partial \nu}|_\Sigma \equiv \frac{\partial \Delta w}{\partial \nu}|_\Sigma \equiv 0$. The new definition of $\mathcal{A}$ incorporates, of course, these boundary conditions, and it is still true that the damping operator is precisely $\mathcal{A}^{1/2}$, so that A now has the same form (3.5) as before. The main difference is that the present $\mathcal{A}$ is non-negative self-adjoint and has $\mu = 0$ as an eigenvalue with corresponding one-dimensional eigenspace, spanned by the constant functions. Thus, the new operator A has $\lambda = 0$ as an eigenvalue with corresponding eigenfunction $\Phi = [\Phi_1,\Phi_2]$, $\phi_1 = \text{const}$, $\phi_2 = 0$. Then, Remark 3.2 applies to stabilize the system, as the condition $\phi(x^0) \neq 0$ is satisfied. (With no harm, one may choose to work on the space $Y = \mathcal{D}(\mathcal{A}^{1/2})\times L_2^0(\Omega)$, where $L_2^0(\Omega)$ is the quotient space $L_2(\Omega)/\mathcal{N}(\mathcal{A})$, the null space of $\mathcal{A}$.)

Example 3.2: <u>The case</u> $\alpha = 1$ [C-T.1-2]. The Kelvin-Voigt model for a plate equation in the deflection $w(t,x)$ is

$$w_{tt}+\Delta^2 w+\Delta^2 w_t = \delta(x-x^0)u(t) \quad \text{in } (0,T]\times\Omega = Q; \tag{3.11a}$$

$$w(0,\cdot) = w_0;\ w_t(0,\cdot) = w_1 \quad \text{in } \Omega; \tag{3.11b}$$

$$\Delta w|_\Sigma+(1-\mu)B_1 w \equiv 0 \quad \text{in } (0,T]\times\Gamma = \Sigma; \tag{3.11c}$$

$$\frac{\partial \Delta w}{\partial \nu}|_\Sigma+(1-\mu)B_2 w \equiv 0 \quad \text{in } \Sigma; \tag{3.11d}$$

with $0 < \mu < 1/2$ the Poisson modulus. The boundary operators B_1 and B_2 are zero for $n = 1$, and [Lag.2] for $n = 2$:

$$B_1 w = 2\nu_1\nu_2 w_{xy}-\nu_1^2 w_{yy}-\nu_2^2 w_{xx};$$

$$B_2 w = \frac{\partial}{\partial r}[(\nu_1^2-\nu_2^2)w_{xy}+\nu_1\nu_2(w_{yy}-w_{xx})], \tag{3.12}$$

where again x^0 is an interior point of the open bounded $\Omega \subset R^n$, $n \leq 2$. Regularity results for problem (3.11) are given in [T.4]. Consistantly with these, we take the cost functional to be the same as (3.2) with $\{w_0, w_1\} \in H^2(\Omega)\times L_2(\Omega)$.

Abstract setting. We introduce the non-negative self-adjoint operator

$$\mathcal{A}h = \Delta^2 h,\ \mathcal{D}(\mathcal{A}) = \{h \in H^4(\Omega):\ \Delta h+(1-\mu)B_1 h|_\Gamma = 0;\ \frac{\partial \Delta h}{\partial \nu} + (1-\mu)B_2 h|_\Gamma = 0\}, \tag{3.13}$$

and select the spaces

$$Y = \mathcal{D}(\mathcal{A}^{1/2})\times L_2(\Omega) = H^2(\Omega)\times L_2(\Omega);\quad U = \mathbb{R}^1, \tag{3.14}$$

and finally, define the operators

$$A = \begin{vmatrix} 0 & I \\ -\mathcal{A} & -\mathcal{A} \end{vmatrix};\ Bu = \begin{vmatrix} 0 \\ \delta(x-x^0)u \end{vmatrix};\ R = I \tag{3.15}$$

to obtain the abstract model (1.1), (1.2).

Assumption (1.3) $(-A)^{-\gamma}B \in \mathcal{L}(U,Y)$. Again, it is straightforward to verify that assumption (1.3) is satisfied with $\gamma = 1$: From (3.15), we require that

$$(-A)^{-1}Bu = \begin{vmatrix} I & \mathcal{A}^{-1} \\ -I & 0 \end{vmatrix} \begin{vmatrix} 0 \\ \delta(x-x^0)u \end{vmatrix} = \begin{vmatrix} \mathcal{A}^{-1}\delta(x-x^0) \\ 0 \end{vmatrix} \in Y, \tag{3.16}$$

i.e., from (3.14) we require that $\mathcal{A}^{-1/2}\delta(x-x^0)$. The same argument below (3.6) then applies yielding that (3.16) holds true if $n \leq 3$.

However, in order to verify assumption (H.1) = (1.5) which requires that γ should be < 1, the most elementary way is to check that assumption (1.3) holds in fact true with $\gamma = 1/2$. In this case, we can in fact rely on the direct computation of $(-A)^{-1/2}$

$$(-A)^{-1/2} = \begin{vmatrix} (1) & \mathcal{A}^{-3/4}(2I+\mathcal{A}^{1/2})^{-1/2} \\ (2) & \mathcal{A}^{-1/4}(2I+\mathcal{A}^{1/2})^{-1/2} \end{vmatrix} \tag{3.17}$$

(where the entries $(1) = \mathcal{A}^{-1/4}(2I+\mathcal{A}^{1/2})^{-1/2}(I+\mathcal{A}^{1/2})$ and $(2) = -\mathcal{A}^{1/4}(2I+\mathcal{A}^{1/2})^{-1/2}$ do not really count in the present analysis), and avoid the domain of fractional powers as in [C-T.4].

We need to compute

$$(-A)^{-\frac{1}{2}}Bu = \begin{vmatrix} A^{-\frac{3}{4}}(2I+A^{\frac{1}{2}})^{-\frac{1}{2}}\delta(x-x^0)u \\ A^{-\frac{1}{4}}(2I+A^{\frac{1}{2}})^{-\frac{1}{2}}\delta(x-x^0)u \end{vmatrix}. \qquad (3.18)$$

From (3.18), we then readily see that $(-A)^{-\frac{1}{2}}Bu \in Y = \mathcal{D}(A^{\frac{1}{2}})\times L_2(\Omega)$ provided (#): $A^{-\frac{1}{2}}\delta(x-x^0) \in L_2(\Omega)$. But $\mathcal{D}(A^{\frac{1}{2}}) = H^2(\Omega)$ (and, in fact, only $\mathcal{D}(A^{\frac{1}{2}}) \subset H^2(\Omega)$ suffices for the present analysis) so taht condition (#) is satisfied provided $\delta(x-x^0) \in [H^2(\Omega)]'$ (duality with respect to $L_2(\Omega)$); i.e., provided $H^2(\Omega) \subset C(\bar{\Omega})$, i.e., by Sobolev embedding provided $2 > \frac{n}{2}$, or $n < 4$, as desired.

We have shown: <u>Assumption (1.3)</u> $(-A)^{-\gamma}B \in \mathcal{L}(U,Y)$ <u>holds true for problem (3.11) with</u> $n \le 3$, <u>and</u> $\gamma = \frac{1}{2}$. The above argument shows some 'leverage.' Indeed, $\gamma = \frac{1}{2}$ is not the least γ for which assumption (1.3) holds true. To obtain the least γ for which assumption (1.3) holds true, we proceed as in the case of problem (3.1) above, in the argument which begins with (3.7) and uses the domains of fractional powers $\mathcal{D}((-A)^{\gamma})$. As below (3.7) and ff., we need to show that (#): $Bu \in [\mathcal{D}((-A)^{\gamma})]'$, duality with respect to Y. But for $0 < \gamma \le \frac{1}{2}$, we have from [C-T.4, with $\alpha = 1$] that

$$\mathcal{D}((-A^{\gamma})) = \mathcal{D}(A^{\frac{1}{2}})\times\mathcal{D}(A^{\gamma}). \qquad (3.19)$$

Thus, from B as in (3.15), we see that condition (#) above holds true, provided $\delta(x-x^0) \subset [\mathcal{D}(A^{\gamma})]'$, duality with respect to $L_2(\Omega)$; i.e., provided $\delta(x-x^0) \subset [H^{4\gamma}(\Omega)]'$, since $\mathcal{D}(A^{\gamma}) \subset H^{4\gamma}(\Omega)$ for the fourth order operator in (3.13), i.e., provided $H^{4\gamma}(\Omega) \subset C(\bar{\Omega})$, which in turn is the case, provided $4\gamma > \frac{n}{2}$, or $\frac{1}{2} \ge \gamma > \frac{n}{8}$. We conclude: <u>Assumption (1.3)</u> $(-A)^{-\gamma}B \in \mathcal{L}(U,Y)$ <u>holds true for problem (3.11) provided</u> $\frac{n}{8} < \gamma \le \frac{1}{2}$, $n \le 3$.

<u>Assumption (H.1) = (1.5)</u>. The operator A in (3.15) generates an s.c. contraction semigroup e^{At} on Y, which moreover is analytic here for $t > 0$. (This is a special case of a much more general result [C-T.2].) This, along with the requirement $\gamma < 1$ proved above guarantees that problem (3.1) satisfies assumption (H.1) = (1.5).

<u>Remark 3.4</u>. Since the semigroup e^{At} is analytic on Y, we have by the just-verified property (1.3), in the norm of $\mathcal{L}(Y,U)$

$$\|B^* e^{A^* t}\| = \|B^*(-A^*)^{-\gamma}(-A^*)^{\gamma} e^{A^* t}\| \le O\left(\frac{1}{t^{\gamma}}\right),$$
$$0 < t, \qquad (3.20)$$

with $\frac{n}{8} < \gamma < \frac{1}{2}$, $n \le 3$, where we can take all $t > 0$ as e^{At} is also uniformly stable [C-T.2]. Thus, for $n \le 3$, we obtain that assumption (H.2) = (1.6) holds as well.

Finite Cost Condition (1.8). With A as in (3.15), the semigroup e^{At} is uniformly (exponentially) stable in $Y/N(A)$, the finite-dimensional nullspace of A [C-T.2], and thus the Finite Cost Condition (1.8) is automatically satisfied on this space with $u \equiv 0$. For the eigenvalue $\lambda = 0$, Remark 3.2 applies also to problem (3.11).

Detectability Condition (2.10). This is satisfied since in our case $R = I$, see (3.15).

Conclusion. Theorem 2.1 applies to problem (3.11) for $n \le 3$. (Theorem 2.2 would also apply for $n \le 3$, but the conclusions of Theorem 2.1 are stronger.)

Example 3.3. (A structurally damped plate with boundary control.) We consider the plate problem.

$$w_{tt}+\Delta^2 w-\Delta w_t = 0 \quad \text{in } (0,T]\times\Omega \equiv Q, \qquad (3.20a)$$
$$w(0,\cdot) = w_0;\ w_t(0,\cdot) = w_1 \quad \text{in } \Omega, \qquad (3.20b)$$
$$w|_{\Sigma} \equiv 0 \quad \text{in } (0,T]\times\Gamma \equiv \Sigma, \qquad (3.20c)$$
$$\Delta w|_{\Sigma} \equiv u \quad \text{in } \Sigma, \qquad (3.20d)$$

which is the same model as the one in (3.1), except that it is acted upon by a boundary control $u \in L_2(0,T;L_2(\Gamma)) \equiv L_2(\Sigma)$, rather than a point control as in (3.1a). We take the same functional J as in (3.2) except that now u is penalized in the $L_2(\Gamma)$-norm. Following [L-T.14], we introduce the Green map G_2 defined by

$$y = G_2 v \iff \{\Delta^2 y = 0 \text{ in } \Omega;\ y|_{\Gamma} = 0;\ \Delta y|_{\Gamma} = v\}. \qquad (3.21)$$

Then, if $\mathcal{A}$ is the same operator defined in (3.3), it is rather straightforward to see that the abstract representation of problem (3.20) is given by the equation

$$w_{tt}+\mathcal{A}w+\mathcal{A}^{1/2}w_t = \mathcal{A}G_2 u. \qquad (3.22)$$

(Indeed, problem (3.20) can be rewritten first as $w_{tt}+\Delta^2(w-G_2u)-\Delta w_t = 0$ in Q; $(w-G_2v)|_{\Sigma} = \Delta(w-G_2u)|_{\Sigma} = 0$ by (3.21); hence abstractly by $w_{tt}+\mathcal{A}(w-G_2u)+\mathcal{A}^{1/2}w_t = 0$ because of the B.C. since now $A^{1/2}h = -\Delta h$, $\mathcal{D}(A^{1/2}) = H^2(\Omega) \cap H_0^1(\Omega)$. From here, (3.22) follows by extending the original A in (3.3), as usual, by isomorphism to, say, A: $L_2(\Omega) \to [\mathcal{D}(A)]'$. It can be shown [L-T.14] that the Green map G_2 can be expressed in terms of the Dirichlet map D defined below, as follows:

$$G_2 = -A^{-1/2}D, \text{ where } y = Dv \Leftrightarrow \{\Delta y = 0 \text{ in } \Omega;\ y|_\Gamma = v\}, \tag{3.23}$$

where D satisfies

$$D: \text{continuous } L_2(\Gamma) \to H^{1/2}(\Omega) \subset H^{1/2-2\varepsilon}(\Omega) \equiv \mathcal{D}(A^{1/4-\varepsilon/2}), \quad \varepsilon > 0. \tag{3.24}$$

<u>**Abstract setting**</u>. Thus, (3.22) becomes the abstract equation

$$w_{tt} + Aw + A^{1/2}w_t = -A^{1/2}Du, \tag{3.25}$$

or

$$\frac{d}{dt}\begin{vmatrix} w \\ w_t \end{vmatrix} = A\begin{vmatrix} w \\ w_t \end{vmatrix};\ A = \begin{vmatrix} 0 & I \\ -A & -A^{1/2} \end{vmatrix};\ Bu = \begin{vmatrix} 0 & \\ -A^{1/2} & Du \end{vmatrix} \tag{3.26}$$

on the spaces $Y = \mathcal{D}(A^{1/2}) \times L_2(\Omega)$; $U = L_2(\Gamma)$.

<u>**Assumption (1.3)**</u>. $(-A)^{-\gamma}B \in \mathcal{L}(U,Y)$. Again, it is elementary to verify that assumption (1.3) is satisfied with $\gamma = 1$: Indeed, from (3.26) we require

$$(-A)^{-1}Bu = \begin{vmatrix} A^{-1/2} & A^{-1} \\ -I & 0 \end{vmatrix} \begin{vmatrix} 0 & \\ -A^{1/2} & Du \end{vmatrix} = \begin{vmatrix} -A^{-1/2} & Du \\ & 0 \end{vmatrix} \in Y$$

$$= \mathcal{D}(A^{1/2}) \times L_2(\Omega), \tag{3.27}$$

which certainly holds true by (3.24). We may also verify that the value $\gamma = 1/2$ fails: from direct computations (as in (3.17)) or from [T.4], we obtain

$$(-A)^{-1/2}Bu = \frac{-1}{\sqrt{3}}\begin{vmatrix} A^{-3/4} & A^{1/2}Du \\ A^{-1/4} & A^{1/2}Du \end{vmatrix} = \frac{-1}{\sqrt{3}}\begin{vmatrix} A^{-1/4}Du \\ A^{1/4}Du \end{vmatrix}, \tag{3.28}$$

and from (3.24) we see that $(-A)^{-1/2}Bu$ in (3.28) fails by $1/4 + \varepsilon/2$, to be in Y.

Indeed, we have that: <u>Assumption (1.3) holds true for all</u> $3/4 < \gamma < 1$.

The above claim can be verified by an argument similar to the ones of the preceding two examples, based on the domain of fractional power [C-T.4, with $\alpha = 1/2$]

$$\mathcal{D}((-A)^{\gamma}) = \mathcal{D}(A^{\frac{1}{2}+\gamma/2})\times\mathcal{D}(A^{\gamma/2}), \quad \tfrac{1}{2} \le \gamma < 1 \qquad (3.29)$$

(only the second component is needed in our argument), whereby the usual condition $Bu \in [\mathcal{D}((-A)^{\gamma})]'$, duality with respect to Y, is satisfied provided, from (3.26) and (3.29), $A^{\frac{1}{2}}Du \in [\mathcal{D}(A^{\gamma/2})]'$, duality with respect to $L_2(\Omega)$; i.e., from (3.24) provided $\frac{3}{8} + \frac{\varepsilon}{2} \le \frac{\gamma}{2}$ or $\gamma = \frac{3}{4} + \varepsilon$, $\forall\ \varepsilon$, and the claim is proved.

__Assumption (H.1) = (1.5)__. This is satisfied as the requirement $\gamma < 1$ was proved above and e^{At} is an s.c. contraction, analytic semigroup on Y [C-T.1-2].

__Remark 3.5__. Since $\frac{1}{2} < \gamma$, a computation as in (3.9) shows that assumption (H.2) = (1.6) is not satisfied.

__Finite Cost Condition (1.8)__. With A as in (3.26), the semigroup e^{At} is uniformly (exponentially) stable on Y [C-T.2], and thus the Finite Cost Condition (1.8) is automatically satisfied with $u \equiv 0$ (see also Remark 3.2).

__Conclusion__. Theorem 2.1 with R = I applies to problem (3.20) for any n. (Theorem 2.2 is not applicable.)

__Remark 3.6__. A similar analysis applies if the B.C. (3.20c-d) are replaced by $\frac{\partial w}{\partial \nu}|_{\Sigma} \equiv 0$, $\frac{\partial \Delta w}{\partial \nu}|_{\Sigma} = u$; refer to Remark 3.3.

So far, we have considered examples of damped plates where the damping operator is equal to the α-th power of the original elastic differential operator, $\alpha = \frac{1}{2}$ in Examples 3.1 and 3.3, and $\alpha = 1$ in Example 3.2. This was due to the special choice of boundary conditions. In the next example, we return to the same Eq. (3.1a), which we now complement with boundary conditions that make the damping operator only 'comparable,' in a technical sense [C-T.1-2], to the α-th power of the elastic operator, $\alpha = \frac{1}{2}$.

__Example 3.5__. On some smooth Ω, dim $\Omega = n \le 3$, consider the plate model:

$$\begin{cases} w_{tt}+\Delta^2 w-\Delta w_t = \delta(x-x^0)u(t) & \text{in } (0,T]\times\Omega = Q, \quad (3.30a)\\ w(0,\cdot) = w_0;\ w_t(0,\cdot) = w_1 & \text{in } \Omega, \quad (3.30b)\\ w|_{\Sigma} \equiv \frac{\partial w}{\partial \nu}|_{\Sigma} \equiv 0 & \text{in } (0,T]\times\Gamma \equiv \Sigma, \quad (3.30c)\end{cases}$$

Regularity results for problem (3.30) are given in [T.4]. Consistently with these, we associate with (3.30) the same cost functional (3.2).

Abstract setting. Now, however, we introduce the positive self-adjoint operators

$$\mathcal{A}h = \Delta^2 h, \ \mathcal{D}(\mathcal{A}) = \{h \in H^4(\Omega): h|_\Gamma = \frac{\partial h}{\partial\nu}|_\Gamma = 0\} \qquad (3.31)$$

$$\mathcal{B}h = -\Delta h, \ \mathcal{D}(\mathcal{B}) = \{h \in H^2(\Omega): h|_\Gamma = \frac{\partial h}{\partial\nu}|_\Gamma = 0\} = H_0^2(\Omega) = \mathcal{D}(\mathcal{A}^{1/2}), \qquad (3.32)$$

where the equality with $\mathcal{D}(\mathcal{A}^{1/2})$ in (3.2) (equivalent norms) is standard [G.1]. Thus, problem (3.30) admits the abstract version

$$w_{tt}+\mathcal{A}w+\mathcal{B}w_t = \delta(x-x^0)u$$

which fits the abstract model (1.1), (1.2) with

$$A = \begin{vmatrix} 0 & I \\ -\mathcal{A} & -\mathcal{B} \end{vmatrix}; \quad Bu = \begin{vmatrix} 0 \\ \delta(x-x^0)u \end{vmatrix}; \quad R = I \qquad (3.33)$$

on the spaces $Y = \mathcal{D}(\mathcal{A}^{1/2})\times L_2(\Omega)$, $U = \mathbb{R}^1$. From (3.32), we have

$$\mathcal{B}^2 h = \Delta^2 h, \ \mathcal{D}(\mathcal{B}^2) = \{h \in H^4(\Omega): h|_\Gamma = \frac{\partial h}{\partial\nu}|_\Gamma = \Delta h|_\Gamma = \frac{\partial\Delta h}{\partial\nu}|_\Gamma = 0\} \subset \mathcal{D}(\mathcal{A}). \qquad (3.34)$$

By Green's second theorem, we have

$$(\mathcal{B}^2 f, f) = (\Delta^2 f, f) = \int_\Gamma \frac{\partial\Delta f}{\partial\nu} f\, d\Gamma - \int_\Gamma \Delta f \frac{\partial f}{\partial\nu} d\Gamma + \int_\Omega (\Delta f)^2 d\Gamma \qquad (3.35)$$

$$= \int_\Omega (\Delta f)^2 d\Omega, \quad f \in \mathcal{D}(\mathcal{B}^2); \qquad (3.36)$$

$$(\mathcal{A}f, f) = (\Delta^2 f, f) = \int_\Gamma \frac{\partial\Delta f}{\partial\nu} f\, d\Gamma - \int_\Gamma \Delta f \frac{\partial f}{\partial\nu} d\Gamma + \int_\Omega (\Delta f)^2 d\Omega \qquad (3.37)$$

$$= \int_\Omega (\Delta f)^2 d\Omega, \quad f \in \mathcal{D}(\mathcal{A}); \qquad (3.38)$$

where the boundary terms on the right hand side of (3.35) and (3.37) still vanish if f is only in $\mathcal{D}(\mathcal{B})$, or $\mathcal{D}(\mathcal{A}^{1/2})$, respectively, see (3.32). Thus, by extension, we get

$$(\mathcal{B}^2 f, f) = (\mathcal{A}f, f) = \int_\Omega (\Delta f)^2 d\Omega, \quad f \in \mathcal{D}(\mathcal{B}) = \mathcal{D}(\mathcal{A}^{1/2}), \qquad (3.39)$$

and thus *a fortiori* the results of [C-T.1-2] apply: These

in particular establish that the operator A in (3.33) <u>generates an s.c., analytic semigroup</u> e^{At} <u>on</u> Y.

<u>Assumption (1.3)</u>: $(-A)^{-\gamma}B \in \mathcal{L}(U,Y)$. Again, it is immediate to see that assumption (1.3) holds true for $\gamma = 1$. In fact, as in the argument below (3.6), we find that

$$(-A)^{-1}Bu = \begin{vmatrix} \mathcal{A}^{-1}\mathcal{B} & \mathcal{A}^{-1} \\ -I & 0 \end{vmatrix} \begin{vmatrix} 0 \\ \delta(x-x^0) \end{vmatrix} = \begin{vmatrix} \mathcal{A}^{-1}\delta(x-x^0) \\ 0 \end{vmatrix} \in Y. \tag{3.40}$$

In effect, <u>assumption (1.3) holds true for problem</u> (3.30) <u>for all</u> γ <u>with</u> $\frac{n}{4} < \gamma < 1$, $n \le 3$, exactly as in the case of problem (3.1). To see this, with $\mathcal{A}$ and $\mathcal{B}$ as in (3.31), (3.32), we denote for convenience

$$A_{\mathcal{B}} = \begin{vmatrix} 0 & I \\ -\mathcal{A} & -\mathcal{B} \end{vmatrix}; \qquad A_{\mathcal{B}}^* = \begin{vmatrix} 0 & -I \\ \mathcal{A} & -\mathcal{B}^* \end{vmatrix};$$

$$A_{\frac{1}{2}} = \begin{vmatrix} 0 & I \\ -\mathcal{A} & -\mathcal{A}^{\frac{1}{2}} \end{vmatrix}; \qquad A_{\frac{1}{2}}^* = \begin{vmatrix} 0 & -I \\ \mathcal{A} & -\mathcal{A}^{\frac{1}{2}} \end{vmatrix}; \tag{3.41}$$

where the adjoints are with respect to Y. Since $\mathcal{B}$ in (3.32) is self-adjoint on $L_2(\Omega)$, we have

$$\mathcal{D}(A_{\mathcal{B}}^*) = \mathcal{D}(A_{\mathcal{B}}) = \mathcal{D}(A_{\frac{1}{2}}) = \mathcal{D}(A_{\frac{1}{2}}^*). \tag{3.42}$$

As a consequence of (part of) (3.34), we have in our present case [C-T.4]

$$\mathcal{D}((-A_{\frac{1}{2}})^{\gamma+\varepsilon}) \subset \mathcal{D}((-A_{\mathcal{B}}^*)^{\gamma}) \subset \mathcal{D}((-A_{\frac{1}{2}})^{\gamma-\varepsilon}),$$
$$0 < \gamma < 1 \tag{3.43}$$

and $\gamma+\varepsilon < 1$. Then, to obtain $(-A_{\mathcal{B}})^{-\gamma}B \in \mathcal{L}(U,Y)$, $\frac{n}{4} < \gamma < 1$, as desired, i.e., $Bu \in [\mathcal{D}((-A_{\mathcal{B}}^*)^{\gamma})]'$, duality with respect to Y, it suffices via the right hand side of (3.44) to have $Bu \in [\mathcal{D}((-A_{\frac{1}{2}})^{\gamma-\varepsilon})]'$. But this was shown to be true in Example 3.1, precisely for $\frac{n}{4} < \gamma-\varepsilon < 1$, in the argument below (3.7).

Alternatively, we may write, using Example 3.1, that

$$(-A_{\mathcal{B}})^{-\gamma}B = (-A_{\mathcal{B}})^{-\gamma}(-A_{\frac{1}{2}})^{\gamma-\varepsilon}(-A_{\frac{1}{2}})^{-(\gamma-\varepsilon)}B \in \mathcal{L}(U,Y), \tag{3.44}$$

since $(-A_{\mathcal{B}})^{-\gamma}(-A_{1/2})^{\gamma-\varepsilon}$ is bounded, for $(-A_{1/2}^*)^{\gamma-\varepsilon}(-A_{\mathcal{B}}^*)^{-\gamma}$ is bounded by (3.42) and the closed graph theorem.

Assumption (H.1) = (1.5). This assumption holds true for problem (3.30) since it was already observed below (3.39) that e^{At} is a s.c. analytic, semigroup on Y, while γ was shown above to be < 1 for $n \leq 3$.

The Finite Cost Condition (1.8) and **The Detectability Condition (2.10)**: These also hold true, since e^{At} is uniformly stable [C-T.2] and R = I, respectively.

Conclusion. Theorem 2.1 applies to problem (3.30).

3.2 Class (H.2): Second order hyperbolic equations with Dirichlet boundary control

We consider the following problem:

$$\begin{cases} w_{tt} = \Delta w & \text{in } (0,T]\times\Omega = Q, & (3.45a) \\ w(0,\cdot) = w_0, \quad w_t(0,\cdot) = w_1 & \text{in } \Omega, & (3.45b) \\ w|_\Sigma \equiv u & \text{in } (0,T]\times\Gamma \equiv \Sigma, & (3.45c) \end{cases}$$

where we take the boundary control $u \in L_2(\Sigma)$. (In (3.45a) we may replace $-\Delta$ by any second order, elliptic operator with time independent, symmetric coefficients of its principal part, with minimal changes in the analysis below.) Consistently with established (optimal) regularity theory [Lio.1], [L-T.2], [LLT], we take $\{w_0, w_1\} \in L_2(\Omega)\times H^{-1}(\Omega)$ and the cost functional

$$J(u,w) = \int_0^\infty \|w(t)\|^2_{L_2(\Omega)} + \|w_t(t)\|^2_{H^{-1}(\Omega)} + |u(t)|^2_{L_2(\Omega)}\,dt. \tag{3.46}$$

Abstract setting. To put problem (3.45), (3.46) into the abstract model (1.1), (1.2), we introduce the positive self-adjoint operator $\mathcal{A}h = -\Delta h$, $\mathcal{D}(\mathcal{A}) = H^2(\Omega) \cap H_0^1(\Omega)$ and define the operators

$$A = \begin{vmatrix} 0 & I \\ -\mathcal{A} & 0 \end{vmatrix}; \quad Bu = \begin{vmatrix} 0 \\ \mathcal{A}Du \end{vmatrix}; \; R = I, \tag{3.47}$$

where D is the Dirichlet map encountered before in Example 3.3, Eq. (3.23),

$$Dv = y \iff \{\Delta y = 0 \text{ in } \Omega, \; y|_\Gamma = v\}, \tag{3.48}$$

and the spaces

$$Y = L_2(\Omega)\times H^{-1}(\Omega); \quad U = L_2(\Gamma). \tag{3.49}$$

The Dirichlet map satisfies the regularity property (3.24).

Assumption (1.3). $(-A)^{-\gamma}B \in \mathcal{L}(U,Y)$. From (3.47) with $u \in L_2(\Gamma)$, we obtain

$$(-A)^{-1}Bu = \begin{vmatrix} 0 & \mathcal{A}^{-1} \\ -I & 0 \end{vmatrix} \begin{vmatrix} 0 \\ \mathcal{A}Du \end{vmatrix} = \begin{vmatrix} Du \\ 0 \end{vmatrix} \in Y, \tag{3.50}$$

a fortiori, by the regularity (3.24) of D, and assumption (1.3) holds true with $\gamma = 1$.

Assumption (H.2) = (1.6). From (3.47) we calculate

$$B^*\begin{vmatrix} z_1 \\ z_2 \end{vmatrix} = D^*z_2 = \frac{\partial}{\partial\nu}\mathcal{A}^{-1}z_2,$$
$$\text{since } D^*\mathcal{A} = \frac{\partial}{\partial\nu} \text{ [L-T.2]}. \tag{3.51}$$

Moreover, we have

$$B^*e^{A^*t}\begin{vmatrix} z_1 \\ z_2 \end{vmatrix} = \frac{\partial\phi(t)}{\partial\nu}, \quad [z_1,z_2] \in Y, \tag{3.52}$$

where $\phi(t) = \phi(t,\phi_0,\phi_1)$ solves the corresponding homogeneous problem

$$\begin{cases} \phi_{tt} = \Delta\phi & \text{in } (0,T]\times\Omega \equiv Q, & \text{(3.53a)} \\ \phi(0,\cdot) = \phi_0,\ \phi_t(0,\cdot) = \phi_1 & \text{in } \Omega, & \text{(3.53b)} \\ \phi|_\Sigma \equiv 0 & \text{in } (0,T]\times\Gamma \equiv \Sigma, & \text{(3.53c)} \end{cases}$$

with

$$\phi_0 = -\mathcal{A}^{-1}z_2 \in \mathcal{D}(\mathcal{A}^{1/2}) = H_0^1(\Omega); \quad \phi_1 = z_1 \in L_2(\Omega). \tag{3.54}$$

Thus, by (3.52), (3.54), an equivalent formulation of assumption (H.2) = (1.6) is the inequality

$$\int_\Sigma \left(\frac{\partial\phi}{\partial\nu}\right)^2 d\Sigma \le c_T\|\{\phi_0,\phi_1\}\|^2_{H_0^1(\Omega)\times L_2(\Omega)} \tag{3.55}$$

for the trace of the solution to problem (3.53). It should be noted that inequality (3.55) does NOT follow from *a priori* (optimal) interior regularity $\phi(t) \in C([0,T];H_0^1(\Omega))$ of the solution to problem (3.53), (3.54). Inequality (3.55) is an independent trace regularity result. It was first established in [L-T.1], [L-T.2]: In these references it was first proved, by means of pseudo-differential operator techniques, that the following interior regularity result holds true: $\{w,w_t\} \in L_2(0,T;L_2(\Omega)\times H^{-1}(\Omega))$ for problem (3.45) with $u \in L_2(\Sigma)$, $\{w_0,w_1\} \in L_2(\Omega)\times H^{-1}(\Omega)$. Next, via a duality argument, it was proved by a purely operator technique, that indeed (3.55) holds true, and that, in fact, $\{w,w_t\} \in C([0,T];\, L_2(\Omega)\times H^{-1}(\Omega))$. Inequality (3.55) was proved independently and directly also in [Lio.1], by a multiplier technique; see also [LLT] for a comprehensive treatment.

Finite Cost Condition (1.8). Sufficient conditions which would imply that the F.C.C. (1.8) is satisfied are:

(i) (exponential) uniform stabilization of problem (3.45) on the space $Y = L_2(\Omega)\times H^{-1}(\Omega)$ by means of an $L_2(0,\infty;L_2(\Gamma))$-feedback u;

(ii) exact controllability of problem (3.45) (to or, equivalently, from the origin) over a finite interval $[0,T]$, on the state space $Y = L_2(\Omega)\times H^{-1}(\Omega)$, within the class of $L_2(0,T;L_2(\Gamma))$-controls u.

A solution to the uniform stabilization problem (i), and consequently, via a known result of D. Russell (1973) of the exact controllability problem (ii) was first obtained in [L-T.12], under some additional geometrical condition on Ω (which includes the class of strictly convex Ω). Later, exact controllability, this time without geometrical conditions on Ω, except for smooth Γ, if u is applied to all of Γ, was established in [Lio.2] via the so-called Hilbert Uniqueness Method, by relying on a lower bound inequality, inequality (3.55) with the reversed inequality sign, if T is sufficiently large

$$\int_\Sigma \left(\frac{\partial\phi}{\partial\nu}\right)^2 d\Sigma \ge c_T\|\{\phi_0,\phi_1\}\|^2_{H_0^1(\Omega)\times L_2(\Omega)}. \qquad (3.56)$$

This latter inequality (3.56) was explicitly obtained in [H] by using the same multiplier methods that had been used in [Lio.1], [L-L-T] for inequality (3.55), and in [L-T.12] to solve the uniform stabilization problem; indeed, such inequality (3.56) is essentially contained also in the estimates of this work [L-T.12], albeit in a less transparent form. A direct approach to exact controllability based on the surjectivity of the input-solution operator and multiplier methods to show the key inequality, in the case where u acts only on a portion of the boundary Σ_0 is given in [T.3]. Later, geometric optics methods--first introduced in [Lit.] for exact controllability questions--provided essentially necessary and sufficient conditions for inequality (3.56) to hold true, with Σ replaced by a subportion $\Sigma_0 \subset \Sigma$ [B-L-R]. In any case, the validity of the Finite Cost Condition (1.8) for problem (3.45), or a more general version thereof, is firmly established.

Detectability Condition (2.17)-(2.19). This holds true since R = I in our case, see (3.47).

Conclusion. Theorem 2.2 applies to problem (3.45).

3.3 Class (H.2): Euler-Bernoulli equations with boundary control

We consider on any smooth bounded $\Omega \subset R^n$:

$$\begin{cases} w_{tt}+\Delta^2 w = 0 & \text{in } (0,T]\times\Omega = Q, & (3.57a) \\ w(0,\cdot) = w_0,\ w_t(0,\cdot) = w_1 & \text{in } \Omega, & (3.57b) \\ w|_\Sigma \equiv 0 & \text{in } (0,T]\times\Gamma = \Sigma, & (3.57c) \\ \frac{\partial w}{\partial \nu}|_\Sigma \equiv u & \text{in } \Sigma, & (3.57d) \end{cases}$$

with boundary control $u \in L_2(\Sigma)$. Consistently with regularity theory, the cost functional to be minimized is

$$J(u,w) = \int_0^\infty \|w(t)\|^2_{L_2(\Omega)} + \|w_t(t)\|^2_{H^{-2}(\Omega)} + \|u(t)\|^2_{L_2(\Gamma)} dt. \tag{3.58}$$

Abstract setting. To put problem (3.57), (3.58) into the abstract model (1.1), (1.2), we introduce the positive self-adjoint operator

$$\mathcal{A}h = \Delta^2 h,\ \mathcal{D}(\mathcal{A}) = \{h \in H^4(\Omega): h|_\Gamma = \frac{\partial h}{\partial \nu}|_\Gamma = 0\}. \tag{3.59}$$

and define the operators

$$A = \begin{vmatrix} 0 & I \\ -\mathcal{A} & 0 \end{vmatrix};\quad Bu = \begin{vmatrix} 0 \\ \mathcal{A}G_2 u \end{vmatrix};\quad R = I \tag{3.60}$$

where G_2 is the appropriate Green map:

$$y = G_2 v \iff \{\Delta^2 y = 0 \text{ in } \Omega;\ y|_\Gamma = 0,\ \frac{\partial y}{\partial \nu}|_\Gamma = v\}, \tag{3.61}$$

and the spaces

$$Y \equiv L_2(\Omega)\times H^{-2}(\Omega),\quad U \equiv L_2(\Gamma). \tag{3.62}$$

Assumption (1.3). $(-A)^{-\gamma}B \in \mathcal{L}(U,Y)$. Since G_2 is certainly bounded $L_2(\Gamma) \to L_2(\Omega)$, we readily obtain from (3.60) with $u \in L_2(\Gamma)$:

$$(-A)^{-1}Bu = \begin{vmatrix} 0 & -\mathcal{A}^{-1} \\ -I & 0 \end{vmatrix} \begin{vmatrix} 0 \\ \mathcal{A}G_2u \end{vmatrix} = \begin{vmatrix} G_2u \\ 0 \end{vmatrix} \in Y, \qquad (3.63)$$

and assumption (1.3) holds true for problem (3.57) with $\gamma = 1$.

Assumption (H.2) = (1.6). One can show that [L-T]

$$B^*e^{A^*t}\begin{vmatrix} z_1 \\ z_2 \end{vmatrix} = \Delta\phi(t), \qquad (3.64)$$

where $\phi(t) = \phi(t,\phi_0,\phi_1)$ solves the corresponding homogeneous problem

$$\begin{cases} \phi_{tt}+\Delta^2\phi = 0 & \text{in } (0,T]\times\Omega = Q, & (3.65a) \\ \phi(0,\cdot) = \phi_0,\ \phi_t(0,\cdot) = \phi_1 & \text{in } \Omega, & (3.65b) \\ \phi|_\Sigma \equiv \frac{\partial\phi}{\partial\nu}|_\Sigma \equiv 0 & \text{in } (0,T]\times\Gamma \equiv \Sigma, & (3.65c) \end{cases}$$

with

$$\phi_0 = -\mathcal{A}^{-1}z_2 \in \mathcal{D}(\mathcal{A}^{1/2}) = H_0^2(\Omega);\ \phi_1 = z_1 \in L_2(\Omega). \qquad (3.66)$$

Thus, by (3.64), (3.66), an equivalent formulation of assumption (H.2) = (1.6) is the inequality

$$\int_\Sigma |\Delta\phi|^2 d\Sigma \le c_T\|\{\phi_0,\phi_1\}\|^2_{H_0^2(\Omega)\times L_2(\Omega)} \qquad (3.67)$$

for the trace of the solution to problem (3.65). As in the case of the wave equation of section 3.2, it should be noted that inequality (3.67) does NOT follow from *a priori* (optimal) interior regularity $\phi(t) \in C([0,T];H_0^2(\Omega))$ of the solution to the problem (3.65), (3.66). It is an independent regularity result which holds indeed true [Lio.2], for any general smooth Ω. Thus, assumption (H.2) = (1.6) holds true for problem (3.57).

Finite Cost Condition (1.8). The same considerations apply now as in the case of the preceding wave equation problem. Exact controllability of problem (3.57) holds true, for any $T > 0$, in fact, on the state space $Y = L_2(\Omega)\times H^{-2}(\Omega)$ within the class of $L_2(\Sigma)$-controls u, with no geometrical conditions on Ω [Lio.2], [L-T].

Uniform stabilization can also be established, likewise without geometrical conditions [O-T].

__Detectability Conditions (2.17)-(2.19)__. This holds true since $R = I$ in our case, see (3.60).

__Conclusion__. Theorem 2.2 applies to problem (3.57), (3.58).

__Remark 3.7__. Theorem 2.2 is also applicable to problem (3.57a-b) with B.C. (3.57c-d) there replaced now by

$$w|_\Sigma \equiv u; \quad \frac{\partial w}{\partial \nu}\Big|_\Sigma \equiv 0, \tag{3.68}$$

and $u \in L_2(\Sigma)$ on a suitable state space. See [L-T.9], [FLT, Appendix C], [L-T.11]. Details are omitted here for lack of space. ∎

3.4 __Class (H.2): Kirchoff plate with boundary control__

In $\Omega \subset R^n$, we consider the Kirchoff plate

$$\begin{cases} w_{tt} - \rho\Delta w_{tt} + \Delta^2 w = 0 & \text{in } (0,T]\times\Omega \equiv Q, & (3.69a) \\ w(0,\cdot) = w_0,\ w_t(0,\cdot) = w_1 & \text{in } \Omega, & (3.69b) \\ w|_\Sigma \equiv 0 & \text{in } (0,T]\times\Gamma, & (3.69c) \\ \Delta w|_\Sigma = u & \text{in } \Sigma, & (3.69d) \end{cases}$$

with $\rho > 0$ a constant, and with just one boundary control u which we take in $L_2(\Sigma)$. Optimal regularity theory of problem (3.69) is given in [L-T.16]. Consistently with these results, we take the following cost functional

$$J(u,w) = \int_0^\infty \|w(t)\|^2_{H^2(\Omega)} + \|w_t(t)\|^2_{H^1(\Omega)} + \|u(t)\|^2_{L_2(\Gamma)} dt \tag{3.70}$$

with initial data $\{w_0, w_1\} \in [H^2(\Omega) \cap H^1_0(\Omega)]\times H^1_0(\Omega)$.

__Abstract setting__. To put problem (3.69), (3.70) into the abstract model (1.1), (1.2), we introduce the positive self-adjoint operators

$$\mathcal{A}h = \Delta^2 h, \quad \mathcal{D}(\mathcal{A}) = \{h \in H^4(\Omega);\ h|_\Gamma = \Delta h|_\Gamma = 0\}$$

$$\mathcal{A}^{1/2}h = -\Delta h, \quad \mathcal{D}(\mathcal{A}^{1/2}) = H^2(\Omega) \cap H^1_0(\Omega)$$

the same as in (3.3) of Example 3.1, and define the operators

$$\mathbf{A} = \begin{vmatrix} 0 & I \\ -\mathcal{A} & 0 \end{vmatrix}; \quad Bu = \begin{vmatrix} 0 \\ \mathcal{A}G_2u \end{vmatrix}; \quad \mathcal{A} = (I+\rho A^{1/2})^{-1}A; \quad R = I \tag{3.71}$$

and G_2 is the same Green map defined in (3.21) in **Example** 3.3, which satisfies Eq. (3.23) in terms of the Dirichlet map D: $G_2 = -A^{-1/2}D$, with D as in (3.24). We also define the spaces

$$Y = [H^2(\Omega) \cap H_0^1(\Omega)]\times H_0^1(\Omega) = \mathcal{D}(A^{1/2})\times\mathcal{D}(A^{1/4});$$
$$U = L_2(\Gamma). \tag{3.72}$$

<u>Assumption (1.3)</u>: $(-\mathbf{A})^{-\gamma}B \in \mathcal{L}(U,Y)$. By (3.71) and (3.72) with $u \in L_2(\Gamma)$, we plainly have

$$(-\mathbf{A})^{-1}Bu = \begin{vmatrix} 0 & \mathcal{A}^{-1} \\ -I & 0 \end{vmatrix} \begin{vmatrix} 0 \\ \mathcal{A}G_2u \end{vmatrix} = \begin{vmatrix} G_2u \\ 0 \end{vmatrix} = \begin{vmatrix} -A^{-1/2}Du \\ 0 \end{vmatrix} \in Y \tag{3.73}$$

and assumption (1.3) holds true for problem (3.69).

<u>Assumption (H.2) = (1.6)</u>. One can show that [L-T.16]

$$B^*e^{\mathbf{A}^*t}\begin{vmatrix} z_1 \\ z_2 \end{vmatrix} = \frac{\partial\Delta\phi(t)}{\partial\nu}\Big|_\Sigma, \tag{3.74}$$

where $\phi(t) = \phi(t,\phi_0,\phi_1)$ solves the corresponding homogeneous problem

$$\begin{cases} \phi_{tt}-\rho\Delta\phi_{tt}+\Delta^2\phi = 0 & (3.75a) \\ \phi(0,\cdot) = \phi_0, \ \phi_t(0,\cdot) = \phi_1 & (3.75b) \\ \phi|_\Sigma \equiv \Delta\phi|_\Sigma \equiv 0 & (3.75c) \end{cases}$$

with

$$\phi_0 = (I+\rho A^{1/2})^{-1}z_2 \in \mathcal{D}(A^{3/4});$$
$$\phi_1 = -(I+\rho A^{1/2})^{-1}A^{1/2}z_1 \in \mathcal{D}(A^{1/2}). \tag{3.76}$$

Thus, by (3.74), (3.76), an equivalent formulation of assumption (H.2) = (1.6) is the inequality

$$\int_\Sigma \left[\frac{\partial\Delta\phi}{\partial\nu}\right]^2 d\Sigma \le c_T\|\{\phi_0,\phi_1\}\|^2_{\mathcal{D}(A^{3/4})\times\mathcal{D}(A^{1/2})}. \tag{3.77}$$

This inequality holds indeed true, as recently shown in [L-T.16] by multiplier methods. Thus, assumption (H.2) = (1.6) holds true for problem (3.69).

Finite Cost Condition (1.8). The same considerations as in section 3.2 for the wave equation apply. It was recently proved that problem (3.69) is exactly controllable for sufficiently large T > 0 on the state space $Y = [H^2(\Omega) \cap H_0^1(\Omega)] \times H_0^1(\Omega)$ within the class of $L_2(\Sigma)$-controls u, with no geometrical conditions on Ω (except Γ smooth), if u is applied to all of Γ [L-T.16]. As a consequence, the Finite Cost Condition (1.8) holds true. (Problem (3.69) is also uniformly stabilizable under some geometrical conditions on Ω, e.g., strict convexity [L-T.16].)

Detectability Condition (2.17)-(2.19). This holds true since R = I, see (3.71).

Conclusion. Theorem 2.2 applies to problem (3.69).

References

[B.1] A. V. Balakrishnan, Applied Functional Analysis, Springer-Verlag, 2nd ed., 1981.

[B.2] A. V. Balakrishnan, Boundary control of parabolic equations: L-Q-R theory, in Non Linear Operators, Proc. 5th Internat. Summer School, Akademie-Berlin, 1978.

[B-K] T. H. Banks and K. Kunish, The Linear Regulator Problem for Parabolic Systems, SIAM J. on Control, Vol. 22, No. 5 (1984), 684-699.

[B-L-R] C. Bardos, G. Lebeau, and R. Rauch, Controle et stabilisation de l'equation des ondes, to appear.

[C-L] S. Chang and I. Lasiecka, Riccati equations for non-symmetric and non-dissipative hyperbolic systems with L_2-boundary controls, J. Math. Anal. and Appl., Vol. 116 (1986), 378-414.

[C-P] R. Curtain and A. Pritchard, Infinite dimensional linear systems theory, LNCS 8, Springer-Verlag, 1978.

[C-R] G. Chen and D. L. Russell, A mathematical model for linear elastic systems with structural damping, Quarterly of Applied Mathematics, January (1982), 433-454.

[C-T.1] S. Chen and R. Triggiani, Proof of two conjectures of G. Chen and D. L. Russell on structural damping for elastic systems: The case $\alpha = \frac{1}{2}$, Proceedings of the Seminar in Approximation and Optimization held at University of Havana, Cuba, January 12-14, 1987, Lecture Notes in Mathematics, 1354, Springer-Verlag, 234-256.

[C-T.2] S. Chen and R. Triggiani, Proof of extension of two conjectures on structural damping for elastic systems: The case $\frac{1}{2} \leq \alpha \leq 1$, Pacific J. Mathematics, Vol. 136, N1 (1989), 15-55.

[C-T.3] S. Chen and R. Triggiani, Differentiable semigroups arising from elastic systems with gentle dissipation: The case $0 < \alpha < \frac{1}{2}$, 1988, submitted. Announcement appeared in Proceedings of International Conference held at Baton Rouge, Louisiana, October 1988.

[C-T.4] S. Chen and R. Triggiani, Characterization of domains of fractional powers of certain operators arising in elastic systems, and applications, preprint 1989.

[D-I] G. Da Prato and A. Ichikawa, Riccati equations with unbounded coefficients, Ann. Matem. Pura e Appl. 140 (1985), 209-221.

[DaP-L-T.1] G. Da Prato, I. Lasiecka, and R. Triggiani, A direct study of Riccati equations arising in boundary control problems for hyperbolic equations, J. Diff. Eqns., Vol. 64, No. 1 (1986), 26-47.

[D-S] M. Delfour and Sorine, The linear-quadratic optimal control problem for parabolic systems with boundary control through the Dirichlet condition, 1982, Tolouse Conference, I.13-I.16.

[F.1] F. Flandoli, Riccati equation arising in a boundary control problem with distributed parameters, SIAM J. Control and Optimiz. 22 (1984), 76-86.

[F.2] F. Flandoli, Algebraic Riccati equation arising in boundary control problems, SIAM J. Control and Optimiz. 25 (1987), 612-636.

[F.3] F. Flandoli, A new approach to the LQR problem for hyperbolic dynamics with boundary control, Springer-Verlag, LINCIS 102 (1987), 89-111.

[F.4] F. Flandoli, Invertibility of Riccati operators and controllability of related systems, Systems and Control Letters 9 (1987), 65-72.

[F-L-T.1] F. Flandoli, I. Lasiecka, and R. Triggiani, Algebraic Riccati equations with non-smoothing observation arising in hyperbolic and Euler-Bernouli equations, Ann. Matem. Pura a Appl., Vol. CLiii (1988), 307-382.

[Gib] J. S. Gibson, The Riccati integral equations for optimal control problems on Hilbert spaces, SIAM J. Control and Optimiz. 17 (1979), 537-565.

[Gr] P. Grisvard, Caracterization de qualques espaces d'interpolation, Arch. Rational Mechanics and Analysis 25 (1967), 40-63.

[H.1] L. F. Ho, Observabilite frontiere de l'equation des ondes, CRAS, Vol. 302, Paris (1986), 443-446.

[L.1] I. Lasiecka, Unified theory for abstract parabolic boundary problems; a semigroup approach, Appl. Math. and Optimiz., 6 (1980), 283-333.

[Lag.1] J. Lagnese, Infinite horizon linear-quadratic regular problem for beams and plates, Lecture Notes LNCIS, Springer-Verlag, to appear.

[Lag.2] J. Lagnese, Uniform boundary stabilization of homogeneous, isotropic plates, Proc. of the 1986 Vorau Conference on Distributed Parameter Systems, Springer-Verlag, 204-215.

[Lio.1] J. L. Lions, Controle des Systems Distribues Singuliers, Gauthier Villars, 1983.

[Lio.2] J. L. Lions, Exact controllability, stabilization and perturbations, SIAM Review 30 (1988), 1-68, to appear in extended versions by Masson.

[Lit] W. Littman, Near optimal time boundary controllability for a class of hyperbolic equations, Springer-Verlag Lecture Notes LNCIS #97 (1987), 307-312.

[L-L-T] I. Lasiecka, J. L. Lions, and R. Triggiani, Non-homogeneous boundary value problems for second order hyperbolic operators, J. Mathem. Pure et Appl., 65 (1986), 149-192.

[L-R] D. Lukes and D. Russell, The quadratic criterion for distributed systems, SIAM J. Control 7 (1969), 101-121.

[L-T.1] I. Lasiecka and R. Triggiani, A cosine operator approach to modeling $L_2(0,T;L_2(\Gamma))$-boundary input hyperbolic equations, Appl. Math. and Optimiz., Vol. 7 (1981), 35-83.

[L-T.2] I. Lasiecka and R. Triggiani, Regularity of hyperbolic equations under boundary terms, Appl. Math. and Optimiz., Vol. 10 (1983), 275-286.

[L-T.3] I. Lasiecka and R. Triggiani, A lifting theorem for the time regularity of solutions to abstract equations with unbounded operators and applications to hyperbolic equations, Proc. Amer. Math. Soc. 103, 4 (1988).

[L-T.4] I. Lasiecka and R. Triggiani, Dirichlet boundary control problem for parabolic equations with quadratic cost: Analyticity and Riccati's feedback synthesis, SIAM J. Control and Optimiz. 21 (1983), 41-68.

[L-T.5] I. Lasiecka and R. Triggiani, An L_2-Theory for the Quadratic Optimal Cost Problem of Hyperbolic Equations with Control in the Dirichlet B.C., Workshop on Control Theory for Distributed Parameter Systems and Applications, University of Graz, Austria (July 1982); Lecture Notes, Vol. 54, Springer-Verlag (1982), 138-153.

[L-T.6] I. Lasiecka and R. Triggiani, Riccati equations for hyperbolic partial differential equations with $L_2(0,T;L_2(\Gamma))$-Dirichlet boundary terms, SIAM J. Control and Optimiz., Vol. 24 (1986), 884-924.

[L-T.7] I. Lasiecka and R. Triggiani, The regulator problem for parabolic equations with Dirichlet boundary control; Part I: Riccati's feedback synthesis and regularity of optimal solutions, Appl. Math. and Optimiz., Vol. 16 (1987), 147-168.

[L-T.8] I. Lasiecka and R. Triggiani, The regulator problem for parabolic equations with Dirichlet boundary control; Part II: Galerkin approximation, Appl. Math. and Optimiz., Vol. 16 (1987), 198-216.

[L-T.9] I. Lasiecka and R. Triggiani, Infinite horizon quadratic cost problems for boundary control problems, Proceedings 20th CDC Conference, Los Angeles (December 1987), 1005-1010.

[L-T.10] I. Lasiecka and R. Triggiani, Hyperbolic equations with non-homogeneous Neumann boundary terms; quadratic cost problem with $L_2(\Sigma)$-boundary observation, to appear.

[L-T.11] I. Lasiecka and R. Triggiani, Exact controllability of the Euler-Bernoulli equation with controls in Dirichlet and Neumann boundary conditions: A non-conservative case, SIAM J. Control and Optimiz., 27 (1989), 330-373.

[L-T.12] I. Lasiecka and R. Triggiani, Uniform exponential energy decay of wave equations in a bounded region with $L_2(0,\infty;L_2(\Gamma)$-feedback control in the Dirichlet boundary conditions, J. Diff. Eqns. 66 (1987), 340-390.

[L-T.13] I. Lasiecka and R. Triggiani, Exact controllability of the wave equation with Neumann boundary control, Applied Math. and Optimiz. 19 (1989), 243-290.

[L-T.14] I. Lasiecka and R. Triggiani, Regularity theory for a class of nonhomogeneous Euler-Bernoulli equations: A cosine operator approach, Bollett. Unione Mathem. Italiana UMI (7), 3-B(1989), 199-228.

[L-T.15] I. Lasiecka and R. Triggiani, Exact controllability of the Euler-Bernoulli equation with boundary controls for displacement and moment, J. Math. Anal. & Appl., to appear 1989.

[L-T.16] I. Lasiecka and R. Triggiani, Regularity, exact controllability and uniform stabilization of Kirchoff plates, preprint 1989.

[M-T] A. Manitius and R. Triggiani, Function space controllability of linear retarded systems: A derivation from abstract operator conditions, SIAM J. Control, Vol. 16 (1978), 599-645.

[O-T] N. Ourada and R. Triggiani, Uniform stabilization of the Euler-Bernoulli equation with feedback only on the Neumann B.C., preprint 1989.

[P-P] J. Pollock and A. Pritchard, The infinite time quadratic cost problem for distributed systems with unbounded control action, J. Inst. Math. Appl. 25 (1980), 287-309.

[P-S] A. Pritchard and D. Solomon, The linear quadratic control problem for infinite dimensional systems with unbounded input and output operators, SIAM J. Control and Optimiz., Vol. 25 (1987), 121-144.

[R] D. Russell, Quadratic performance criteria in boundary control of linear symmetric hyperbolic systems, SIAM J. Control 11 (1973), 475-509.

[Sal] D. Salomon, Infinite dimensional linear systems with unbounded control and observation: A functional analytic approach, Trans. Am. Math. Soc., 1987, 383-431.

[Sor] M. Sorine, Une resultat d'existence et unicité pour l'equation de Riccati stationnaire, report INRIA, No. 55, 1981.

[T.1] R. Triggiani, On the stabilizability problem in Banach space, J. Math. Anal. and Appl., Vol. 52 (1975), 383-403.

[T.2] R. Triggiani, A cosine operator approach to modeling $L_2(0,T;L_2(\Gamma))$-boundary input problems for hyperbolic systems, Proceedings of 8th IFIP Conference, University of Wurzburg, W. Germany, July 1977, Lecture Notes CIS, Springer-Verlag #6 (1978), 380-390.

[T.3] R. Triggiani, Exact boundary controllability on $L^2(\Omega)\times H^{-1}(\Omega)$ for the wave equation with Dirichlet control acting on a portion of the boundary, and related problems, Appl. Math. and Optimiz. 18 (1988), 241-277.

[T.4] R. Triggiani, Regularity of structurally damped systems with point/boundary control, preprint 1989.

[Z] J. Zabczyk, Remarks on the algebraic Riccati equation in Hilbert space, Appl. Math. and Optimiz. 2 (1976), 251-258.

Algebraic Riccati equations arising in boundary/point control: A review of theoretical and numerical results
Part II: Approximation theory

Irena Lasiecka and Roberto Triggiani
Department of Applied Mathematics
Thornton Hall
University of Virginia
Charlottesville, Virginia 22903

The present Part II is a continuation of Part I of the paper bearing the same title. Part I referred to the continuous case. The present Part II refers to the corresponding discrete problem. To emphasize the continuity of the two parts, we keep a progressive numbering of the sections.

4. Numerical approximations of the solution to the abstract Algebraic Riccati Equation

The main goal of this section is twofold: (i) to formulate a numerical algorithm for the computation of the solution to the Algebraic Riccati Equation (ARE) (2.1); (ii) to present the relevant convergence results.

To begin with, we introduce a family of approximating subspaces $V_h \subset Y \cap \mathcal{D}(B^*)$, where h, $0 < h \le h_0 < \infty$, is a parameter of discretization which tends to zero. Let Π_h be the orthogonal projection of Y onto V_h, with the usual approximating property

$$\|\Pi_h y - y\|_Y \to 0, \quad y \in Y. \tag{4.1}$$

Let A_h: $V_h \to V_h$ and B_h: $U \to V_h$ be approximations of A, respectively B, which satisfy the usual, natural requirements:

(i) $\quad \Pi_h \hat{A}^{-1} - \hat{A}_h^{-1}\Pi_h \to 0$, strongly in Y; $\quad$ (4.2a)

(ii) $\quad \|\hat{A}^{-1}(B_h - B)u\|_Y \to 0$, $u \in U$. $\quad$ (4.2b)

We consider the following approximation of the ARE (2.1):

$$(A_h^* P_h x_h, y_h)_Y + (P_h A_h x_h, y_h)_Y + (Rx_h, Ry_h)_Z$$
$$= (B_h^* P_h x_h, B_h^* P_h y_h)_U, \quad x_h, y_h \in V_h. \qquad (\text{ARE})_h$$

Our main goal is to prove that, under natural assumptions which are the discrete counterpart of the hypothesis (H.1) = (1.5) or (H.2) = (1.6) of the continuous case, we have (among other things):

(i) $\quad P_h \to P,$ strongly in Y; (4.3)

(ii) $\quad B_h^* P_h \to B^* P,$ in a technical sense to be made precise; (4.4)

(iii) $\quad \left\| e^{(A_h - B_h B_h^* P_h)t} \right\|_{\mathcal{L}(Y)} \le C\, e^{-\omega t}, \ \omega > 0.$ (4.5)

Although there are a number of papers in the literature which deal with the problem of approximating ARE, most of these works [G.1], [B-K.1], [K-S.1], [I-T.1], treat the case where the input operator B is bounded. When instead B is genuinely unbounded, an array of new difficulties arise. Some of them are the same which are already encountered in the continuous case treatment; some others are new, and are intrinsically connected with the approximating schemes. We list a few.

(a) Open loop approximation. Consider the input → solution operator

$$(Lu)(t) = \int_0^t e^{A(t-\tau)} Bu(\tau)\,d\tau. \qquad (4.6)$$

Under either hypothesis (H.1) = (1.5), or else hypothesis (H.2), the operator L is continuous: $L_2(0,T;U) \to L_2(0,T;Y)$ (indeed $\to C([0,T];Y)$ in the case of assumption (H.2), and also in the case of assumption (H.1) with $\gamma < \frac{1}{2}$, or with $\gamma = \frac{1}{2}$, when A has a Riesz basis on Y). In the corresponding approximation theory, the question arises whether the discrete map

$$(L_h u)(t) = \int_0^t e^{A_h(t-\tau)} B_h u(\tau)\,d\tau \qquad (4.7)$$

which is continuous: $L_2(0,T;U) \to L_2(0,T;V_h)$, is also continuous as an operator $L_2(0,T;U) \to L_2(0,T;Y)$, uniformly with respect to the parameter h. (For instance, one may take $B_h = \Pi_h B$, where we note that $\Pi_h B$ is well defined since $V_h \in \mathcal{D}(B^*)$ by assumption:

$$(\Pi_h Bu, v_h)_Y = (Bu, v_h)_Y = (u, B^* v_h)_U). \qquad (4.8)$$

In the case where B is bounded, this stability requirement is true if the approximation of A is consistent, i.e., subject to (4.2), as it follows via Trotter-Kato theorem. Instead, in the case where B is unbounded, special care must be exercised to select a suitable approximation scheme, which guarantees the above stability requirement.

(b) <u>Approximation of gain operators</u> B^*P_h. The problem here is that even if (4.3) holds true, it is far from clear to (4.4) will also follow. Thus, special care must be given in selecting the approximating schemes, as to obtain convergence (4.4) for the gain operators.

Thus, a theory of approximations in the case where the operator B is genuinely unbounded, such as it arises in boundary control and point control for partial differential equations, offers new challenges which are not present in the B-bounded case. In order to cope with these difficulties, we need--as in the continuous case--distinguish between dynamics which satisfy assumption (H.1) = (1.5) and dynamics which satisfy assumption (H.2) = (1.6).

4.1 Approximation for the (H.1)-class

4.1.1 Approximation assumptions

Approximation of A. Let $A_h: V_h \to V_h$ be an approximation of A which satisfies the following requirements:

(uniform analyticity)

$$|A_h e^{A_h t}|_{\mathcal{L}(Y)} \le \frac{C}{t} e^{(\omega+\varepsilon)t}; \quad t > 0 \qquad (A.1)$$

discrete analog of (H.1), where the constant C is uniform with respect to h;

$$|\Pi_h \hat{A}^{-1} - \hat{A}_h^{-1} \Pi_h|_{\mathcal{L}(Y)} \le Ch^s \quad \text{for some } s > 0. \qquad (A.2)$$

Approximation of B. We take $B_h = \Pi_h B$, which is well defined, see comments leading to (4.8). We shall assume that the operator $B: U \to [\mathcal{D}(A^*)]'$ satisfies the following "approximation" properties

$$|B^* x_h|_U \le Ch^{-s+\varepsilon_0} |x_h|_Y \quad \text{for some } \varepsilon_0 > 0; \qquad (A.3)$$

$$|B^*(\Pi_h x - x)|_{\mathcal{L}(\mathcal{D}(\hat{A}^{*1-\varepsilon_0}), U)} \to 0 \quad \text{as } h \downarrow 0. \qquad (A.4)$$

Remark 4.1. Notice that (A.2) throughout (A.4) are the standard approximation properties. They are consistent with the regularity of the original operators A and B. Moreover, they are satisfied by typical schemes (finite elements, finite differences, mixed methods, spectral approximations). The property of uniform analyticity (A.1) is not a standard assumption and needs to be verified in each case. However, to our knowledge, it is satisfied for most of the schemes and examples which arise from analytic semigroup problems. For instance, sufficient condition for (A.1) to hold is the uniform coercitivity of the bilinear form associated with A_h (see Lemma 4.2 in [L-1]). There are, however, a number of significant physical examples (e.g., damped elastic systems) where the bilinear form is not coercive, while the underlying semigroups $e^{A_h t}$ are uniformly analytic (see section 6).

4.1.2 Consequences of approximating assumptions on A and B

From (A.2) and (A.1), the following "rough" data estimates follow (see [L-1, Appendix]).

$$\text{(i)} \qquad |e^{A_h t}\Pi_h - \Pi_h e^{At}|_{\mathcal{L}(Y)} \leq \frac{Ch^{s\Theta}(\omega+\varepsilon)t}{t^{\Theta}}, \qquad (4.9)$$

where $0 \leq \Theta \leq 1$ and $\varepsilon > 0$ can be arbitrarily small;

$$\text{(ii)} \qquad |\Pi_h R(\lambda,A) - R(\lambda,A_h)\Pi_h|_{\mathcal{L}(Y)} \leq C\, h^s,\ s > 0$$

uniformly in $\lambda \in \Sigma^{co}_{app}(A)$, where $\Sigma_{app}(A)$ = closed triangular sector containing the axis $[-\infty, a]$ and delineated by the two rays $a+\rho^{\pm i\phi}$ for some $\pi/2 < \phi < 2\pi$; $a = \omega+\varepsilon$;

$$\text{(iii)} \qquad |e^{A_h^* t}\Pi_h - \Pi_h e^{A^* t}|_{\mathcal{L}(\mathcal{D}(A^*),Y)} \leq C\, h^s$$

uniformly in $t > 0$ on compact subintervals. From (A.4) and the assumptions imposed on B, we obtain

$$|B^*\Pi_h x|_U \leq C|\hat{A}^{*\,1-\varepsilon_0} x|_Y. \qquad (4.10)$$

4.1.3 Approximation of dynamics and of control problems. Related Riccati equation

We now introduce an approximation of the control problem and of the corresponding ARE.

Control problem. Given the approximating dynamics $y_h(t) \subset V_h$ such that

$$\dot{y}_h(t) = A_h y_h(t) + \Pi_h B u(t);\ y(0) = \Pi_h y_0$$

minimize (4.11)

$$J(u,y_h(u)) \equiv \int_0^\infty [|Ry_h(t)|_Z^2+|u(t)|_U^2]dt.$$

The optimal solution to (4.11) (which we shall see later to exist) will be denoted by $\{u_h^0 = u(y_h^0);y_h^0\}$.

Riccati Equation. The approximation of the Riccati Equation is defined by equation $(ARE)_h$ below (4.2) with $B_h = \Pi_h B$.

4.1.4 Main results of approximating schemes

For the present 'analytic' class, subject to assumption (H.1) = (1.5), the Finite Cost Condition will be guaranteed by the following Stabilizability Condition (S.C.)

(S.C.) $\exists\ F \in \mathcal{L}(Y,U)$ such that the s.c. analytic semigroup $e^{(A+BF)t}$ (as guaranteed by (1.3)) is exponentially stable on Y:

$$\|e^{(A+BF)t}\|_{\mathcal{L}(Y)} \le M_F\, e^{-\omega_F t}$$

for some $\omega_F > 0$.

Our main results are formulated in the theorems below.

Theorem 4.1 [L-T.2]. Assume:

I. The continuous hypothesis (H.1) = (1.5), the above Stabilization Condition, the Detectability Condition (2.10), and, in addition

$$\begin{cases} \text{(i)} & \text{either } R > 0 \\ \text{(ii)} & \text{or } \hat{A}^{-1}KR\colon Y \to Y \text{ compact;} \end{cases} \tag{4.12}$$

$$\begin{cases} \text{(i)} & \text{either } B^*\hat{A}^{*-1}\colon Y \to U \text{ compact} \\ \text{(ii)} & \text{or } F\colon Y \to U \text{ compact.} \end{cases} \tag{4.13}$$

II. The approximation properties (4.1), (A.1)-(A.4). Then there exists $h_0 > 0$ such that for all $h < h_0$, the solution P_h to the equation $(ARE)_h$ exists, is unique and the following convergence properties hold:

$$|e^{-A_{h,p}t}|_{\mathcal{L}(Y)} \le C\, e^{-\bar{\omega}_p t}, \quad \bar{\omega}_p > 0, \tag{4.14}$$

where $A_{h,p} \equiv A_h - \Pi_h BB^* P_h$

$$|\hat{A}_h^{*1-\rho} P_h|_{\mathcal{L}(Y)} + |\hat{A}_h^{*\frac{1}{2}-\rho} P_h A_h^{\frac{1}{2}-\rho}|_{\mathcal{L}(Y)} \le C$$
$$\text{for any } 0 < \rho \le 1. \tag{4.15}$$

$$|[P_h \Pi_h - P]x|_Y \to 0 \text{ for all } x \in Y \text{ as } h \downarrow 0. \tag{4.16}$$

$$|B^* [P_h \Pi_h - P]x|_U \to 0 \text{ for all } x \in Y \text{ as } h \downarrow 0. \tag{4.17}$$

$$|u_h^0 - u^0|_{L_2(0,\infty;U)} + |u_h^0 - u^0|_{C([0,\infty];Y)} \to 0 \text{ as } h \downarrow 0. \tag{4.18}$$

$$|y_h^0 - y^0|_{L_2(0,\infty;U)} + |y_h^0 - y^0|_{C([0,\infty];Y)} \to 0 \text{ as } h \downarrow 0. \tag{4.19}$$

$$J(u_h^0, y_h^0) \to J(u^0, y^0). \tag{4.20}$$

If, in addition, $V_h \in \mathcal{D}(\hat{A}^\alpha)$ and

$$|\hat{A}^\alpha x_h| \sim |\hat{A}_h^\alpha x_h| \tag{4.21}$$

then

$$|\hat{A}^{*\alpha} [P_h \Pi_h - P]x|_Y \to 0; \ x \in Y; \ 0 \le \alpha < 1. \tag{4.22}$$

If (4.21) holds with $\alpha = \frac{1}{2}$, then

$$|\hat{A}^{*\frac{1}{2}-\rho} (P_h \Pi_h - P) \hat{A}^{\frac{1}{2}-\rho} x|_Y \to 0, \quad x \in Y,$$
$$\text{for any } 0 < \rho \le \tfrac{1}{2}. \tag{4.23}$$

Remark 4.2. Eq. (4.22) typically holds with $\alpha = \frac{1}{2}$. This is certainly the case when A is coercive and A_h is a standard Galerkin approximation of A; i.e., $(A_h u_h, y_h) = (Au_h, y_h)$. ∎

Remark 4.3. If A is self-adjoint (or "slightly non-self-adjoint," i.e., $A = A_1 + A_2$, where A_1 is self-adjoint and A_2: $\mathcal{D}(A_1^{1-\varepsilon}) \to Y$ is bounded), one can take $\rho = 0$ in (4.23). ∎

Remark 4.4. If $\hat{A}^{-1}$ is compact, the convergence in (4.22), (4.23), and (4.17) is uniform. ∎

Theorem 4.2 [L-T.2]. If, in addition to the assumptions I and II of Theorem 4.1, we assume

$$B^*\hat{A}^{*-\sigma}: Y \to U \text{ is compact for some } 0 \le \sigma < 1, \tag{4.24}$$

then

$$|B^*(P_h\Pi_h - P)|_{\mathcal{L}(Y;U)} \to 0 \quad \text{as } h \downarrow 0; \tag{4.25}$$

$$|e^{(A-BB^*P_h)t}|_{\mathcal{L}(Y)} \le C\, e^{-\hat{\omega}_p t}; \quad \hat{\omega}_p > 0. \tag{4.26}$$

Remark 4.5. Assumption (4.24) is obviously satisfied if $\hat{A}^{-1}: Y \to Y$ is compact. ■

Theorem 4.1 provides a basic convergence result for the optimal solutions of the approximating problem (4.11), the corresponding Riccati operators, and the gain operators to the same quantities associated with the original problem (1.1), (1.2). The advantage of Theorem 4.2 is this: Under the additional hypothesis (4.24) (which is satisfied for all the applications which we have in mind), it states that the original system, once acted upon by the discrete feedback control law given by $u_h^*(t) = -B^*P_h y_h^*(t)$ yields (uniformly) exponentially stable solutions. ■

Remark 4.6. By tracing the estimates established in the proof of Theorem 4.1, one could derive the rates of convergence of approximations expressed in terms of the parameter s and ε_0. ■

Remark 4.7. Instead of the original inner product $(x_h, y_h)_Y$, one can introduce an equivalent inner product $(x_h, y_h)_{Y_h}$, where

$$c_1|x_h|_Y \le |x_h|_{Y_h} \le c_2|x_h|_Y.$$

In some situations, it is more convenient to work with a discrete inner product $(\ ,\ )_{Y_h}$ as to simplify the computations for the adjoint operators for the discrete problem. ■

Remark 4.8. The literature on approximating schemes of optimal control problems and related Riccati equations generally assumes

(i) convergence properties of the 'open loop' solutions, i.e., of the maps $u \to y$ of the continuous problem;

(ii) "uniform stabilizability/detectability" hypotheses for the approximating problems.

In contrast, our basic assumptions are:

(a) stabilizability/detectability hypotheses (S.C.)/(D.C.) of the continuous system;

(b) a "uniform analyticity" hypothesis (A.1) on the approximations.

Starting from (a) and (b), we then derive both the convergence properties of the open loop and the uniform stabilizability/detectability hypotheses--(i) and (ii) above--which are taken as assumptions in other treatments. Thus, the theory presented here is "optimal," in the sense that it assumes only what is strictly needed. Indeed, it can be shown that assumptions (A.1), (S.C.)/(D.C.) are not only sufficient, but also necessary, for the main theorems presented here.

These considerations are an important aspect of the entire theory since, in the case where B is an unbounded operator, the requirement of convergence $L_h \to L$ of the open loop solutions is a very strong assumption as remarked before. Generally, even when L is bounded, and the scheme is consistent, it may well happen that the scheme is not even stable; i.e., L_h may not be uniformly bounded in h. The properties of the composition $e^{At}B$ may not be retained in the approximation $e^{A_h t}B_h$. Special care must be exercised in approximating B.

4.1.5 Literature

There is rather extensive literature which is concerned with the general issue of approximation schemes for Algebraic Riccati Equations in infinite-dimensional spaces. Here we shall concentrate only on works which focus on the case where the original free dynamics is modelled by an analytic semigroup e^{At}, as in the present section. Reference [B-K] presents approximation results for parabolic problems with distributed controls; i.e., with the operator B bounded. Next, reference [L-T.1] analyzed the case of a parabolic problem with Dirichlet control, via an abstract semigroup approach. The abstract treatment in [L-T.1] of the physically concrete, important problem of a parabolic equation defined on a bounded Ω of R^n, with Dirichlet boundary control--where then B^* is $(\hat{A}^*)^{3/4+\varepsilon}$-bounded, or $(\hat{A}^*)^{-(3/4+\varepsilon)}B$ is bounded--may be viewed as a canonical illustration of the purely abstract situation where one has $\hat{A}^{-\gamma}B$ bounded for $\gamma < 1$, and A has compact resolvent (the latter property being automatically satisfied in the parabolic problem over a bounded domain $\Omega \subset R^n$). Thus, <u>the treatment in</u> [L-T.1] <u>works equally well</u>, <u>*mutatis mutandis*</u>, <u>in the abstract case of an analytic semigroup generator</u> A <u>with compact resolvent</u>, <u>and with</u> $\hat{A}^{-\gamma}B$ <u>bounded</u>, $0 \le \gamma < 1$. There is a natural 'cutting line' in the range of values of γ, which crucially bears on the degree of technical difficulties present in the treatment of the optimal control problem and its Algebraic Riccati approximation; this is given by the special value $\gamma = 1/2$.

Indeed, if $\hat{A}^{-\gamma}B$ is bounded or, equivalently, B^* is $(\hat{A}^*)^{-\gamma}$-bounded, with $\gamma \leq ½$, then the corresponding input → solution operator L is <u>*a priori*</u> continuous to $C([0,T];Y)$, so that all the trajectories of the continuous dynamical system are *a priori* pointwise continuous in time, and the operator B^*P is then <u>*a priori*</u> a bounded operator. Thus, in the case $\gamma \leq ½$, a derivation of the A.R.E. may be given which closely parallels the pattern where B is a bounded operator ($\gamma = ½$, if A has a Riesz basis of eigenvectors).

Instead, if B^* is $(\hat{A}^*)^{\gamma}$-bounded or, equivalently, $\hat{A}^{-\gamma}B$ is bounded, with $½ < \gamma$, the operator L is not continuous into $C([0,T];Y)$; i.e., the open loop trajectories are generally not pointwise continuous in time. Here a main technical difficulty is therefore to show that, nevertheless, the gain operator B^*P is bounded. This is done by carefully analyzing the properties of the <u>optimal solutions</u> $y^0(t)$ (as distinguished from ordinary solutions $y(t)$), and by eventually showing via a bootstrap argument that the <u>optimal</u> solutions $y^0(t)$ are pointwise continuous in time (unlike ordinary solutions $y(t)$ which are only, say, in $L_2(0,T;Y)$. This strategy then succeeds in proving boundedness of the gain operator B^*P.

The strategy outlined above the case $\gamma > ½$ was successfully implemented in [L-T.1] in the canonical case of the parabolic equation with Dirichlet boundary control, where in fact $\gamma = ¾+\varepsilon$; although--as mentioned before--the technique, which was originally designed for this specific problem, works with no essential changes also for the general case $\gamma < 1$, when A has compact resolvent. The treatment in [L-T.1] uses, as a starting point, two sources: On the one hand, the properties of the continuous optimal control problem and related Algebraic Riccati Equation following the variational approach of [L-T.1] (see also [F.1], [D-I] of Part I for an alternative direct approach); and on the other hand, the approximation results for analytic semigroups proved in [L.1].

More recently, [B-I] considered the approximations of a subclass of analytic problems modelled by strictly coercive bilinear forms. The results in [B-I] assume that the operator B is "$A^{½}$" bounded and that the resolvent $R(\lambda,A)$ is compact. The importance of having a theory of approximation valid for $\gamma > ½$ include, in addition to parabolic problems with Dirichlet boundary control, also plate problems with strong damping and point control (section 3.1).

The approximation Theorem 4.1 in the general case $\gamma < 1$ is proved in [L-T.2]. Here the techniques rely on a combination of ideas of the continuous problem [L-T.1], [F.1], [D-I] of Part I, together with approximating properties of analytic semigroups [L.1] and convergence properties for the open loop problem, [L.2].

4.2 Approximation for the (H.2)-class

4.2.1 Approximating assumptions

Approximation of A. We assume as $h \downarrow 0$:

$$\hat{A}_h^{-1}\Pi_h - \Pi_h\hat{A}^{-1} \to 0 \quad \text{in } Y; \qquad \text{(B.1)}$$
$$\hat{A}_h^{*-1}\Pi_h - \Pi_h\hat{A}^{*-1} \to 0 \quad \text{in } Y.$$

$$|e^{A_h t}|_{\mathcal{L}(Y)} \le C\, e^{\omega t}, \quad t > 0. \qquad \text{(B.2)}$$

Approximation of B. We assume as $h \downarrow 0$:

$$\int_0^T |B_h^* e^{A_h^* t}\Pi_h x|_U^2 dt \le C_T |x|_Y^2$$
(discrete analog of (H.2)). (B.3)

(i) $|\hat{A}^{-1}(B_h - B)u|_Y \to 0; \quad u \in U;$

(ii) $|(\hat{A}_h^{-1} - \hat{A}^{-1})B_h u|_Y \to 0; \quad u \in U;$

(B.4)

(iii) $|(B_h^*\Pi_h - B^*)\hat{A}^{*-1}x|_U \to 0; \quad x \in Y;$

(iv) $|B_h^*(\hat{A}_h^{*-1}\Pi_h - \Pi_h\hat{A}^{*-1})x|_U \to 0; \quad x \in Y.$

4.2.2 Approximation of dynamics and of control problem. Related Riccati Equation

Control problem. Given the approximating dynamics $y_h(t) \in V_h$ such that

$$\dot{y}_h(t) = A_h y_h(t) + B_h u(t), \quad y_h(0) = \Pi_h y(0)$$

minimize

$$J(u, y_h(u)) = \int_0^\infty [\,|Ry_h(t)|_Y^2 + |u(t)|_U^2]dt.$$

Riccati Equation. The approximating Riccati Equation is given in equation $(\text{ARE})_h$ below (4.2).

4.2.3 Approximating results

Theorem 4.3 [L.3]. Assume

I. the continuous hypotheses (H.2), (S.C.)(D.C.).

II. The approximation properties (B.1)-(B.3) and, in addition:

$(F.C.C.)_h$ (uniform Finite Cost Condition):

$$\exists\ \alpha > 0;\ \forall\ y_0 \in Y,\ \exists\ u_h \in L_2(0,\infty;U)$$
$$\text{such that } J(u, y_h(u)) < \alpha|y_0|_Y^2.$$

$(D.C.)_h$ (uniform Detectability Condition): There exist $K_h: Z \to V_h$ such that

$$|K_h^* x_h|_Z \leq C[|B_h^* x_h|_U + |x_h|_Y],$$

and

$$|e^{A_{K_h} t}|_{\mathcal{L}(Y)} \leq C\ e^{-\omega_1 t},$$

where $A_{K_h} = A_h - K_h R$. Then:

I. (convergence of Riccati operators)

$$|P_h \Pi_h x - Px|_Y \to 0, \quad x \in Y, \tag{4.27}$$

$$|e^{A_{P,h} t} x_n|_Y \leq C\ e^{-\bar{\omega}_0 t} |x_h|_Y, \tag{4.28}$$

where $A_{P,h} = A_h - B_h B_h^* P_h$.

II. (convergence of optimal solutions)

$$|u_h^0 - u^0|_{L_2(0,\infty;U)} \to 0; \tag{4.29}$$

$$|y_h^0 - y^0|_{L_2(0,\infty;Y)} \to 0; \tag{4.30}$$

$$|y_h^0 - y^0|_{C[0,\infty;Y]} \to 0. \tag{4.31}$$

Hence

$$|e^{A_{P,h} t} \Pi_h x - e^{A_P t} x|_{L_2(0,\infty;Y)} \to 0; \tag{4.32}$$

$$|e^{A_{P,h} t} \Pi_h x - e^{A_P t} x|_{C(0,\infty;Y)} \to 0. \tag{4.33}$$

III. (convergence of "gain" operators)

(i) $$|B_h^* P_h \Pi_n x|_U \to |B^* Px|_U, \quad x \in \mathcal{D}(A). \tag{4.34}$$

(ii) For each $x \in \mathcal{D}(A_F)$ there exists a sequence $x_h \in Y_h$ such that $x_h \to x$ in Y and

$$|B_h^* P_h x_h - B^* P x| \to 0. \tag{4.35}$$

The proof of the above theorem is given in [L.3]. The above theorem provides us with the convergence theory for the Riccati operators and the gain feedbacks with minimal assumptions imposed on the model. We notice that the convergence of the gain operator holds on a dense set in Y, and not on the whole space Y. This is consistent with the continuous theory, where the gain operator B^*P is only densely defined.

Remark 4.9. Note that in both cases (general abstract C_0 semigroup and analytic semigroup) the above results are optimal, as they reconstruct numerically the properties of the solution which are present in the continuous case. This means that in the case of general C_0 semigroups and unbounded operators B, we obtain strong convergence of the Riccati operators on the full space Y, while the gain operators converge as unbounded operators on some dense set. In the analytic case, the uniform convergence of both Riccati operators and gain operators hold on the entire space Y. This, again, is in agreement with the regularity of the continuous theory (i.e., our approximation reconstructs "optimally" the properties of the continuous problem).

4.2.4 Discussion on the assumptions

(i) Note that in Theorem 4.3--in contrast with the analytic situation of section 4.1--we need to assume the "uniform Finite Cost Condition" $(F.C.C.)_h$ and the "uniform Detectability Condition" $(D.C.)_h$. In the analytic case subject to assumption (H.1) = (1.5), these properties can be deduced from the "uniform analyticity condition" (A.1). Instead, in the case of an arbitrary s.c. semigroup, these properties need to be established independently for a specific choice of the approximation scheme. Indeed, these conditions are in general rather sensitive and scheme dependent. They may fail, even in the case of B bounded, if an inappropriate approximating scheme is selected. Negative examples are known, even in the case of retarded (delay) equations, with spline approximations [L-M], [M.1], [P]. These conditions are related to the following fact: The spectrum of the original operator A should be "faithfully" approximated by the chosen approximating scheme.

(ii) All the remaining assumptions (B.1)-(B.4) are very natural and, in fact, minimal ones. They are consistent with the hypotheses imposed on the continuous problem. Indeed, (B.1) and (B.2) are the usual requirements of consistency of the approximation of the original semigroup and its adjoint.

Hypothesis (B.3) is a discrete counterpart of (H.2), while the assumptions grouped in (B.4) are in line with the continuous property that $\hat{A}^{-1}B$ be bounded (and, in fact, they are satisfied if one takes $B_h = \Pi_h B$).

4.2.5 Literature

Most of the literature dealing with approximation schemes for Riccati Equations for arbitrary C_0-semigroup treats the case of the input operator B bounded, see e.g., [G], [I.1], [KS]. In the B-unbounded case and with arbitrary C_0-semigroups, we are aware, in addition to [L.3], of only one paper [I-T] where the approximations of ARE are discussed subject to the condition (H.2) and the additional requirement that the observation R is smoothing like in [P-S] of Part I. Since, as already discussed, the framework of [P-S] is not applicable to all the examples of sections 3.2, 3.3, 3.4, the treatment of [I-T] cannot be applied either.

5. Examples of numerical approximation for the classes (H.1) and (H.2)

In this section we illustrate the applicability of the approximation Theorem 4.1 (class (H.1)) and of the approximation Theorem 4.2 (class (H.2)) in a few examples taken from the continuous section 3. For a full treatment of the case of the heat equation with Dirichlet boundary control, we refer to [L-T.1].

5.1 Class (H.1): The structurally damped plate problem in Example 3.1

We return to Example 3.1, model (3.1) in section 3, with $n = \dim \Omega \leq 3$.

Choice of V_h. We shall select the approximating space $V_h \subset H^2(\Omega) \cap H_0^1(\Omega)$ to be a space of splines (e.g., cubic splines), which comply with the usual approximation properties

$$|Q_h z - z|_{H^{\ell}(\Omega)} \leq C h^{s-\ell} |z|_{H^s(\Omega)}, \quad z \in H^s(\Omega) \cap H_0^1(\Omega),$$

$$0 \leq \ell \leq 2; \ \ell \leq s < r; \qquad (5.1)$$

$$|z_h|_{H^{\alpha}(\Omega)} \leq C h^{-s} |z_h|_{D^{\alpha-s}(\Omega)}, \quad 0 \leq \alpha \leq 2, \qquad (5.2)$$

where Q_h is the orthogonal projection of $L_2(\Omega)$ onto V_h and where r is the order of approximation.

Choice of $\mathcal{A}_h$. We let

$$\mathcal{A}_h = Q_h \mathcal{A} Q_h : V_h \rightarrow V_h;$$

i.e.,

$$(A_h\phi_h,\psi_h)_\Omega = (\Delta\phi_h,\Delta\psi_h)_\Omega = (A^{1/2}\phi_h,A^{1/2}\psi_h)_\Omega,$$

$$\phi_h,\psi_h \in V_h, \quad (5.3)$$

$$|A_h^{1/2}\phi_h|_{L_2(\Omega)} = |A^{1/2}\phi_h|_{L_2(\Omega)}; \quad |A_h^{1/2}\phi_h| \sim |\phi_h|_{H^2(\Omega)},$$

$$\phi_h \in V_h, \quad (5.4)$$

where (5.4) is a consequence of (5.3). From elliptic estimates

$$\begin{cases} |(A^{-1}-A_h^{-1}Q_h)z|_{H^2(\Omega)} \le C\, h^2|z|_{L_2(\Omega)}; \\ |(A^{-1/2}-A_h^{-1/2}Q_h)z|_{H^2(\Omega)} \le C\, h^2|z|_{H^2(\Omega)}, \quad z \in \mathcal{D}(A^{1/2}). \end{cases} \quad (5.5)$$

Choice of A_h **and** B_h. To begin with, we let $Y_h \equiv V_{h1}\times V_{h2}$, where V_{h1} consists of the elements of V_h equipped with norm $|v_h|_{V_h^1} = |A_h^{1/2}v_h|_{L_2(\Omega)}$ and V_{h2} consists of the elements of V_h equipped with the $L_2(\Omega)$-norm. We shall write $x_h = [x_{h1},x_{h2}] \in Y_h$. Next, we define

$$A_h\colon Y_h \to Y_h\colon A_h = \begin{vmatrix} 0_1 & Q_h \\ -A_h & -A_h^{1/2} \end{vmatrix}; \quad (5.6)$$

$$B_h\colon L_2(\Gamma) \to Y_h\colon B_h u = \begin{vmatrix} 0 \\ \mathcal{B}_h u \end{vmatrix}, \quad (\mathcal{B}_h u, v_h)_\Omega = v_h(x^0)u.$$

$$(5.7)$$

Finally, we let $\Pi_h\colon Y \to Y_h$ be defined as

$$\Pi_h = \begin{vmatrix} Q_h & 0 \\ 0 & Q_h \end{vmatrix}.$$

Computation of adjoints A_h^* **and** B_h^*. To compute the adjoints of A_h and B_h, we use the inner products generated by the topology on V_{h1} and V_{h2}. We find, as in the continuous case,

$$A_h^* = \begin{vmatrix} 0 & -Q_h \\ \mathcal{A}_h & -\mathcal{A}_h^{1/2} \end{vmatrix}; \quad B_h^* x_h = x_{h2}(x^0)$$

as it follows from $(A_h x_h, y_h)_{Y_h} = (x_h, A_h^* y_h)_{Y_h}$ and $(B_h u, x_h)_{Y_h}$ and $(u, B_h^* x_h)_U$, respectively.

Approximating control problem. With the above notation, the approximating version of the dynamics (4.11) is

$$\begin{cases} (\ddot{y}_h, \phi_h) + (\mathcal{A}_h y_h, \phi_h) + (\mathcal{A}^{1/2} y_h, \phi_h) = \phi_{h2}(x^0) u; \\ (\mathcal{A}_h y_h, \phi_h) = (\Delta y_n, \Delta \phi_h), \quad \text{all } \phi_h \in V_h; \\ (y_h(0), \phi_h) = (y_0, \phi_h); \; (\dot{y}_h(0), \phi_h) = (y_1, \phi_h); \end{cases} \tag{5.8}$$

where all inner products are in $L_2(\Omega)$.

The optimal feedback control for the finite-dimensional problem is given by $u_h^0(t) \equiv -[\overline{P}_{h2} \Psi_h(t)][x^0]$, where

$$P_h y_h \equiv \begin{cases} P_{h1} y_{h1} + P_{h2} y_{h2} \equiv \overline{P}_{h1} y_h; \\ P_{h3} y_{h1} + P_{h4} y_{h2} \equiv \overline{P}_{h2} y_h, \end{cases} \tag{5.9}$$

and P_h satisfies the following algebraic equation with $L_2(\Omega)$-inner products

$$-(\mathcal{A}_h x_{h2}, \overline{P}_{h1} y_h) + (\mathcal{A}_h x_{h1} - \mathcal{A}_h^{1/2} x_{h2}, \overline{P}_{h2} y_h) - (\mathcal{A}_h \overline{P}_{h1} x_h, y_{h2})$$

$$+ (\overline{P}_{h2} x_h, \mathcal{A}_h y_{h1} - \mathcal{A}_h^{1/2} y_{h2}) + (\mathcal{A}_h x_{h1}, y_{h1}) + (x_{h2}, y_{h2})$$

$$= (\overline{P}_{h2} x_h)(x^0)(\overline{P}_{h2} y_h)(x^0). \tag{5.10}$$

Verification of assumptions of Theorem 4.1. In order to apply Theorem 4.1, we need to verify the approximating assumptions (A.1)-(A.4), as well as assumptions (4.12), (4.13). Indeed, the last two are plainly satisfied: (4.12) since R = I, while (4.13) follows from (3.6) and the argument below it (in essence, $\mathcal{A}^{-(1/2-\varepsilon)} \delta \in L_2(\Omega)$, while $\mathcal{A}^{-\varepsilon}$ is compact on $L_2(\Omega)$.

Assumption (A.1). This follows by applying the arguments of [C-T.2] of the continuous case to the finite-dimensional operator given by (5.6).

Assumption (A.2). By (5.5), we have that (A.2) with s = 2 holds true:

$$|(A_h^{-1}-A^{-1})x_h|_Y = |-(A^{-\frac{1}{2}}-A_h^{-\frac{1}{2}})x_{h1}+(A^{-1}-A_h^{-1})x_{h2}|_{H^2(\Omega)}$$

$$\leq C\, h^2[|x_{h1}|_{H^2(\Omega)}+|x_{h2}|_{L_2(\Omega)}]$$

$$= C\, h^2[|x_h|_Y].$$

The same result holds for the adjoint A^*, in view of its definition.

Assumption (A.3). By Sobolev embedding and the inverse approximation property (5.2), we have for any $\varepsilon > 0$,

$$|B^* x_h|_U = |x_{h2}(x^0)| \leq C|x_{h2}|_{H^{\frac{3}{2}+\varepsilon}(\Omega)}$$

$$\leq C\, h^{-\frac{3}{2}-\varepsilon}|x_{h2}|_{L_2(\Omega)}$$

$$\leq C\, h^{-2+\varepsilon_0}|x_h|_Y,$$

where ε_0 can be taken $0 < \varepsilon_0 < \frac{1}{2}$.

Assumption (A.4). By (5.1) we compute

$$|B^*(\Pi_h x-x)|_U = |(Q_h x_2)(x^0)-x_2(x^0)|_{R^1}$$

$$\leq C|Q_h x_2-x_2|_{H^{\frac{3}{2}+\varepsilon}(\Omega)}$$

$$\leq C\, h^{\varepsilon}\ |x_2|_{H^{\frac{3}{2}+2\varepsilon}(\Omega)}$$

$$\leq C\, h^{\varepsilon}\ |x|_{\mathcal{D}(\hat{A}^{*^{1-\varepsilon_0}})},$$

with $0 < \varepsilon_0 < \frac{1}{2}$, by the fractional powers results in [C-T.4], Part I, reported in (3.8), Part I.

Thus, we have verified all the assumptions of Theorem 4.1. Thus, Theorem 4.1 applies to our problem, Example 3.1, Eq. (3.1) of Part I, and yields the following convergence results:

$$P_h \Pi_h x \to Px \text{ in } H^{4-\varepsilon}(\Omega)\times H^{2-\varepsilon}(\Omega),$$
$$\text{for all } x \in H^2(\Omega)\times L_2(\Omega).$$

In particular,

$$B^* P_h \Pi_h x \to B^* Px \quad \text{for all } x \in H^2 \times L_2(\Omega),$$

where P_h is computed from (5.10). Also, if we use the feedback law given by

$$\hat{u}_h(t) = -[\overline{P}_{h2} y(t)][x^0]$$

inserted into the __original__ dynamics

$$w_{tt} + \Delta^2 w - \Delta w = \delta(x-x^0)\hat{u}_h; \; w|_\Gamma = \Delta w|_\Gamma = 0,$$

then the corresponding feedback system is uniformly (in h) exponentially stable in the topology of $H^2(\Omega)\times L_2(\Omega)$. This means that the numerical algorithm provides a feedback control which yields uniform (in h) stability results for the original system.

We conclude this section by pointing out that the other examples of section 3.1 dealing with structurally damped plate problems can be dealt with by a similar approximating scheme. For other p.d.e. examples in the 'analytic' class (H.1)-parabolic equations with $L_2(\Sigma)$-boundary control in the Dirichlet (or Neumann) B.C., we refer to the original treatment in [L-T.1].

5.2 __Class (H.2): The wave equation with Dirichlet boundary control__

In this subsection, we verify the applicability of the approximating Theorem 4.2 to the wave equation with $L_2(\Sigma)$-Dirichlet boundary control, treated in section 3.2, Part I, Eq. (3.45).

__Choice of__ V_h. Let $V_h \subset H_0^1(\Omega)$ be an approximating subspace of $L_2(\Omega)$. Let Q_h be the L_2-orthogonal projection of $L_2(\Omega)$ onto V_h. We assume that V_h enjoys the following approximating properties:

$$\begin{cases} \text{(i)} & |Q_h z - z|_{H_0^\alpha(\Omega)} \to 0, \ z \in H_0^\alpha(\Omega), \ |\alpha| \le 1; \\ \text{(ii)} & |\frac{\partial}{\partial \nu} z_h|_{L_2(\Omega)} \le C\, h^{-1/2-\varepsilon} |z_h|_{H_0^1(\Omega)}; \\ \text{(iii)} & |z_h|_{H_0^\alpha(\Omega)} \le C\, h^{-s} |z_h|_{H_0^{\alpha-s}(\Omega)}, \quad 0 \le \alpha \le 1, \ \alpha - s \ge 0; \\ \text{(iv)} & |\frac{\partial}{\partial \nu} (Q_h z - z)|_{L_2(\Omega)} \le C\, h^{1/2} |z|_{H^2(\Omega)}. \end{cases} \tag{5.11}$$

It is well known that the above approximation properties are satisfied for, say, spline approximations (of order r, $r \ge 1$) defined on a uniform mesh. Also, modal (eigenfunction) approximations are typical examples of schemes which satisfy the requirements in (5.11).

__Choice of__ $\mathcal{A}_h$. We let

$$\mathcal{A}_h = Q_h \mathcal{A} Q_h : V_h \to V_h; \text{ i.e., } (\mathcal{A}_h \phi_h, \psi_h) = \int_\Omega \nabla\psi_h \cdot \nabla\psi_h d\Omega.$$

From the definition of $\mathcal{A}_h$, it follows immediately that

$$|\mathcal{A}_h^{1/2} z_h|_{L_2(\Omega)} = |\mathcal{A}^{1/2} z_h|_{L_2(\Omega)},$$

and $\mathcal{A}_h^{1/2}$ is an isomorphism $H_0^1(\Omega) \to L_2(\Omega) \cap V_h$ (with a norm uniform in h). Since $\mathcal{A}_h^{1/2}$ is self-adjoint on $L_2(\Omega)$, we have that $\mathcal{A}_h^{1/2}$ is also an isomorphism $L_2(\Omega) \to H^{-1}(\Omega) \cap V_h$. Moreover, elliptic estimates give

$$|(\mathcal{A}_h^{-1} Q_h - \mathcal{A}^{-1}) z|_{H_0^s(\Omega)} \le C\, h^{2-s} |z|_{L_2(\Omega)}, \qquad 0 \le s \le 1. \tag{5.12}$$

__Choice of__ A_h __and__ B_h. We introduce the space $Y_h = V_{h1} \times V_{h2}$, where V_{h1} (resp. V_{h2}) is the space V_h equipped with norm $L_2(\Omega)$ (resp. $\mathcal{A}_h^{-1/2}$-norm):

$$|v_h|_{V_{h1}} = |v_h|_{L_2(\Omega)}; \quad |v_h|_{V_{h2}} = |\mathcal{A}_h^{-1/2} v_h|_{L_2(\Omega)}.$$

By (5.12), Y_h is topologically equivalent to Y. The operators $B_h: U \to Y_h$ are defined by

$$B_h u = \begin{vmatrix} 0 \\ \mathcal{B}_h u \end{vmatrix} = \begin{vmatrix} 0 \\ Q_h \mathcal{A} D u \end{vmatrix}, \tag{5.13}$$

well defined since

$$(\mathcal{B}_h u, \phi_h)_{L_2(\Omega)} = (\mathcal{A} Du, \phi_h)_{L_2(\Omega)} = \left(u, \frac{\partial \phi_h}{\partial \nu}\right)_{L_2(\Omega)}. \quad (5.14)$$

Finally, we let $\Pi_h: Y_h \to Y_h$ be as before, below (5.7): a diagonal matrix with Q_h on the main diagonal.

<u>**Computation of adjoints**</u> A_h^* <u>**and**</u> B_h^*. These are computed with respect to the topologies of V_{h1}, V_{h2}. As in the continuous case, one readily obtains

$$A_h^* = -A_h \text{ and } B_h^* x_h = \frac{\partial}{\partial \nu} \mathcal{A}_h^{-1} x_{h2}. \quad (5.15)$$

<u>**Approximating control problem**</u>. With the above notation, the discrete version of the state equation is equivalent to

$$\begin{cases} (\ddot{y}_h(t), \phi_h)_\Omega + \int_\Omega \nabla y_h \cdot \nabla \phi_h d\Omega = \left(u(t), \frac{\partial}{\partial \nu} \phi_h\right)_\Gamma; \\ (y_h(0), \phi_h)_\Omega = (y_0, \phi_h)_\Omega, \ (\dot{y}_h(0), \phi_h)_\Omega = (y_1, \phi_h)_\Omega, \end{cases} \quad (5.16)$$

for all $\phi_h \in V_h$, in the L_2-inner products of Ω and Γ, where

$$y_h(t) = \sum_{\ell=1}^{h} y_\ell \phi_\ell(x) = y_{h1}; \ y_{h2}(t) = \dot{y}_h(t).$$

The optimal feedback control for finite-dimensional problem (CP_h) is given by $u_h^0(t) = -\frac{\partial}{\partial \nu} \mathcal{A}_h^{-1} \overline{P}_{h2}(y_h(t))$ where

$$P_h \begin{bmatrix} x_{h1} \\ x_{h2} \end{bmatrix} = \begin{bmatrix} P_{h1} x_{h1} + P_{h2} x_{h2} \\ P_{h3} x_{h1} + P_{h4} x_{h2} \end{bmatrix} \equiv \begin{bmatrix} \overline{P}_{h1}(x_h) \\ \overline{P}_{h2}(x_h) \end{bmatrix},$$

and P_h satisfies the Riccati equation

$$-(\overline{P}_{h1} x_h, y_{h2})_\Omega + (\overline{P}_{h2} x_h, y_{h1})_\Omega - (x_{h2}, \overline{P}_{h1} y_h)_\Omega$$

$$+ (x_{h1}, \overline{P}_{h2} y_h)_\Omega + (x_{h1}, y_{h1})_\Omega + (\mathcal{A}_h^{-1} x_{h2}, y_{h2})_\Omega$$

$$= \left[\frac{\partial}{\partial \nu} A_h^{-1}\overline{P}_{h2}, \frac{\partial}{\partial \nu} A_h^{-1}\overline{P}_{h2}y_h\right]_\Gamma ,$$

$$\text{for all } x_h = \begin{bmatrix} x_{h1} \\ x_{h2} \end{bmatrix}; \quad y_h = \begin{bmatrix} y_{h1} \\ y_{h2} \end{bmatrix} \subset V_{h1} \times V_{h2} = Y_h. \tag{5.18}$$

Thus, if one can apply the theory presented in Theorem 4.2, then the feedback law (5.16) computed with the aid of Riccati operator P_h (from (5.18)), once inserted into the dynamics (5.16) yields a feedback semigroup which is uniformly exponentially stable, and it is convergent to the original, infinite-dimensional solution.

It remains to verify the assumptions of Theorem 4.2; i.e., conditions (B.1)-(B.4), $(F.C.C.)_h$, and $(D.C.)_h$, since the other assumptions (H.2), (F.C.C.), and (D.C.) have already been verified in Part I, section 3.2.

Assumptions (B.1), (B.2), (B.4), $(D.C.)_h$. By using the approximating properties (5.11), together with the elliptic estimates (5.12), one can show ([L.3]) that hypotheses (B.1), (B.2), (B.4) are indeed satisfied for our dynamics of the wave equation with Dirichlet control (3.45). Moreover, assumption $(D.C.)_h$ is automatically satisfied, since R = I in our case.

Discussion on hypotheses (B.3) **and** $(F.C.C.)_h$. More delicate is the issue of validity of hypotheses (B.3) and $(F.C.C.)_h$ in the case of the wave equation problem with Dirichlet control (3.45) under study. Although both conditions are very natural--or they are discrete counterpart of properties which hold true in the continuous case--their validity in approximating problem (3.45) may well depend on the specific selection of the numerical scheme adopted. This fact, or pathology, should not--really--be surprising. Indeed, in the case, say, of delay differential equations which are uniformly stabilizable to begin with, it is known that the validity of the uniform Finite Cost Condition $(F.C.C.)_h$ depends on the particular numerical scheme selected (see [L-M], [M.1], [P]). Before we analyze further the validity of conditions (B.3) and $(F.C.C.)_h$ in our present case of the wave equation (3.45), it will be expedient to re-write them in an explicit, equivalent form, as they apply to our case. It can be shown [L.3] that

Lemma 5.1. Let $\psi_h(t) \in V_h$ be a semi-discrete solution of the following ODE problem

$$(\ddot{\psi}_h(t), \phi_h)_\Omega + \int_\Omega \nabla\psi_h(t)\cdot\nabla\phi_h dr = 0, \ \forall\ \phi_h \in V_h. \tag{5.19}$$

Then:

(i) Condition (B.3) is equivalent to the following inequality:

$$\left|\frac{\partial}{\partial\nu}\psi_h\right|_{L_2(\Sigma_T)} \le C_T\left[|\dot{\psi}_h(0)|_{L_2(\Omega)}+\|\nabla\psi_h(0)\|_{L_2(\Omega)}\right]. \tag{5.20}$$

(ii) Condition $(F.C.C.)_h$ is satisfied, provided

$$\left|\frac{\partial}{\partial\nu}\psi_h\right|_{L_2(\Sigma_T)} \ge C_T\left[|\dot{\psi}_h(0)|_{L_2(\Omega)}+\|\nabla\psi_h(0)\|_{L_2(\Omega)}\right]. \tag{5.21}$$

Remark 5.1. Notice that both condition (5.20), (5.21) are satisfied for the continuous solutions of the homogeneous wave equation. In fact, in this case, they are equivalent to, respectively, the regularity assumption (H.2) and the exact controllability of the wave equation with boundary control as in (3.45d), [L-T.1-3], [Lio.1-2], [H.1], [T.3].

To question whether these estimates (5.20), (5.21) are always satisfied for schemes which comply only with (5.11) is far from being obvious (see our earlier remark), and we are not in a position to give a full and general answer. It is clearly a technical issue which depends very much on the specific algorithm employed. Although very natural (in fact, necessary) for our problem, and although they have a continuous counterpart for the original system, their validity for a specific numerical scheme is, in general, an open problem of numerical analysis. Below, we shall provide an affirmative answer, but under an additional hypothesis (satisfied, for instance, by modal approximations). The general case, say of spline approximations, is still an open question. (Here, we report some promising numerical computations [G-L-L], [D-G-K-W] which confirm numerically the validity of these conditions in the case of finite differences and mixed finite elements.)

Lemma 5.2 [L.3]. Assume that the following commutativity property holds true

$$Q_h A = A Q_h. \tag{5.22}$$

Then, any approximation scheme which complies with (5.11) and (5.22) satisfies the inequalities (5.20) and (5.21) as well.

Remark 5.2. As noted above, examples of such an approximation are given, for instance, by modal approximations.

6. Conclusions

The present two-part paper attempts to collect--within obvious space limitations--the most significant results concerning the existence (and uniqueness) of solutions to abstract operator Algebraic Riccati Equations as well as the numerical analysis of their numerical approximation for actual computations. We have considered a rather general dynamics (1.1) with unbounded operators B, subject to two different (but not mutually exclusive) regularity hypotheses, which were labeled (H.1) = (1.5) and (H.2) = (1.6).

6.1 Theoretical aspects

(H.1) class. This class comprises pairs $\{A,B\}$ of operators in (1.1), where A generates an s.c. analytic semigroup and B is "almost as unbounded as" A, in the technical sense of assumption (1.3).

Class (H.2). The regularity requirement (H.2) has been shown in recent years--by purely p.d.e. techniques--to hold true for a variety of hyperbolic problems as well as of plate problems with certain boundary conditions, see the examples of sections 3.2, 3.3, 3.4, and of Appendix C in [FLT], Part I. However, (H.2) imposes certain restrictions even within these dynamics.

In the case of wave equations (or more generally, second order hyperbolic equations) with Neumann boundary control (as opposed to the Dirichlet control of section 3.2), the full theory of Theorem 2.2 applies in dim Ω = 1 on the space $H^1(\Omega)\times L_2(\Omega)$. It is in relation to this space that assumption (H.2) holds true: This is the space of optimal regularity in this case of dim Ω = 1, where exact controllability/uniform stabilization hold likewise true. However, for dim $\Omega \geq 2$, the situation is not so satisfactory. The solution to $L_2(\Sigma)$-controls lives pointwise in time in a space definitely larger than (with weaker topology than) the space $H^1(\Omega)\times L_2(\Omega)$ of finite "energy," see the sharp regularity results in [L-T.3-4], which moreover depend on the geometry of Ω. Thus, (H.2) is not true in relation to $H^1(\Omega)\times L_2(\Omega)$, if dim $\Omega \geq 2$. However, results of exact controllability/uniform stabilization--needed to verify the Finite Cost Condition--are presently available only on $H^1(\Omega)\times L_2(\Omega)$ with $L_2(\Sigma)$-controls, not in the spaces of sharp regularity where (H.2) instead is true.

In the case of plates problems, we have seen in section 3.3 that assumption (H.2) holds true for lower order boundary conditions (Dirichlet/Neumann), in relations to explicitly identified spaces of optimal regularity, where exact controllability/uniform stabilization results are also available to verify here the Finite Cost Condition. Thus, for these fourth order problems (in space), the situation is as complete as for wave equations with Dirichlet control. However, for plates with higher order boundary conditions, the situation is, instead, rather the counterpart of wave equations with Neumann control. As in the latter case of waves, we have at present that assumption (H.2) is not true for plates with, say, shear forces and bending moments as controls in relation to the "energy" space $H^2(\Omega)\times L_2(\Omega)$, where instead exact controllability/uniform stabilization results are available [LL].

Thus, there is a need to further relax assumption (H.2), as to incorporate these foregoing classes of problems and, consequently, to develop a theory for the ARE for them.

6.2 Numerical aspects

As to the numerical results discussed in this paper, we have shown that a fully consistent numerical theory can be developed for dynamical models which satisfy either assumption (H.1) = (1.5), or else (H.2) = (1.6).

Class (H.1). In the case of models which comply with assumption (H.1) ('analytic' class), the numerical theory is optimal, as it provides optimal convergence results (which are in line with the properties of the original continuous solutions) under minimal assumptions imposed on the approximating subspaces. These are just basic convergence properties which are satisfied by all spline approximations, modal or spectral approximations, etc.

Class (H.2). In the case od dynamics which comply with assumption (H.2), the numerical theory is less satisfactory. The basic theory of Theorem 4.2 is, yes, 'optimal,' in the sense that optimal convergence conclusions are obtained under only 'natural' approximating assumptions. However, the question remains as to whether or not a few of these 'natural' approximating assumptions are actually satisfied by various approximating subspaces (algorithms). This is, in fact, a delicate issue, and our Lemma 5.1 provides just one answer. The issue is purely technical: whether a given approximating subspace will guarantee the desired (and 'natural') stability results as expressed by inequalities (5.20), (5.21). In full generality beyond our Lemma 5.1, this is presently an open question.

References

[B-I] H. T. Banks, K. Ito, Approximation techniques for parabolic control systems: A variational approach. SIAM Conference on Control in the 90's. San Francisco, 1989.

[B-K] H. T. Banks, K. Kunisch. The linear regulator problem for parabolic systems. SIAM J. Control, Vol. 22, No. 5 (1984), 684-699.

[D-G-K-W] T. Dupont, R. Glowinski, W. Kinton, M. Wheeler, Mixed finite element methods for time dependent problems: Application to control, Research Report UM/MD-54, Univ. of Houston, 1989.

[G.1] J. S. Gibson. The Riccati integral equations for optimal control problems on Hilbert spaces. SIAM J. on Control, Vol. 17, No. 4 (1979), 537-565.

[G-L-L] R. Glowinski, C. H. Li, J. L. Lions, A numerical approach to the exact boundary controllability of the wave equation, Japan J. of Applied Mathematics, to appear.

[I.1] K. Ito. Strong convergence and convergence rate of approximating solutions for Algebraic Riccati equations in Hilbert spaces, Lecture Notes Control Information Sciences, 102 (1987), Springer-Verlag.

[I-T] K. Ito, M. T. Tran, Linear quadratic control problem for linear systems with unbounded input and output operators: Numerical approximations, Proceedings of the Vorau Conference, 1988, to appear.

[K-S] F. Kappel and D. Salomon, An approximation theorem for the algebraic Riccati equation. Proceedings of the 27th IEEEC D.C. Conference in Austin, Texas, 1988.

[L-L.1] J. Lagnese and J. L. Lions, Modeling, analysis and control of thin plates, Collection Recherches en Mathematiques Appliquees, Vol. 6, Masson, Paris, 1988.

[L-M.1] I. Lasiecka and A. Manitius, Differentiability and convergence of approximating semigroups for retarded functional differential equations, SIAM J. Numerical Anal. 25 (1988), 883-907.

[L-T.1] I. Lasiecka and R. Triggiani, The regulator problem for parabolic equations with Dirichlet boundary control, Part II: Galerkin approximation, Appl. Math. and Optimiz., Vol. 16 (1987), 198-216.

[L-T.2] I. Lasiecka and R. Triggiani, Approximation theory for Algebraic Riccati equations with unbounded input operators: The case of analytic semigroups, to appear.

[L-T.3] I. Lasiecka, and R. Triggiani, Sharp regularity results for mixed second order hyperbolic equations of Neumann type: The L_2-boundary case, Ann. di Matem. Pura e Appl., to appear.

[L-T.4] I. Lasiecka and R. Triggiani, Trace regularity of the solutions of the wave equation with homogeneous Neumann boundary conditions and compactly supported data, J. of Math. Anal. and Appl.

[M.1] A. Manitius, G. Tran, G. Payre, and R. Roy, Computation of eigenvalues associated with functional differential equations, SIAM J. Sci. Statist. Comp., to appear.

[L.1] I. Lasiecka, Convergence estimates for semidiscrete approximations of nonselfadjoint parabolic equations, SIAM J. Num. Anal., Vol. 21 (1984), 894-909.

[L.2] I. Lasiecka, Galerkin approximations of abstract parabolic boundary value problems with rough boundary data; L_p theory, Math. Comp., Vol. 47, No. 175 (1986), 55-75.

[L.3] I. Lasiecka, Approximations of the solutions of infinite dimensional Algebraic Riccati Equations with unbounded input operators, to appear.

[P] G. Propst, Numerische Untersuchung der Spectra Endlichmensionaler Approximation für Differenzen-Differentialgleichungen, Technical Report No. 40, 1984, Univ. of Graz.

SOME CONTROL PROBLEMS FOR TIME-DELAY SYSTEMS

V. M. Marchenko

Abstract

The paper deals with the problem of canonical representations and their applications for linear stationary systems with aftereffect. We introduce non-singular functional transformations which induce the corresponding equivalence relations on the set of considered systems. Then the problem of determination of canonical forms is interpreted as a problem of universality for each class of transformations. Generalizations of the natural canonical forms of Frobenius as well as some other canonical forms are given for time-delay systems. Applications of the obtained results to such problems of qualitative control theory as controllability, modal control, stabilization, reconstruction and others are also presented.

Introduction.

One of the basic problems of qualitative theory of control for linear dynamic systems is the problem of construction of canonical forms. For linear ordinary stationary systems this problem is well understood [1–3]. For systems with aftereffect this problem turns out to be much more complicated. The first result is associated with canonical decomposition of Shimanov [4, 5]. Then Kappel [6] considered the theory of Jordan canonical forms for time-delay systems. The history of generalizations of Frobenius' forms to the retarded argument systems can be found in [7–11]. The obtained results are based on algebraic properties of delay operator. Similar ideas are used by Lee, Neftci and Olbrot [12] to obtain some other canonical representations.

The canonical forms were used to study different problems of qualitative control theory for systems with aftereffect, e.g.: problems of modal control [13, 14], controllability and stabilization [7], reconstruction [15], classification [8], decoupling [16] and others.

The aim of the paper is to present some results on qualitative control theory obtained by the author and his collaborators.

Transformations of systems with time-delays.

Let us consider the class F of linear stationary control systems

$$\dot{x}(t) = \sum_{j=0}^{l} \big(A_j x(t-h_j) + B_j u(t-h_j)\big), \quad t > 0, \tag{1}$$

with initial conditions

$$x(\tau) = \varphi(\tau), \quad u(\tau) = 0, \quad \tau \in [-h, 0], \quad x(+0) = \varphi_0 \in R^n,$$

where $A_j \in R^{n\times n}$, $B_j \in R^{n\times r}$, $j = 0,\dots,l$; $R^{n\times 1} = R^n$; $0 = h_0 < h_1 < \cdots < h_l = h$ are constant time-delays; $\varphi(\cdot)$ is a piecewise continuous vector function in $[-h, 0]$ with values in R^n. Here by the symbol $R^{m\times d}$ we denote the space of $m \times d$ matrices.

Every system (1) is uniquely defined by matrix couple $\big(A(\cdot), B(\cdot)\big)$ where

$$A(m) = \sum_{j=0}^{l} A_j m^{h_j}, \quad B(m) = \sum_{j=0}^{l} B_j m^{h_j}, \quad m \geq 0.$$

Then the class F of the systems (1) can be identified with a class $\mathcal{F}(\cdot)$ of the matrix couples $\big(A(\cdot), B(\cdot)\big)$.

For the class $\mathcal{F}(\cdot)$ we shall consider the following transformations

$$\mathcal{R}_1(\cdot) = \Big\{ T(\cdot) \mid T(m) = \sum_{j=0}^{k} T_j m^{s_j}, \ k \in N, \ T_j \in R^{n\times n}, \\ 0 = s_0 < s_1 < \cdots < s_k; \ \det T(m) \equiv \text{const} \neq 0, \ m \geq 0 \Big\}$$

$$\mathcal{R}_2(\cdot) = \Big\{ T(\cdot) \mid T(m) = \sum_{j=0}^{k} T_j m^{s_j}, \ k \in N, \ T_j \in R^{n\times n}, \\ 0 = s_0 < s_1 < \cdots < s_k; \ \det T(m) \not\equiv 0, \ m \geq 0 \Big\},$$

where N denotes the set of natural numbers.

The classes $\mathcal{R}_1(\cdot)$ and $\mathcal{R}_2(\cdot)$ induce two classes of non-singular transformations on F:

$$\mathcal{R}_1: \quad \begin{cases} x(t) = \sum_{j=0}^{k} T_j\, y(t - s_j) = T\big(\exp(-p)\big) y(t), \\ y(t) = T^{-1}\big(\exp(-p)\big)\, x(t), \quad t > 0, \\ x(\tau) = 0 \quad \text{for} \quad \tau < -h; \end{cases}$$

and

$$\mathcal{R}_2: \quad \begin{cases} x(t) = T\big(\exp(-p)\big)\, y(t), \quad \det T\big(\exp(-p)\big)\, y(t) = \\ = \det T\big(\exp(-p)\big) T^{-1}\big(\exp(-p)\big)\, x(t) = \sum_{j=0}^{\bar{k}} \overline{T}_j m^{\bar{s}_j}, \\ t \in R; \quad x(\tau) = 0, \quad y(\tau - \beta h) = 0 \quad \text{for} \quad \tau < -h, \end{cases}$$

where β is a natural number; $\exp(-p)$ is a delay operator $\big(\exp(-ps_j)\, y(t) = y(t - s_j)\big)$.

The class F is invariant with respect to transformations of the class $\mathcal{R}_1$, that allows to interpret the problem of finding the canonical forms in the class F as the problem of construction of universal elements.

Relation with the problem of universality.

Let us introduce the following equivalence relation $\mathcal{Z}$ on $F(\cdot)$

$$\begin{aligned}\big(A(\cdot),B(\cdot)\big)\mathcal{Z}\big(A^*(\cdot),B^*(\cdot)\big) \iff \exists T(\cdot)\in\mathcal{R}_1(\cdot),\ & B^*(m)=T^{-1}(m)B(m),\\ & A^*(m)=T^{-1}(m)A(m)T(m),\\ & m\geq 0.\end{aligned}$$

Let S be an arbitrary set of properties of the class $F(\cdot)$ which is invariant with respect to $\mathcal{R}(\cdot)$. We denote by $W(S)$ the set of all mappings f: $F(\cdot)\to S$ which are invariant with respect to $\mathcal{Z}$, i.e.

$$f\big(A^*(\cdot),\,B^*(\cdot)\big)=f\big(A(\cdot),\,B(\cdot)\big)\quad\text{if}\quad\big(A^*(\cdot),\,B^*(\cdot)\big)\mathcal{Z}\big(A(\cdot),\,B(\cdot)\big).$$

Then the problem of universality is to find a set C and a mapping $\xi\in W(C)$ such that for every set S and every mapping f there exists the unique mapping η: $C\to S$ such that $f=\eta\times\xi$, i.e. the following diagram

$$\begin{array}{ccc} F(\cdot) & \xrightarrow{\ \xi\ } & C\\ f\searrow & & \swarrow\eta\\ & S & \end{array}\tag{2}$$

is commutative.

The couple $(\xi,\,C)$ is an universal element for the problem. The problem of universality has a standard solution $(\xi,\,F(\cdot)/\mathcal{Z})$ where ξ maps $F(\cdot)$ on the factor set $F(\cdot)/\mathcal{Z}$. The solution is unique up to an isomorphism. If $C\subset F(\cdot)$ then the mapping ξ from the universal couple is canonical, i.e.

$$D(\cdot)\mathcal{Z}\xi\big(D(\cdot)\big)\quad\text{for}\quad D(\cdot)\in F(\cdot),$$

$$D(\cdot)\mathcal{Z}D^*(\cdot)\iff\xi\big(D(\cdot)\big)=\xi\big(D^*(\cdot)\big),\ \text{for any } D(\cdot),\ D^*(\cdot)\in F(\cdot).$$

and the set C is called the set of canonical forms for $\mathcal{Z}$ on $F(\cdot)$. This set is unique up to an isomorphism. Thus the set of canonical forms of the systems of class F for the transformations $\mathcal{R}_1$ can be given as the set C from the universal couple $(\xi,\,C)$ where $C\subset F$. Analogously one can investigate the problem of canonical forms for $\mathcal{R}_2$ on F.

The problem of universality was solved for systems without delays by Popov [3]. A generalization to systems with retarded argument and transformations from $\mathcal{R}_2$ is given by Kirillova and Marchenko [7]. The problem for the class $\mathcal{R}_1$ is still open.

Canonical forms.

We give canonical forms for delay-differential systems similar to the well-known canonical form of Brunovsky [1] for systems without delay. Consider the set of column vectors

$$\begin{array}{ccccc} \big(B(m)\big)_1, & \big(B(m)\big)_2, & \ldots, & \big(B(m)\big)_r, & (3_0)\\ A(m)\big(B(m)\big)_1, & A(m)\big(B(m)\big)_2, & \ldots, & A(m)\big(B(m)\big)_r, & (3_1)\\ \vdots & \vdots & \vdots & \vdots & \\ \big(A(m)\big)^{n-1}\big(B(m)\big)_1, & \big(A(m)\big)^{n-1}\big(B(m)\big)_2, & \ldots, & \big(A(m)\big)^{n-1}\big(B(m)\big)_r, & (3_{n-1}) \end{array}$$

where by the symbol $(D)_i$ we denote the i-th column of the matrix D.

Let $l_1(m)$ be the largest number of linearly independent vectors in (3_0); $l_2(m)$ be the largest number of vectors in (3_1) linearly independent mod (3_0) and so on. Assume that for some $m \geq 0$ the condition

$$\operatorname{rank}\big[B(m),\, A(m)B(m), \ldots, \big(A(m)\big)^{n-1}B(m)\big] = n \tag{4}$$

holds. Then for some natural μ we have

$$l_1(m) \geq l_2(m) \geq \cdots \geq l_\mu(m); \quad \sum_{i=1}^{\mu} l_i(m) = n.$$

Denote by$M(\cdot)$ the set of linearly independent vectors in (3) chosen in the following way: in every row we go from the left to the right and then we go from the top to the bottom. For every column vector $\big(B(m)\big)_i$ we find the number $n_i(m)$ as the largest non-negative integer such that

$$A^{n_i(m)-1}(m)\big(B(m)\big)_i \in M(m), \quad m \geq 0.$$

Taking into account (4) we conclude that

$$n_1(m) + \cdots + n_{l_1(m)}(m) = n, \quad l_1(m) \leq r, \quad m \geq 0.$$

It is possible to show that there exists a set Ω, of the zero Lebesgue measure, outside of which the relations

$$\begin{aligned} A^{n_i(m)}(m)\big(B(m)\big)_i = &-\sum_{j=1}^{i-1} \sum_{k=0}^{\min\{n_i(m),\, n_j(m)-1\}} a_{ijk}(m)A^k(m)\big(B(m)\big)_j \\ &-\sum_{j=i}^{l_1(m)} \sum_{k=0}^{\min\{n_i(m),\, n_j(m)\}-1} a_{ijk}(m)A^k(m)\big(B(m)\big)_j, \end{aligned}$$

$$i \geq 1, \quad m \notin \Omega, \quad m \geq 0,$$

hold.

Consider a non-singular transformation of class $\mathcal{R}_2$ represented by the matrix $T(\cdot) = \big[t_1(\cdot),\, t_2(\cdot), \ldots,\, t_n(\cdot)\big]$ the columns of which are defined as follows

$$
\begin{aligned}
t_{s_{i-1}+\nu}(m) = A^{n_i+\nu}(m)\big(B(m)\big)_i &+ \sum_{j=1}^{i-1}\ \sum_{k=\nu}^{\min\{n_j-1,\,n_i\}} a_{ijk}(m)A^{k-\nu}(m)\big(B(m)\big)_j \\
&+ \sum_{j=i}^{l_1(m)}\ \sum_{k=\nu}^{\min\{n_j,\,n_i\}-1} \alpha_{ijk}(m)A^{k}(m)\big(g(m)\big)_j,
\end{aligned}
$$

where $s_0 = 0$, $s_i = \sum\limits_{j=1}^{i} n_j$, $n_j = n_j(m)$, $m \notin \Omega$, $m \geq 0$. Using this transformation we rewrite the system (1) in the form

$$
\dot y(t) = \left[\begin{array}{cccccccccccccc}
0 & 1 & & & & & & & & & & & \\
 & & \ddots & & & & & & & & & & \\
 & & & 1 & & & & & & & & & \\
* & * & \cdots & * & * & * & \cdots & * & \cdots & * & * & \cdots & * \\
 & & & & 0 & 1 & & & & & & & \\
 & & & & & & \ddots & & & & & & \\
 & & & & & & & 1 & & & & & \\
* & * & \cdots & * & * & * & \cdots & * & \cdots & * & * & \cdots & * \\
 & & & & & & & & \ddots & & & & \\
 & & & & & & & & & 0 & 1 & & \\
 & & & & & & & & & & & \ddots & \\
 & & & & & & & & & & & & 1 \\
* & * & \cdots & * & * & * & \cdots & * & \cdots & * & * & \cdots & *
\end{array}\right] y(t) +
$$

$$
+\left[\begin{array}{cccccccc}
0 & & & & & * & \cdots & * \\
\vdots & & & & & \vdots & & \vdots \\
0 & & & & & * & \cdots & * \\
1 & * & * & \cdots & * & * & \cdots & * \\
 & 0 & & & & * & \cdots & * \\
 & \vdots & & & & \vdots & & \vdots \\
 & 0 & & & & * & \cdots & * \\
 & 1 & * & \cdots & * & * & \cdots & * \\
 & & \ddots & & & \vdots & & \vdots \\
 & & & & 0 & * & \cdots & * \\
 & & & & \vdots & \vdots & & \vdots \\
 & & & & 0 & * & \cdots & * \\
 & & & & 1 & * & \cdots & *
\end{array}\right] v(t)\,, \tag{5}
$$

where $v(t) = Q\big(\exp(-p)\big)u(t)$, $Q(\cdot)$ is a quasipolynomial matrix with $\det Q(m) \equiv 1$, $m \geq 0$, and the symbol $*$ denotes the elements which may belong to the set of functions represented as a quotient of quasi-polynomials. Let us assume that

$$
B_j = b_j \in R^n, \quad j = 0, 1, \ldots, l: \quad B(m) = b(m) \in R^n, \quad \forall\, m \geq 0, \tag{6}
$$

and

$$\det W(m) = \det \left[b(m),\ A(m)b(m), \ldots, \left(A(m)\right)^{n-1} b(m) \right] \equiv \text{const} \neq 0, \quad m \geq 0. \tag{7}$$

Then the representation (5) can be rewritten as follows [9]

$$\dot{y}(t) = \begin{bmatrix} 0 & 1 & 0 & \ldots & 0 \\ 0 & 0 & 1 & \ldots & 0 \\ & & & \ddots & \\ 0 & 0 & 0 & \ldots & 1 \\ * & * & * & \ldots & * \end{bmatrix} y(t) + \begin{bmatrix} 0 \\ 0 \\ \vdots \\ 0 \\ 1 \end{bmatrix} u(t). \tag{8}$$

Here the symbol $*$ means some quasipolynomials with respect to the differential operator $p = \dfrac{d}{dt}$.

Other canonical forms and their applications to the classification of systems with infinite number of delays are considered in [7, 8]. The reduction of the system with delays to the corresponding canonical form allows to simplify the investigation of pointwise and complete controllability, modal control by dynamical and discrete regulators [13] and so on. Below we give some applications of canonical forms to such problems of qualitative control theory as modal control, controllability, stabilization and reconstruction.

Modal control.

Let the system (1) be connected with a linear difference regulator of the form

$$u(t) = \sum_{j=0}^{k} Q_j x(t - s_j), \quad Q_j \in R^{r \times n}, \tag{9}$$

where $j = 0, 1, \ldots, k$; $0 = s_0 < s_1 \cdots < s_k$, $k \in N$.

The system (1) is said to be modally controllable by the regulator (9) if and only if for any real numbers r_{ij}, $\beta_j, \ldots$, there exists a regulator of the form (9) such that

$$\det \left[\lambda I_n - A\big(\exp(-\lambda)\big) - B\big(\exp(-\lambda)\big) \sum_{j=0}^{k} Q_j \exp(-\lambda s_j) \right] \equiv$$

$$\equiv \lambda^n + \sum_{i=0}^{n-1} \sum_{j=0}^{k} r_{ij} \lambda^i \exp(-\lambda \beta_j)$$

for all complex numbers $\lambda \in C$.

Using the canonical form (8) we obtain [13]:

Proposition 1. The system (1), (6) is modally controllable by the regulator (9) if and only if the conditions (7) hold.

The regulator (9) is suitable for its technical realization since it requires the measurement of the output signal at the present moment and at a finite number of previous moments. However, the condition (7) of the existance of the regulator is very restrictive. To weaken this condition, linear integral regulators are used in [13].

Below we propose another way to weaken the requirements concerning regulators. Namely, we introduce control delays [14]. Let $h_j = j\gamma$, $j = 0, 1, \ldots, l$, for some $\gamma > 0$ and

$$\sum_{i=0}^{M} a_i u(t - i\gamma) = \sum_{j=0}^{N} q_j' x(t - j\gamma), \tag{10}$$

where $a_i \in R$, $q_j \in R^n$; $i = 0, 1, \ldots, M$; $j = 0, 1, \ldots, N$ and ($'$) denotes transposition.

Definition 1. For a given polynomial s, $s(m) = \sum_{j=0}^{M} a_j m^j$, the system (1), (6) with $h_j = j\gamma, \ldots$, is said to be s-modally controllable by the regulator (10) if for any real numbers $r_{ij}, \ldots$, there exists a regulator of the form (10) such that

$$\begin{aligned} &\det\Big[s\big(\exp(-\lambda\gamma)\big)\Big(\lambda I_n - \sum_{j=0}^{l} A_j \exp(\lambda j\gamma)\Big) \\ &\quad - \sum_{j=0}^{l} b_j \exp(-\lambda j\gamma) \sum_{j=0}^{N} q_j' \exp(-\lambda j\gamma)\Big] \\ &\equiv \Big(s\big(\exp(-\lambda\gamma)\big)\Big)^n \Big(\lambda^n + \sum_{i=1}^{n}\sum_{j=0}^{\theta} r_{ij}\lambda^{n-i}\exp(-\lambda j\gamma)\Big), \qquad \forall\lambda \in C. \end{aligned}$$

The system is strongly modally controllable if it is s-modally controllable for every polynomial s and weakly modally controllable if it is s-modally controllable for at least one polynomial s.

Using the canonical form (8) we obtain [14]:

Proposition 2. The system (1), (6) with $h_j = j\gamma, \ldots$, is s-modally controllable by the regulator (10) if and only if the quotient $s\big(\exp(-\lambda\gamma)\big) : \det W\big(\exp(-\lambda\gamma)\big)$ is an entire function with respect to λ.

The system is strongly modally controllable if and only if the condition (7) holds.

The system is weakly modally controllable if and only if the condition

$$\det w(m) \not\equiv 0, \qquad m > 0, \tag{11}$$

is valid.

It is known [17] that the condition (11) gives a criterion of pointwise controllability. Then the system (1), (6) with $h_j = j\gamma, \ldots$, is weakly modally controllable by the regulator (10) if and only if it is pointwise controllable. This statement is a generalization of well-known result of Wonham [18].

A detailed consideration of the problem of modal control in case of incomplete information can be found in [13, 14].

Controllability.

One of the most difficult problems of controllability for time-delay systems is the problem of complete controllability to zero function [19].

The system (1) is said to be completely controllable if for any admissible initial data there exists a number t_1 and a piecewise continuous control $u(\cdot)$ for which the corresponding solution of the system possesses the property $x(t) = 0$, $u(t) = 0$ for $t \geq t_1$.

Proposition 3. If the condition (7) holds then the system (1), (6) is completely controllable.

Stabilization.

For the simplest system with delay

$$\dot{x}(t) = A_0 x(t) + A_1 c(t-\gamma) + bu(t), \tag{12}$$

where $b \in R^n$, the problem of stabilization is stated as follows: find a regulator of the form (10) such that the closed-loop system

$$\sum_{i=0}^{M} a_i \dot{x}(t-i\gamma) = \sum_{i=0}^{M} a_i \big(A_0 x(t-i\gamma) + A_1 x(t-i\gamma-h)\big) + \sum_{j=0}^{N} bq_j' x(t-j\gamma) \tag{13}$$

is asymptotically stable.

Proposition 4. If all roots of the equation

$$\det \left[b,\ (A_0 + \lambda a_1)b,\ \ldots,\ (A_0 + \lambda A_1)^{n-1} b \right] = 0$$

are located outside of the disc $|\lambda| < 1$ then the system (12) is stabilizable by the regulator (13).

Reconstruction of systems with infinite number of delays by difference regulators.

Let us consider a class $\widetilde{\mathcal{F}}$ of systems of the form

$$\dot{x}(t) = \sum_{j=0}^{+\infty} \big(A_j x(t-h_j) + B_j u(t-h_j)\big), \qquad t > 0, \tag{14}$$

with initial conditions

$$x(\tau) \equiv 0, \qquad u(\tau) \equiv 0 \quad \text{for} \quad \tau < \mu$$

where μ is a non-negative number; $0 = h_0 < h_1 < \cdots$

Introduce the class $\mathcal{R}_3$ of the following transformations

$$
\begin{aligned}
x(t) &= \sum_{j=0}^{+\infty} D_j y(t-\gamma_j) = D\big(\exp(-p)\big)y(t), \quad t>0, \\
y(t) &= D^{-1}\big(\exp(-p)\big)x(t) = \sum_{j=0}^{+\infty} C_j x(t-\beta_j), \quad t>0,
\end{aligned}
\tag{15}
$$

$D(m)D^{-1}(m) = I_n$ for almost all $m \geq 0$; $\det D(0) \neq 0)$; $p = \dfrac{d}{dt}$, $\exp(-p)$ is a delay operator; $0 = \gamma_0 < \gamma_1 < \cdots$, $0 = \beta_0 < \beta_1 < \cdots$.

Consider the regulator

$$
u(t) = \sum_{j=0}^{+\infty} Q_j x(t-\delta_j) \tag{16}
$$

Definition 2. The system (14) is said to be reconstructed to the system

$$
\dot{y}(t) = \sum_{j=0}^{+\infty} \widetilde{A}_j y(t-\eta_j), \qquad t>0, \tag{17}
$$

if there exists a regulator (16) and a transformation $D(\cdot)$ of class $\mathcal{R}_3$ (see (15)) such that $y(t) = D^{-1}\big(\exp(-p)\big)x(t)$ where

$$
\dot{x}(t) = \sum_{j=0}^{+\infty}\Big(A_j x(t-h_j) + B_j \sum_{i=0}^{+\infty} Q_i x(t-h_j-\delta_i)\Big).
$$

By analogy with the above we can identify the system with a couple $\big(A(\cdot),\, B(\cdot)\big)$ of formal matrix series:

$$
A(m) = \sum_{j=0}^{+\infty} A_j m^{h_j}, \qquad B(m) = \sum_{j=0}^{+\infty} B_j m^{h_j}.
$$

Further, for simplicity, we assume that there exists a number m_* such that the numerical series $\sum_{j=0}^{+\infty} \|A_j\|\,(m_*)^{h_j}$ and $\sum_{j=0}^{+\infty} \|B_j\|\,(m_*)^{h_j}$ are convergent. Then for all $k = 1, 2, \ldots$ and $g \in R^n$ the matrix series $\sum_{j=0}^{+\infty}(A_j m^{h_j})g$ converges uniformly in the interval $[0,\, m_*]$. It permits to introduce numbers $n_j(m)$, $j = 1, \ldots, l_1(m)$, in the usual way (see section: Canonical forms).

Let $K_1(m), \ldots, K_{\nu(m)}(m)$ denote the degrees of invariant non-trivial polynomials of the matrix $\widetilde{A}(m)$ from (17).

Proposition 5. If we assume that $n_i(m) \equiv n_i = \text{const}$, $K_j(m) \equiv k_j = \text{const}$; $i = 1, \ldots, l_1(m) \equiv l_1$; $j = 1, \ldots, \nu(m) \equiv \nu$; $m \geq 0$ $(n_1 \geq n_2 \geq \cdots \geq n_{l_1},$

$K_1 \geq K_2 \geq \cdots \geq K_\nu$) then the system (14) is reconstructed to (17) if and only if the conditions

$$\sum_{i=1}^{j} K_i \geq \sum_{i=1}^{j} n_i, \qquad j = 1, 2, \ldots, \min\{l_1, \nu\},$$

hold.

The results of this section are due to Yanovich [15]. In his paper also the case of finite number of delays has been extensively studied.

Other canonical forms and their applications.

The problem of interest is to find conditions under which the system (12) with $b = 0$ can be reduced to a diagonal form by using the transformations of the class $\mathcal{R}_1$. We have

Proposition 6. A necessary and sufficient condition for solvability of the above problem is that

$$\operatorname{rank} \mathcal{R}_n\big(a_{i0} I_n - A_0,\, a_{i1} I_n - A_1\big) < n^2 \tag{18}$$

for $a_{ij} \in U_j$; $j = 0, 1$; $i = 1, 2, \ldots, n$; where U_j is the set of eigenvalues of the matrix A_j for every $j = 0$, 1 and

$$\mathcal{R}_n(D, Q) = \begin{bmatrix} D & 0 & \ldots & 0 & 0 \\ Q & D & \ldots & 0 & 0 \\ & & & & \\ 0 & 0 & \ldots & Q & D \\ 0 & 0 & \ldots & 0 & Q \end{bmatrix} \quad \text{is } n(n+1) \times n^2\text{–matrix.}$$

This result can be applied to the problem of pointwise degeneracy, namely, if the system (12) with $b = 0$ is pointwise degenerated then the condition (18) holds for some $a_{i0} \in U_0$ and $a_{i1} = 0$.

Note that a time-delay system is pointwise degenerated if for some time moments its solutions do not fill out the whole space R^n.

The results of this section have been obtained by V. L. Merezsa.

Conclusion.

In this paper we have presented some methods of construction of canonical representations for linear stationary systems with retarded argument as well as examples of their applications to study some problems of qualitative control theory for such systems. It should be pointed out that the universality problem is solved for the $\mathcal{R}_2$ transformations class only. However the class $\mathcal{R}_1$ is more important in theory and applications. It preserves [7] such important qualitative properties of time-delay systems as pointwise and complete controllability, modal control, stabilization. The universality problem for the transformations of class $\mathcal{R}_1$ has not been solved yet. Also the problems of modal control and stabilization by difference multi-input regulators still remain open.

To conclude, we illustrate the obtained results by some examples.

Example 1. Consider the systems

$$\dot{x}(t) = \begin{bmatrix} 0 & 1 & 0 & 0 \\ 1 & 0 & 0 & 0 \\ 0 & 0 & 0 & 1 \\ 0 & 0 & 1 & 0 \end{bmatrix} x(t) + \begin{bmatrix} 0 & 1 & 0 & 0 \\ 0 & 0 & 0 & 0 \\ 1 & 0 & 0 & 0 \\ 0 & 0 & 1 & 0 \end{bmatrix} x(t-h) + \begin{bmatrix} 1 & 0 \\ 0 & 0 \\ 0 & 1 \\ 0 & 0 \end{bmatrix} u(t), \tag{19}$$

$$\dot{y}(t) = \begin{bmatrix} -1 & 1 & 0 & 0 \\ 0 & -1 & 1 & 0 \\ 0 & 0 & -1 & 1 \\ 0 & 0 & 0 & -1 \end{bmatrix} y(t). \tag{20}$$

If we close the system (19) by the regulator

$$\begin{aligned} u(t) = &\begin{bmatrix} 0 & -1 & 0 & 1 \\ -4 & -1 & -4 & -7 \end{bmatrix} x(t) \\ &+ \begin{bmatrix} 0 & -1 & 0 & -1 \\ -1 & 0 & 0 & 6 \end{bmatrix} x(t-h) \\ &+ \sum_{k=2}^{+\infty} \begin{bmatrix} 0 & 0 & 0 & 1 \\ 0 & 0 & 0 & -6 \end{bmatrix} \cdot (-1)^k x(t-kh) \end{aligned}$$

then the closed-loop system has a unique non-trivial polynomial $(\lambda+1)^4$. It is possible to show, that using the transformation

$$x(t) = \begin{bmatrix} -3 & 1 & 1 & 1 \\ 3 & 2 & 1 & 0 \\ -3 & 7 & -4 & 0 \\ 3 & -4 & 0 & 0 \end{bmatrix} y(t) + \begin{bmatrix} 0 & 0 & 0 & 0 \\ 0 & 0 & 0 & 0 \\ 0 & 0 & 0 & 0 \\ 3 & -4 & 0 & 0 \end{bmatrix} y(t-h)$$

the closed-loop system is reduced to the system (20), i.e. the system (19) is reconstructed to the system (20).

Example 2. Consider stabilization problem for the system (12) with matrices

$$A(m) = \begin{bmatrix} 0 & 0 & -1 & 0 \\ 2+m & 0 & 0 & 1 \\ 0 & 0 & 0 & 1 \\ -2 & 0 & 0 & -1 \end{bmatrix}, \qquad b = \begin{bmatrix} 0 \\ 0 \\ 0 \\ 1 \end{bmatrix} \tag{21}$$

with the characteristic equation

$$\lambda^4 + \lambda^3 - 2\lambda = 0.$$

Using $\mathcal{R}_2$-transformations with matrices

$$T(m) = \begin{bmatrix} 0 & -1 & 0 & 0 \\ -(2+m) & 0 & 1 & 0 \\ 0 & 0 & 1 & 0 \\ 0 & 0 & 0 & 1 \end{bmatrix},$$

$$T^{-1}(m) = \frac{1}{2+m}\begin{bmatrix} 0 & 1 & -1 & 0 \\ 2+m & 0 & 0 & 0 \\ 0 & 0 & -(2+m) & 0 \\ 0 & 0 & 0 & -(2+m) \end{bmatrix}$$

we obtain the system

$$\dot{y}(t) = \begin{bmatrix} 0 & 1 & 0 & 0 \\ 0 & 0 & 1 & 0 \\ 0 & 0 & 0 & 1 \\ 0 & 2 & 0 & -1 \end{bmatrix} y(t) + \begin{bmatrix} 0 \\ 0 \\ 0 \\ 1 \end{bmatrix} u(t). \tag{22}$$

If we choose a regulator $u = q'y$ for the system (22) such that the characteristic equation of the closed-loop system has the form $(\lambda + 1)^4 = 0$, then we obtain the corresponding regulator of the type (10) in the form

$$\begin{aligned} 2u(t) + u(t-h) = &- 12x_1(t) - x_2(t) + 13x_3(t) + 6x_4(t) - \\ &- 6x_1(t-h) + 6x_3(t-h) + 3x_4(t-h) \end{aligned}$$

which stabilizes the system (12), (21).

Example 3. Consider an automatic reostat regulator of voltage (see [20]) described by the system (12) with matrices

$$A_0 = \begin{bmatrix} 0 & 1 & 0 & 0 \\ * & 0 & * & 0 \\ 0 & 0 & 0 & 1 \\ * & 0 & * & * \end{bmatrix}, \quad A_1 = \begin{bmatrix} 0 & 0 & 0 & 0 \\ * & 0 & 0 & 0 \\ 0 & 0 & 0 & 0 \\ 0 & 0 & 0 & 0 \end{bmatrix}, \quad b = \begin{bmatrix} 0 \\ 1 \\ 0 \\ 0 \end{bmatrix} \tag{23}$$

(the symbol $*$ denotes non-zero elements).

Using the transformations of the class $\mathcal{R}_1$, the system (12), (23) can be transformed to the canonical form (8). Therefore this system is modally and completely controllable.

References

1. Brunovsky P. A classification of linear controllable systems. *Kybernetika*, 6 (1970), p. 173–188.
2. Luenberger D. G. Canonical forms for linear multivariable systems. *IEEE Trans. Automat. Contr.*, 12 (1967), p. 290–293.
3. Popov V. M. Invariant description of linear time-invariant controllable systems. *SIAM J. Contr.*, 10 (1972), p. 252–264.
4. Shimanov S. N. On the theory of linear differential equations with delays. *Differentsial'nye Uravneniya*, 1 (1965), p. 102–116. (in Russian).
5. Hale J. Theory of functional–differential equations. *Springer–Verlag*, New York, 1977.
6. Kappel F. Degenerate difference–differential equations. Algebraic theory. *J. Differential Eqns.*, 24 (1977), p. 99–126.

7. Kirillova F. M., Marchenko V. M. Functional transformations and canonical forms in linear systems with retarded argument. Prepr. No 7 (39). *Institute of Mathematics* AN BSSR, Minsk, 1978. (in Russian).
8. Kirillova F. M., Marchenko V. M. On classification and construction of canonical forms for multiply connected time-delay systems. In B. N. Petrov and M. V. Meyerov (Eds). Issledovaniya po teorii mnogosvyaznykh sistem. *Nauka*, Moskva, 1982, p. 22–30. (in Russian).
9. Marchenko V. M. Transformations of systems with retarded argument. *Differentsial'nye Uravneniya*, 12 (1977), p. 1882–1884. (in Russian).
10. Marchenko V. M. On the theory of canonical forms for control systems with delay. *Mat. Sbornik* 105 (147), (1978), p. 403–412. (in Russian).
11. Marchenko V. M. Some problems of qualitative control theory for linear stationary systems with aftereffect. In Cz. Olech (Ed.) Mathematical Control Theory. *Banach Center Publ.*, 14, PWN, Warsaw, 1985, p. 361–381.
12. Lee E. B., Neftci S., Olbrot A. Canonical forms for linear multivariable systems. *IEEE Trans. Automat. Contr.*, 27 (1982), p. 128–132.
13. Marchenko V. M., Asmykovich I. K. On the problem of modal control in linear systems with delay. In S. G. Tzafestas (Ed.) *Simulation of Distributed-Parameter and Large-Scale Systems.* North-Holland Publ. Company, Amsterdam, 1980, p. 73–78.
14. Marchenko V. M. Modal control for linear time-delay retarded-feedback objects. In Y. Sunahara, S. G. Tzafestas, T. Futagami (Eds.) Modelling and Simulation of Distributed-Parameter Systems. *Proc. of the IMACS/IFAC Intern. Symposium*, Hiroshima, Japan (October 6th–9th, 1987), p. 685–690.
15. Yanovich V. I. A reconstruction of systems with delays. *Problems of optimal control*, Minsk, 1981, p. 189–198 (in Russian).
16. Asmykovich I. K. On decoupling of multiinput systems with delays. *Problems of optimal control*, Minsk, 1981, p. 198–207 (in Russian).
17. Marchenko V. M. Pointwise controllability and observability of linear systems with neutral type retarded argument. *Dokl. AN BSSR*, 3 (1986), p. 208–211 (in Russian).
18. Wonham W. M. On pole-assignment in multi-input controllable systems. *IEEE Trans. Automat. Contr.*, 12 (1967), p. 660–665.
19. Krasovsky N. N. Theory of Control of Movement. *Nauka*, Moskva, 1968 (in Russian).
20. Vikker D. A. *Elektrichestvo*, 9 (1934), p. 26–30 (in Russian).

Vladimir M. Marchenko
220630, Minsk, USSR,
ul. Sverdlova, 13[a],
BTI im S. M. Kirova
(Dept. of Math.).

STATIC AND DYNAMIC FEEDBACK LINEARIZATION OF NONLINEAR SYSTEMS

R.Marino

Abstract

The necessary and sufficient conditions for linearization via state space change of coordinates and for (static) feedback linearization via nonsingular state feedback and change of coordinates are recalled from[15],[23],[9],[13] and discussed. For those nonlinear systems which are not feedback linearizable, the determination of the largest feedback linearizable subsystem is discussed and the main results from[16] and[18]are recalled. Finally the problem of dynamic feedback linearization via nonsingular dynamic state feedback and extended state space change of coordinates is posed and recent results from[3],[4],[5] are stated and discussed.

1. Introduction

We consider the problem of transforming a nonlinear continuous-time control system ($z \in R^n$, $u \in R^m$)

$$\dot{z} = f(z) + \sum_{i=1}^{m} u_i(t) g_i(z) = f(z) + G(z)u(t) \tag{1.1}$$

with $f(0)=0$ and rank $G(0)=m$ into a linear and controllable one ($x \in R^N$, $v \in R^M$)

$$\dot{x} = Ax + \sum_{i=1}^{m} b_i v_i(t) = Ax + Bv \tag{1.2}$$

The state z is assumed to be available for measurments. We are interested in transformations which preserve the stability properties, so that the stability theory for linear control systems can be used to design state feedback controls for those systems (1.1) which are transformable into (1.2).

State space diffeomorphisms

$$x = \phi(z), \quad \phi(0) = 0, \quad x \in R^n \tag{1.3}$$

were the first transformations to be studied in Krener[15]. State feedback transformations

$$u = \alpha(z) + \beta v, \quad v \in R^m \tag{1.4}$$

with $\alpha(0)=0$ and β nonsingular constant matrix were introduced and studied by Brockett[1]. They were later generalized by Jakubczyk-Respondek[13] and Hunt-Su-Meyer[9] where the feedback transformation

$$u = \alpha(z) + \beta(z), \quad v \in R^m \tag{1.5}$$

allows for the nonsingular matrix β to depend on the state as well. The study of the above transformations led to a successful characterization of those systems (1.1) which can be transformed into (1.2) by (1.3), which will be called linearizable, and of those systems transformable into (1.2) by (1.3) and (1.5), which will be called (static) feedback linearizable. The transformations (1.3) and (1.5) with $\beta(z)$ singular of constant rank have not yet been studied.

Adaptive feedback linearization has been studied in[24] and [14] for systems with unknown constant parameters $(p_1,\ldots,p_q)=p$ appearing linearly

$$\dot{z} = f_0(z) + \sum_{j=1}^{p} f_j(z)p_j + G_0(z)u(t) + \sum_{j=1}^{p} G_j(z)p_j u(t) \tag{1.6}$$

The problem has been solved under the assumption of matching conditions in [24]

$$f_j(z) \in \mathrm{Im}\ G_0(z),\ 1 \leqslant j \leqslant q, \quad G_j(z) \in \mathrm{Im}\ G_0(z),\ 1 \leqslant j \leqslant q \tag{1.7}$$

and under the assumption of extended matching conditions in [14]

$$\begin{aligned} &f_j(z) \in \mathrm{Im}[G_0(z),\ ad_{f_0} G_0(z)], \quad 1 \leqslant j \leqslant q \\ &G_j(z) \in \mathrm{Im}\ G_0(z), \quad 1 \leqslant j \leqslant q \end{aligned} \tag{1.8}$$

where $ad_{f_0} G_0 = [ad_{f_0} g_{01}, \ldots, ad_{f_0} g_{0m}]$. In[24] the problem of robustness of feedback linearization versus unmodeled dynamics is also analysed using singular perturbation techniques.

For those systems which are not feedback linearizable the problem of partial feedback linearization naturally arises, namely the transformation of (1.1) by (1.3) and (1.5) into a partially linear and controllable system

$$\begin{aligned} &\dot{x}^{(1)} = Ax^{(1)} + Bv \\ &\dot{x}^{(2)} = a(x^{(1)}, x^{(2)}) + b(x^{(1)}, x^{(2)})v \end{aligned} \tag{1.9}$$

The problem was solved in the single-input case in[16] and in the multi-input case in[18] (see also[17], [21]). One can compute the dimension of the largest feedback linearizable subsystem and construct the corresponding transformation for any system (1.1).

Partial feedback linearization is related to input-output decoupling. Given m outputs

$$y_j = h_j(z), \quad 1 \leq j \leq m, \tag{1.10}$$

the input-output decoupling problem is to determine a transformation (1.3)-(1.5) which takes the system (1.1)-(1.10) into

$$\begin{aligned} \dot{x}^{(1)} &= A\, x^{(1)} + Bv \\ \dot{x}^{(2)} &= a(x^{(1)}, x^{(2)}) + b(x^{(1)}, x^{(2)})v \\ y &= C\, x^{(1)} \end{aligned} \tag{1.11}$$

with (A,B,C) in prime canonical form. Necessary and sufficient condition have been determined for this problem (see Isidori[10]). For linear systems it is known that those conditions can be weakened if one allows for a dynamic compensator (Morse-Wonham [19]). This motivated the introduction of a nonlinear dynamic state feedback transformation by Singh[22]

$$\begin{aligned} \dot{w} &= \gamma(z,w) + \delta(z,w)v \qquad w \in R^q \\ u &= \alpha(z,w) + \beta(z,w)v \qquad v \in R^m \end{aligned} \tag{1.12}$$

with $\gamma(0,0)=0$, $\alpha(0,0)=0$, which is a generalization of the stateic state feedback (1.5). Necessary and sufficient conditions for input-output decoupling via dynamic feedback transformations (1.12) and extended state space diffeomorphisms

$$x = \phi(z,w), \quad \phi(0,0)=0, \quad x \in R^{n+q} \tag{1.13}$$

were obtained in Descusse-Moog ([6],[7]) and Nijmeijer-Respondek[20]. Isidori[11] and Isidori-De Luca-Moog [12] give sufficient conditions in order to achieve both input-output decoupling and full linearization.

Charlet-Levine-Marino address in[3] and [5] the problem of transforming the system (1.1) into (1.2) with $N=n+q$, $M=m$, via transformations (1.12) and (1.13), which is called the dynamic feedback linearization problem. It is shown in [3]that single-input systems (1.1) which are dynamic feedback linearizable are also static feedback linearizable. Special classes of dynamic feedback linearizable systems (1.1) with $m>1$ are given in [5], showing that dynamic feedback linearization is a multi-input phenomenon. A different approach is taken in [8]by Fliess: transformations (1.12) are considered but the closed loop system is no longer required to be linear and controllable in some new coordinates of the extended state space (z,w) ; the only requirement is the existence of a w-dependent state space change of coordinates

$$x = \Phi(z,w), \qquad x \in R^n \tag{1.14}$$

in which the system is linear and controllable from the new input v. While this is a less restrictive notion of dynamic feedback linearization, the stability properties of the dynamic compensator(1.12) remain to be analysed. This analysis, which could be rather difficult, is not needed with the notion of dynamic feedback linearization introduced in[3] .

Dynamic feedback linearization by a special class of dynamic compensators (1.12) and extended state space diffeomorphisms (1.13) is studied in[5]; compensators (1.12) are restricted to be of the following form:

$$\begin{aligned}
\dot{w}^j_i &= w^j_{i+1} && 1 \leqslant i \leqslant \mu_j - 1 \\
\dot{w}^j_{\mu_j} &= \alpha_j(z,w) + \sum_{l=1}^{m} \beta_{jl}(z,w) v_l(t), && \mu_j \geqslant 1 \\
u_j &= w^j_1 && \mu_j \geqslant 1 \\
u_j &= \alpha_j(z,w) + \sum_{l=1}^{m} \beta_{jl}(z,w) v_l(t), && \mu_j = 0
\end{aligned} \tag{1.15}$$

with $\mu_i \geqslant 0$, $1 < i < m$, $\mu = \mu_1 + \ldots + \mu_m$, $\alpha_j(0,0) = 0$, $1 \leqslant j \leqslant m$, $\beta(z,w)$ non-singular in a neighborhood of the origin in $R^{n+\mu}$. Sufficient conditions for dynamic feedback linearization via the above special class of compensators are given in[5]. They require the involutivity of certain distribution defined on the original state space which are computed on the basis of the vector fields f, $g_1, \ldots, g_m$ and on a set of integers $\mu_1, \ldots, \mu_m$ which characterize the linearizing compensator (1.15). When $\mu_1 = \ldots = \mu_m = 0$, the conditions become the necessary and sufficient conditions for static feedback linearization. The conditions are not necessary, but the serve as a guide in the construction of the linearizing compensator (1.15).

The paper is organized as follows. In Section 2 the necessary and sufficient conditions for linearization and for feedback linearization are discussed both in the single-input case and in the multi-input case; the problem of partial feedback linearization and the determination of the largest feedback linearizable subsystem are also addressed. Section 3 is devoted to dynamic feedback linearization and to the illustration of the sufficient conditions given in[5] .

2. Static feedback linearization

Consider the nonlinear system (1.1) with $f, g_1, \ldots, g_m$ smooth vector fields in a neighborhood of the origin.

Definition 2.1. System (1.1) is said to be locally linearizable if there exists a local diffeomorphism (1.3) such that it is expressed in new coordinates x as a linear and controllable system (1.2) with N=n, M=m, that is

$$f(z) = (d\phi/dz)^{-1}A\phi(z), \quad G(z) = (d\phi/dz)^{-1}B \tag{2.1}$$

The transformation $(f,G)\to((d\phi/dz)f\phi^{-1},(d\phi/dz)G\phi^{-1})$ is the nonlinear generalization of the linear transformation $(A,B)\to(TAT^{-1},TB)$, where x=Tz is the linear change of coordinates.

Definition 2.2. System (1.1) is said to be locally state feedback linearizable if there exists a nonsingular state feedback (1.5), with α_i, $1\leq i\leq m$, β_{ji}, $1\leq i,j\leq m$, smooth functions in a neighborhood of the origin, such that the closed loop system

$$\dot z = f(z) + G(z)\alpha(z) + G(z)\beta(z)v = f(z) + g(z) + G(z)v =$$

$$= f(z) + G(z)v = \tilde f(z) + \sum_{i=1}^{m} \tilde g_i(z)\, v_i \tag{2.2}$$

is locally linearizable.

The transformation $(f,G)\to((d\phi/dz)(f+G\alpha)\phi^{-1},(d\phi/dz)G\beta\phi^{-1})$ is called feedback transformation. Two systems related by a feedback transformation are said to be feedback equivalent. The above tranformation is the nonlinear generalization of the linear transformation $(A,b)\to(T(A+B\alpha)T^{-1},TB\beta)$ studied in [2] (see also [25]), which is made of a linear change of coordinates x=Tz and of a linear feedback $u = \alpha z + \beta v$, with β nonsingular. It is shown in [2] that every controllable linear system can be transformed via a linear feedback transformation into its Brunovski canonical form. No canonical form is known for a linear controllable multi-input system (A,B) under the action of a linear change of coordinates. For a single-input controllable system (A,b) on the other hand it is known that there exists a canonical form (the controller canonical form [25]) under the action of a linear change of coordinates.

Definitions 2.1 and 2.2 pose the problems of identifying those nonlinear systems (1.1) which are linearizable and which are feedback linearizable. We now state and discuss the Theorems which solve those problems.

Theorem 2.1 [13]. System (1.1) is locally feedback linearizable if, and only if,

(i) rank span $\{g_j,\ldots, ad_f^{n-1}g_j : 1\leq j\leq m\} = n$ in U_0;

(ii) span$\{g_j,\ldots,ad_f^i g_j : 1\leq j\leq m\}$ is involutive and of constant rank m_i in U_0 for $i=0,\ldots,n-1$.

A set of m indices $\{k_1,\ldots,k_m\}$, called controllability or Kronecker indices, is uniquely associated with a feedback linearizable system (1.1) as follows:

$$k_j = \text{card}\{ r_i \geq j : r_0 = m_0,\ r_i = m_i - m_{i-1},\ i \geq 0 \} \tag{2.3}$$

Since $m_{n-1} = n$ by assumption (i) it follows that $k_1 + \ldots + k_m = n$. Since by definition $m_0 = m$, (2.3) defines m controllability indices, which are invariant under feedback transformations.

There is some redundancy in the statement of Theorem 2.1 which can be stated as follows.

Theorem 2.2[9]. System (1.1) is locally feedback linearizable if, and only if,

(i) rank span$\{ g_j, \ldots, ad_f^{n-1} g_j : 1 \leq j \leq m \} = n$ in U_0;

(ii) rank span$\{ g_j, \ldots, ad_f^{i} g_j : 1 \leq j \leq m \} = m_i$ in U_0 for every $i = 1, \ldots, n-1$;

(iii) span$\{ g_j, \ldots, ad_f^{k_i - 1} g_j : 1 \leq j \leq m \}$ is involutive for every $i = 1, \ldots, m$, with k_i defined by (2.3).

The linearization problem is solved by the following stronger conditions.

Theorem 2.3([15],[23]). System (1.1) is locally linearizable if, and only if,

(i) rank span$\{ g_j, \ldots, ad_f^{n-1} g_j : 1 \leq j \leq m \} = n$ in U_0;

(ii) rank span$\{ g_j, \ldots, ad_f^{i} g_j : 1 \leq j \leq m \} = m_i$ in U_0 for every $i = \ , \ldots, n-1$;

(iii) $[ad_f^s g_i , ad_f^t g_j] = 0$, $0 \leq s+t \leq 2n-1$; $1 \leq i, j \leq m$.

In the single-input case (m=1) the formulation of Theorem 2.2 is further simplified since the only controllability index is $k_1 = n$.

Theorem 2.4[9]. System (1.1) with m=1 is locally feedback linearizable if, and only if,

(i) rank span$\{ g, \ldots, ad_f^{n-1} g \} = n$ in U_0;

(ii) span$\{ g, \ldots, ad_f^{n-2} g \}$ is involutive.

Since the distribution generated by a single vector field is always involutive, condition (ii) is always satisfied when n=2. From Theorems 2.2 and 2.4 we have the following.

Corollary 2.5. If system (1.1) is locally feedback linarizable then its linear approximation about the origin

$$\dot{z} = df/dz\big|_{z=0}\, z + G(0)u = Fz + Gu \tag{2.4}$$

is controllable, i.e. rank$[G, FG, \ldots, F^{n-1}G] = n$.

Corollary 2.6. System (1.1) with n=2, m=1 is locally feedback linearizable if, and only if, its linear approximation about the origin (2.4) is controllable.

Definition 2.3. System (1.1) is said to be locally partially feedback linearizable with controllability indices $k_1,..,k_p$, $p\leq m$, if there exists a state feedback transformation (1.5) and local coordinates (1.3) in which the closed loop system (2.2) is expressed as (1.9) with (A,B) controllable with controllability indices $k_1,...,k_p$, $q=k_1+...+k_p$ and $x^{(1)} = (x_1,....,x_q)$, $x^{(2)} = (x_{q+1},....,x_n)$.

If system (1.1) is not feedback linearizable one may ask what is the largest feedback linearizable subsystem, i.e. the largest integer q. To this purpose define the distributions

$$Q^j = \operatorname{span}\{\bar{G}^{j-1},\ ad_f^j G^0\} \quad j\geq 0 \tag{2.5}$$

where $\bar{D}$ denotes the involutive closure of the distribution D and

$$G^j = \operatorname{span}\{g_i,...,ad_f^j g_i :\ 1\leq i\leq m\} \tag{2.6}$$

Under the hypothesis of constant rank in U_0 for $\bar{G}^j$ and Q^j, compute the codimension of $\bar{G}^j$ in Q^{i+1} as follows

$$r_0^* = \operatorname{rank} G^0 = m, \qquad r_i^* = \operatorname{rank} Q^i - \operatorname{rank} \bar{G}^{i-1} \tag{2.7}$$

It is shown in [18] that the sequence of integers r_i^*, $i\geq 0$, is a nonincreasing one so that one can define a set of m integers

$$k_j^* = \operatorname{card}\{ r_i^* \geq j\ :\ i\geq 0\}. \tag{2.8}$$

They are invariant under feedback transformation. The integer $q^* = k_1^*+....+k_m^*$ gives the dimension of the largest feedback linearizable subsystem in the following sense.

Theorem 2.7 [18]. System (1.1) is locally partially feedback linearizable with controllability indices $(k_1^*,...,k_m^*)$. If system (1.1) is locally partially feedback linearizable with controllability indices $(k_1,...,k_p)$, $p\leq m$, then for every i, $1\leq i\leq p$, $k_i\leq k_i^*$.

Theorem 2.7 is a generalization of Theorem 2.2, which can be restated as follows: system (1.1) is locally feedback linearizable if, and only if, $k_1^*+...+k_m^*=n$.

If the distribution G^0 is involutive one has a special decomposition.

Corollary 2.8 [18]. System (1.1) is transformable by a nonsingular feedback transformation into (1.9), with $b(x^{(1)},x^{(2)}) = 0$ and (A,B) a controllable system with controllability indices $k_1^*,\dots,k_m^*$ if, and only if, G^0 is an involutive distribution.

Note that the involutivity assumption of Corollary 2.8 is always met in the single-input case.

In the single-input case (m=1), Theorem 2.7 can be stated in a much simpler way. The index k^* defined in (2.8) is given in this case (following [16]) as the smallest integer such that in U_0

$$ad_f^{k^*-1} g \notin \bar{G}^{k^*-2} , \qquad ad_f^{k^*} g \in \bar{G}^{k^*-1} \tag{2.9}$$

Theorem 2.9[16]. System (1.1) with m=1 is locally partially feedback linearizable with controllability index k^* defined by (2.9). If system (1.1) with m=1 is locally partially feedback linearizable with index k, then $k \leq k^*$.

3. Dynamic feedback linearization

In this section we address the problem of determining a dynamic state feedback transformation (1.12) and an extended state space diffeomorphism (1.13) such that the closed-loop system

$$\begin{aligned} \dot{z} &= f(z) + G(z)\alpha(z,w) + G(z)\beta(z,w)v \\ \dot{w} &= \gamma(z,w) + \delta(z,w)v \end{aligned} \tag{3.1}$$

is transformed into a linear and controllable one (1.2) with N=n+q, M=m. If we view the inputs u to system (1.1)

$$u_j = \alpha_j(z,w) + \sum_{i=1}^{m} \beta_{ji}(z,w)v_i , \quad 1\leq j\leq m \tag{3.2}$$

as outputs of the closed loop system (3.1), we can define m characteristic indices $(\nu_1,\dots,\nu_m)$ as

$$\begin{aligned} \nu_j &= 0 \quad \text{if} \quad \beta_{ji}(z,w) \neq 0 \text{ for some } i,\ 1\leq i\leq m \\ \nu_j &= \min\{ r: L_{g'_i} L_{f'}^{r-1} \alpha (z,w) \neq 0 \text{ for some } i,\ 1\leq i\leq m\} \end{aligned} \tag{3.3}$$

where

$$f'(z,w) = \begin{pmatrix} f(z) + G(z)\alpha(z,w) \\ \gamma(z,w) \end{pmatrix} \qquad G'(z,w) = \begin{pmatrix} G(z)\beta(z,w) \\ \delta(z,w) \end{pmatrix}$$

When all ν_j are finite, the mxm matrix

$$D(z,w) = (\sigma_{ji}(z,w)) \tag{3.4}$$

with $\sigma_{ji}(z,w) = \beta_{ji}(z,w)$ if $\nu_j=0$ and $\sigma_{ji}(z,w) = L_{g'_i} L_{f'}^{\nu_j-1} \alpha_j(z,w)$

if $\nu_j > 0$, is called the decoupling matrix of the compensator (1.12) for sytem (1.1).

Definition 3.1. A dynamic compensator (1.12) is said to be regular for system (1.1) if the corresponding decoupling matrix $D(z,w)$ is nonsingular in a neighborhood of the origin of the extended state space (z,w).

The above definition is a generalization of the notion of static nonsingular compensator (1.5).

Definition 3.2. System (1.1) is said to be locally dynamic feedback linearizable if there exists a regular dynamic state feedback compensator (1.12)and an extended state space diffeomorphism (1.13) which transform system (1.1) into system (1.2) with $N=n+q$ and $M=m$.

We first recall a negative result for single-input systems.

Theorem 3.1[3] .The following statements are equivalent.

(i) System (1.1) with $m=1$ is locally(static) feedback linearizable.

(ii) System (1.1) with $m=1$ is locally dynamic feedback linearizable.

A necessary condition easily follows from Definition 3.2.

Theorem 3.2. If system (1.1) is locally dynamic feedback linearizable then its linear approximation about the origin (2.4) is controllable.

Consider the example

$$\dot{x}_1 = x_2 + x_3^2 , \quad \dot{x}_2 = x_3 , \quad \dot{x}_3 = u \tag{3.5}$$

Its linear approximation about the origin

$$\dot{x}_1 = x_2 , \quad \dot{x}_2 = x_3 , \quad \dot{x}_3 = u \tag{3.6}$$

is controllable. On the other hand (3.5) is not feedback linearizable since condition (ii) of Theorem 2.4 fails. According to the above Theorem 3.1, (3.5) is not dynamic feedback linearizable either, which shows that the necessary condition of Theorem 3.2 is far from being sufficient.

Sufficient conditions for the existence of a linearizing compensator of special structure (1.15) can be given in terms of the following distributions defined on the original state space. Given a set of integers $\mu_1, \ldots \mu_m$ define

$$D^0 = \text{span}\{ g_j , \ 1 \leqslant j \leqslant m : \mu_j = 0 \} ;$$

$$D^{i+1} = \text{span}\{ D^i, \ ad_f D^i, \ g_j , \ 1 \leqslant j \leqslant m : \mu_j = i+1\} , \ i > 0.$$

Theorem 3.3 [5]. If for a set of integers $\mu_1, \ldots, \mu_m$, $0 \leqslant \mu_1 \leqslant \ldots \leqslant \mu_m$, $\mu = \mu_1 + \ldots + \mu_m$, the distributions D^i are such that

(i) D^i is involutive and of constant rank in U_0 for $i = 0,\ldots,n+\mu_m-1$;

(ii) rank $D^{n+\mu_m-1} = n$;

(iii) $[g_j, D^i] \subset D^{i+1}$ for all j, $1 \leqslant j \leqslant m$, such that $\mu_j \geqslant 1$ and all i, $0 \leqslant i \leqslant n+\mu_m-1$;

then system (1.1) is locally dynamic feedback linearizable by a dynamic compensator (1.15) with indices $\mu_1,\ldots,\mu_m$ and a local diffeomorphism (1.13) with $x \in R^{n+\mu}$.

Theorems 3.2 and 3.3 allow us to solve completely the problem of dynamic feedback linearization when n=m+1 in (1.1).

Corollary 3.4. Consider system (1.1) with n=m+1. The following statements are equivalent.

(i) System (1.1) is locally dynamic feedback linearizable.

(ii) The linear approximation about the origin (2.4) is controllable.

Corollary 3.4 is a generalization of Corollary 2.6. Theorem 3.1 and Corollary 3.4 show that dynamic feedback linearization is a multi-input phenomenon.

It is shown in [5] that the conditions of Theorem 3.3 (in particular condition (iii)) are not necessary even for dynamic feedback linearization via compensators of special structure (1.15).

If $\mu_i = 0$, $1 \leqslant i \leqslant m$, then $D^j = G^j$ for every $j \geqslant 0$. Conditions (i) of Theorem 3.3 require that the distributions $D^j = G^j$ be involutive and of constant rank for j=0,...,n-1 and therefore conditions (iii) of Theorem 3.3 are always satisfied. It follows that for $\mu_i = 0$, $1 \leqslant i \leqslant m$, the sufficient conditions of Theorems 2.1 and 3.3 coincide.

4. References

[1] R.W.Brockett. Feedback invariants for nonlinear systems. Proc. VII IFAC Congress, Helsinki, p. 1115-1120, 1978.

[2] B.Brunovski. A classification for linear controllable systems. Kybernetica, 6, p. 173-188, 1970.

[3] B.Charlet, J.Levine, R.Marino. Two sufficient conditions for feedback linearization. In Analysis and Optimization of Systems, Lecture Notes in Control and Information Sciences, A.Bensoussan,J.L.Lions Eds, p.181-192,Springer Verlag 1988.

[4] B.Charlet, J.Levine, R.Marino. On dynamic feedback linearization. To appear in Systems and Control Letters.

[5] B.Charlet, J.Levine, R.Marino. New sufficient conditions for dynamic feedback linearization. To appear in Proc. IFAC Symposium Nonlinear Control Systems Design, Capri (Italy) 14-16 June 1989.

[6] J.Descusse, C.H.Moog. Decoupling with dynamic compensation for strong invertible affine nonlinear systems. Inter.

Jour. of Control, 42,6, p. 1387-1398, 1985.

[7] J.Descusse, C.H.Moog. Dynamic decoupling for right invertible nonlinear systems. Systems and Control Letters, 8, 4, p.345-349, 1987.

[8] M.Fliess. Generalisation non lineaire de la forme canonique de commande et linearisation par bouclage. C.R. Acad. Science Paris, 308, I, p. 377-379, 1989.

[9] L.R.Hunt, R.Su, G.Meyer. Design for multi-input nonlinear systems. In Differential Geometric Control Theory, R.Brockett, R.Millmann, H.Sussmann Eds.,p.268-298,Birkhauser, 83.

[10] A.Isidori. Control of nonlinear systems via dynamic state feedback. In Algebraic and Geomtric Methods in Nonlinear Control Theory, M. Hazewinkel, M.Fliess Eds., p. 121-145, Reidel, 1986.

[11] A.Isidori. Nonlinear Control Systems: an Introduction. Lecture Notes in Contro and Information Sciences, 72, Springer, Berlin, 1985.

[12] A.Isidori, C.H.Moog, A. De Luca. A sufficient condition for full linearization via dynamic state feedback. Proc. 25th IEEE CDC, Athens, p.203-207, 1986.

[13] B.Jakubczyk, W.Respondek. On linearization of control systems. Bull. Acad. Pol. Sci . Ser. Sci. Math., 28, 9-10 p. 517-522, 1980.

[14] I.Kanellakopoulos, P.V.Kokotovic, R.Marino. Robustness of adaptive nonlinear control under an extended matching condition. To appear in Proc. IFAC Symposium Nonlinear Control Systems Design, Capri (Italy) 14-16 June 1989.

[15] A.J.Krener. On the equivalence of control systems and the linearization of nonlinear systems. SIAM Jour. of Control, 11,4,p.670-676, 1973.

[16] A.J.Krener, A.Isidori, W.Respondek. Partial and robust linearization by feedback. Proc. 22nd IEEE CDC, p.126-130, 1983.

[17] J.H.Lewis. the Kronecker indices of affine nonlinear control systems. Proc. 24th IEEE CDC, p. 371-374, 1985.

[18] R.Marino. On the largest feedback linearizable subsystem. Systems and Control Letters, 6, p. 345-351, 1986.

[19] A.S.Morse, W.M.Wonham. Decoupling and pole assignment by dynamic compensation. SIAM Jour. Control, 8, 3, p.317-337, 1970.

[20] H.Neijmeijer, W.Respondek. Dynamic input-output decoupling of nonlinear control systems. IEEE Automatic Control, 33, p.1065-1070, 1988.

[21] W.Respondek. Partial linearization, decomposition and fibre linear systems. In Theory and Applications of Nonlinear Control Systems, C.I. Byrnes, A.Lindquist Eds. Elsevier, North-Holland, p. 137-154, 1986.

[22] S.Singh. Decoupling if invertible nonlinear systems with state feedback and precompensation. IEEE Trans. Automatic Control, 25, p. 1237-1239, 1980.

[23]H.J.Sussmann. Lie brackets, real analyticity and geometric control. In Differential Geometric Control Theory, R.W.Brockett, R.Millmann, H.Sussmann Eds., Birkhauser, Boston, Ma. p. 1-116, 1983.

[24]D.G.Taylor, P.V.Kokotovic, R.Marino, I. Kanellakopoulos. Adaptive regulation of nonlinear systems with unmodeled dynamics, IEEE Trans. Automatic Control, 34, p. 405-412, 1989.

[25]W.M.Wonham. Linear Multivariable Control: A Geometric Approach. 2nd Edition, Springer, New York, 1979.

REMARKS ON THE CONTROL OF DISCRETE TIME NONLINEAR SYSTEMS

Henk Nijmeijer
Department of Applied Mathematics
University of Twente, PO Box 217
7500 AE Enschede, The Netherlands

Abstract

In this paper we study the (dynamic) input-output decoupling problem and the problem of right-invertibility for discrete-time nonlinear systems. It is shown that under generic conditions these problems are solvable around an equilibrium point if and only if the same problems are solvable for the linearization of the nonlinear system. The results typically apply to well-known questions in economics and an example stemming from economic modelling is given.

1. Introduction

In this note we study discrete-time nonlinear systems of the form

$$\begin{cases} x(k+1) = f(x(k),\ u(k)) \\ y(k) = h(x(k),\ u(k)) \end{cases} \tag{1.1}$$

where $x = (x_1,\ldots,x_n)^T$ is the n-dimensional state vector, $u = (u_1,\ldots,u_m)^T$ and $y = (y_1,\ldots,y_m)^T$ are the m-dimensional control and output vector respectively. We assume the state x belongs to an open part $\mathcal{X}$ of $\mathbb{R}^n$ and similarly u and y belong to open neighborhoods $\mathcal{U}$ and $\mathcal{Y}$ in $\mathbb{R}^m$. Systems of the form (1.1) naturally appear as the discretization of continuous time nonlinear control systems $\dot{x} = \tilde{f}(x,u)$, $y = \tilde{h}(x,u)$, but also arise for instance in economic and econometric modelling. Usually in an economic context one refers to $u(\cdot)$ as the instrument variables and $y(\cdot)$ as the target variables. In the theory of economic policy making a standard question deals with the ability of guiding a given set of target variables along arbitrary time paths by a suitable selection of the policy instruments at each time instant. This problem of dynamic path controllability has been studied in the economics literature exclusively for linear systems of the form (1.1), see [2,3, 4,6,20,23,24,25] and other references in there.

The linearity assumption is clearly usually not met. To avoid nonlinearities in the model a common approach is that one linearizes the economic model around a fixed equilibrium point and then one applies the linear theory on dynamic path controllability. To what extent this method may be successfully applied, fully depends on the ignored higher order terms in the linearization.

The purpose of this note is firstly to study the question of nonlinear dynamic path controllability or what in engineering often is referred to as right-invertibility or functional reproducibility, see e.g. [5,22,26]. Secondly we investigate to what extent the linearization of the model (1.1) contains essential information on this question for the nonlinear system (1.1) itself.

We study the above problem via the theory on input-output decoupling, in which one requires that each component of the input u influences one and only one output component. Usually such a situation of noninteracting is not met and one may ask for the possibility of adding extra control loops to (1.1) in order to achieve the decoupling requirements. These extra control loops may be either static:

$$u(k) = \alpha(x(k),\ v(k)) \tag{1.2}$$

or dynamic

$$\begin{cases} z(k+1) = \varphi(z(k),\ x(k),\ v(k)) \\ u(k) = \psi(z(k),\ x(k),\ v(k)), \end{cases} \tag{1.3}$$

where $v(\cdot)$ denotes some new reference input. Necessary and sufficient conditions are given which guarantee the local existence of a decoupling controller of the form (1.2) or (1.3). A straightforward observation shows that the forementioned necessary and sufficient conditions exactly yield the dynamic path controllability property. We also show that under fairly general (generic) conditions these conditions on the nonlinar system (1.1) are equivalent to (known) conditions on the linearization of (1.1), thereby demonstrating that the solvability requirements for input-output decoupling of (1.1) or dynamic path controllability of (1.1) are equivalent to the solvability conditions for these problems for the linearization of (1.1). This will be illustrated on an example stemming from economic modelling, cf. [25,19].

Recently several authors have studied decoupling problems for discrete-time nonlinear systems, see [10-13,17] and [18] which forms the starting point of this note. Of separate interest in this respect we also mention the recent contributions on the problem of feedback linearization of discrete-time nonlinear systems, see e.g. [11,15,16] and [14].

We finally note that recently a novel approach to discrete-time nonlinear control problems is proposed in [8] where tools from difference algebra are employed.

2. Local input-output decoupling

Consider the discrete-time nonlinear system (1.1) and assume all data to be analytic. When necessary we denote the components of a vector by using lower indices e.g.

$$x(k) = (x_1(k),\dots,x_n(k))^T,$$

$$h(x,u) = (h_1(x,u),\dots,h_m(x,u))^T.$$

With each component of the output y_i we can associate a characteristic number or relative order ρ_i in the following manner, see also [15]. Given an arbitrary initial state $x^o \in \mathcal{X}$ we can compute for $i \in \underline{m} = \{1,\dots,m\}$ the derivative

$$\frac{\partial h_i}{\partial u}(x^o,u) = \left(\frac{\partial h_i}{\partial u_1}(x^o,u),\dots,\frac{\partial h_i}{\partial u_m}(x^o,u)\right) \tag{2.1}$$

From the analyticity of the system it follows that either the vector in (2.1) is nonzero for all (x^o,u) belonging to an open and dense submanifold O_i of $\mathcal{X} \times \mathcal{U}$, or this vector vanishes for all $(x_o,u) \in \mathcal{X} \times \mathcal{U}$. In the first case we define $\rho_i = 0$, whereas in the latter case we continue by observing that the function $h_i(x^o,u)$ does not depend on u and so we may write $h_i(x^o,u) = h_i^o(x^o)$ for some analytic function h_i^o on $\mathcal{X}$. Next we compute in an analogous fashion $(\partial/\partial u)\, h_i^o(f(x^o,u))$. If this vector is nonzero on an open and dense submanifold O_i in $\mathcal{X} \times \mathcal{U}$ we set $\rho_i = 1$, otherwise we continue with the function $h_i^1(x^o) = h_i^o(f(x^o,u))$. In this way the number ρ_i — if it exists — determines the inherent delay between the inputs and the i-th output. In case none of the iterated functions $h_i^k(f(x,u))$ depends on u we define $\rho_i = \infty$. When $\rho_i = \infty$ the i-th output evolves in time independent from the input sequence applied to the system (1.1). Notice that a finite characteristic number satisfies, cf. [18],

$$\rho_i \le n \tag{2.2}$$

Assuming that each characteristic number ρ_i is finite we introduce a so called decoupling matrix

$$A(x,u) = \begin{bmatrix} \frac{\partial}{\partial u} h_1^{\rho_1 - 1}(f(x,u)) \\ \vdots \\ \frac{\partial}{\partial u} h_m^{\rho_m - 1}(f(x,u)) \end{bmatrix} \tag{2.3}$$

where by definition we set $h_i^{-1}(f(x,u)) = h_i(x,u)$. From the definition of the ρ_i's the rows of the matrix $A(x,u)$ are nonvanishing functions on an open and dense submanifold $O = O_1 \cap O_2 \ldots \cap O_m$ of $\mathcal{X} \times \mathcal{U}$. Note that the matrix given in (2.3) is a generalization of the Falb-Wolovich matrix appearing in linear decoupling theory, cf. [7]. In what follows we will explain the relation between the matrix $A(x,u)$ and the input-output decoupling problem. The system (1.1) is said to be input-output decoupled if — after a possible relabeling of the outputs — the i-th input $u_i(\cdot)$ only influences the i-th output component $y_i(\cdot)$ and none of the other outputs $y_j(\cdot)$, $j \neq i$. In case the system does not possess the above property one may ask whether it is possible to achieve input-output decoupling after adding extra control loops to the system. The simplest case is that where we allow for a regular static state feedback (1.2) where the analytic mapping $\alpha(x,\cdot) : \mathcal{U} \longrightarrow \mathcal{U}$ is a diffeomorphism for each x. In general it is extremely hard to find conditions which guarantee the existence of such a globally defined feedback $\alpha: \mathcal{X} \times \mathcal{U} \longrightarrow \mathcal{U}$. We therefore will concentrate only on local solutions. That is we will derive conditions that guarantee the existence of α on some neighborhood in $\mathcal{X} \times \mathcal{U}$. For simplicity we assume we work around a neighborhood of a fixed point, i.e. let $(x^o, u^o) \in \mathcal{X} \times \mathcal{U}$ be given such that

$$f(x^o, u^o) = x^o \tag{2.4}$$

We will say that the static state feedback decoupling problem for the system (1.1) is locally solvable around (x^o, u^o) if there exists a neighborhood $\tilde{O}$ in $\mathcal{X} \times \mathcal{U}$ of (x_o, u_o) and a regular static state feedback α defined on $\tilde{O}$ which achieves input-output decoupling on $\tilde{O}$. The following theorem solves this problem.

<u>Theorem 2.1</u> Consider the system (1.1) around the equilibrium point (2.4) and assume that the characteristic numbers ρ_i, $i \in \underline{m}$ are finite. The static state feedback input-output decoupling problem is locally solvable if and only if

$$\text{rank } A(x^o, u^o) = m \tag{2.5}$$

The proof of this theorem follows along the same lines as the proof of theorem 3.2 of [18]. The essential observation which enables us to construct a decoupling feedback $u = \alpha(x,v)$ locally is that the equation

$$\begin{bmatrix} v_1 \\ \vdots \\ v_m \end{bmatrix} - \begin{bmatrix} h_1^{\rho_1 - 1}\big(f(x,u)\big) \\ \vdots \\ h_m^{\rho_m - 1}\big(f(x,u)\big) \end{bmatrix} = \begin{bmatrix} 0 \\ \vdots \\ 0 \end{bmatrix} \tag{2.6}$$

possesses a local solution $u = \alpha(x,v)$ by applying the Implicit Function Theorem, because the Jacobian matrix of the left-hand side of (2.6) equals $A(x,u)$, which is nonsingular at (x°,u°) by (2.5). Notice that $\alpha : \tilde{O} \longrightarrow \tilde{V}$ is analytic for some (possibly small) neighborhoods $\tilde{O}$ and $\tilde{V}$ of (x°,u°) in $\mathcal{X} \times \mathcal{U}$ and of u° in $\mathcal{U}$. This implies that when we apply a time sequence $v(k)$, $k \geq 0$, to the feedback modified system

$$\begin{cases} x(k+1) = f(x(k),\ \alpha(x(k),\ v(k))) \\ \quad y(k) = h(x(k),\ \alpha(x(k),\ v(k))) \end{cases} \tag{2.7}$$

which is such that $(x(k),v(k)) \in \tilde{O}$ and $\alpha(x(k),v(k)) \in \tilde{V}$ for all $k \geq 0$ then the input-output behavior of the decoupled system is given as

$$y_i(k+\rho_i) = v_i(k) \quad , \quad k \geq 0, \quad i \in \underline{m} \tag{2.8}$$

Of course the decoupling property is lost if we leave the neighborhoods $\tilde{O}$ resp. $\tilde{V}$. In this respect it is perhaps better to speak about a finite time input-output decoupling as is done in the forementioned reference.

Of course the next question is, what happens if the rank condition (2.5) is not satisfied. Let

$$r_1(x,u) = \operatorname{rank} A(x,u) \tag{2.9}$$

and assume $r_1(\cdot,\cdot)$ is constant, say $r_1(x,u) = r_1$, on a neighborhood $\tilde{O} \subset O$ of (x°,u°). Note that from the analyticity it follows that $r_1(x,u)$ is constant on een open and dense part of O and thus of M. Clearly the case that $r_1 = m$ is treated in Theorem 2.1 and when $r_1 < m$ there does not exist a local solution around (x°,y°) of the static state feedback input output decoupling problem. In this case a dynamic state feedback (1.3) possibly solves the local decoupling problem. The following algorithm is essential in the solution of the dynamic decoupling problem. In the first step, assuming $r_1(x,u)$ is constant around (x°,u°), we reorder the output functions $h_1,\ldots,h_m$ in such a way that the first r_1 rows of $A(x,u)$ are linearly independent. Then construct a local feedback $u = \alpha_1(x,v)$ around (x°,u°) such that the first r_1 input-output channels are decoupled, i.e.

$$y_i(\rho_i) = v_i(0), \qquad i = 1,\dots,r_1 \tag{2.10}$$

whereas for the other outputs we have

$$y_i(\rho_i) = h^{\rho_i -1}\big(f(x^\circ,\alpha_1(x^\circ,v_1(0),\dots,v_m(0))\big), \; i = r_1+1,\dots,m \tag{2.11}$$

Observe that, as in Theorem 2.1, similar expressions are also true for $y_1(k+\rho_i)$ as long as we are working in a sufficiently small neighborhood on which the Implicit Function Theorem applies. The basic observation is that the right-hand side of (2.11) does not depend on $v_{r+1}(0),\dots,v_m(0)$. This follows from the rank condition for $A(x,u)$. In the second step of the algorithm we introduce a precompensator. Define as precompensator a n-fold time delay for the inputs $v_1,\dots,v_{r_1}$, that is, let

$$\begin{cases} z_{i1}(k+1) = z_{i2}(k) \\ z_{i2}(k+1) = z_{i3}(k) \\ \quad\vdots \\ z_{in}(k+1) = w_i(k), \qquad i = 1,\dots,r_1 \end{cases} \tag{2.12a}$$

together with the linking map

$$v_i(k) = z_{i1}(k), \qquad i = 1,\dots,r_1 \tag{2.12b}$$

The system (1.1) together with the feedback $u = \alpha_1(x,v)$ and the precompensator (2.12a,b) yields a new system $\sum_1$ with inputs $(w_1,\dots,w_{r_1},v_{r_1+1},\dots,v_m)$, states $(x,z_{11},\dots,z_{1n},\dots,z_{r_1},\dots,z_{r_1n})$ $= (x,z^1)$ and outputs $y_1,\dots,y_m$. To avoid the introduction of new control variables we again use $(u_1,\dots,u_m)$ to denote the inputs of $\sum_1$. Setting $x_e = (x,z^1)$ the system $\sum_1$ may be written as

$$\begin{cases} x_e(k+1) = f_e(x_e(k),u(k)) \\ y(k) \quad\;\; = h_e(x_e(k),u(k)) \end{cases} \tag{2.13}$$

and this is again a system of the form (1.1). Note that the original equilibrium point (2.4) of the system (1.1) transforms into an equilibrium of the dynamics (2.13):

$$f_e(x_e^\circ,u^\circ) = x_e^\circ \tag{2.14}$$

where x_e° is of the form $x_e^\circ = (x^\circ,z^\circ)$. Now again we define the characteristic numbers $\rho_1^e,\dots,\rho_m^e$ of the system (2.13). Notice that

$$\rho_i^e = \rho_i+n, \qquad i = 1,\dots,r_1 \tag{2.15}$$

Then consider the decoupling matrix of $\sum_1$, which will be written as $A^1(x_e,u) = A^1(x,z^1,u)$. Let

$$r_2(x_e,u) = \text{rank } A^1(x_e,u) \tag{2.16}$$

Clearly $r_2(x_e,u)$ is constant on an open and dense subset of $\mathcal{X}_e \times \mathcal{U}$, say $r_2(x_e,u) = r_2$. Assume (x_e^o,u^o) is an interior point where $r_2(x_e,u)$ is constant, then we may repeat the arguments used in the first step. In this way we obtain after relabeling outputs and adding state feedback to $\sum_1$ r_2 decoupled input-output channels, i.e.

$$y_i(\rho_i^e) = v_i(0), \qquad i = 1,\ldots,r_2 \tag{2.17}$$

and for the other outputs we have

$$y_i(\rho_i^e) = h_e^{\rho_i^e-1}\big(f_e(x_e,\alpha_2(x_e,v_1(0),\ldots,v_m(0))\big), \; i = r_2+1,\ldots,m \tag{2.18}$$

and by (2.16) we know that the right-hand side of (2.18) is independent of $v_{r_2+1}(0),\ldots,v_m(0)$. To continue the algorithm we repeat step 2 as many times as necessary and we clearly have a local solution for the dynamic decoupling problem if $r_k = m$ for a certain k (of course we implicitly assume here that at each step the equilibrium point as in (2.14) is an interior point where the rank of the corresponding decoupling matrix is constant). That the here given procedure indeed works is based on the fact that the number of iterations in the algorithm is finite and the algorithm terminates as soon as $r_{k+1} = r_k$. As $r_k \le m$ for all k there exists an integer k^* for which $r_{k^*+1} = r_{k^*}$, This rank r_{k^*} will be denoted as r_* and determines the maximal number of input-output decoupled channels of the system (1.1) under addition of a dynamic state feedback (1.3). Clearly the integer r_* is the rank of the decoupling matrix of the system $\sum_{k^*}$ and it is therefore constant on an open dense submanifold of the Cartesian product of its state space and input space. The following theorem solves the dynamic decoupling problem around (x^o,u^o).

<u>Theorem 2.2</u> Consider the system (1.1) around the equilibrium point (2.4). Suppose the corresponding equilibrium point of the precompensated system $\sum_{k^*}$ is an interior point of the subset where its decoupling matrix has constant rank. Then the dynamic state feedback input-output decoupling problem is locally solvable around the equilibrium point if and only if

$$r_* = m \tag{2.19}$$

Although this result slightly differs, its proof follows along the same lines as the finite time decoupling result in [18].

3. Local right-invertibility

We now address the problem of right-invertibility (or functional reproducibility or dynamic path controllability) of the system (1.1). We say that the system (1.1) initialized at $x(0) = x^{o}$ is <u>right-invertible</u> at x^{o} if there exists a fixed adjustment time k_0 - depending only on the system (1.1) - such that for all possible time functions $\xi(k)$, $k \geq 0$, in the output space $\mathcal{Y}$ there exists a proper choice of instruments $\bar{u}(k)$, $k \geq 0$, which yield as target variable $y(k) = \xi(k)$, $k \geq k_0$. In the sequel we will establish a result on invertibility which parallels its continuous time counterpart, cf. [21]. This result essentially deals with <u>local</u> (finite-time) right-invertibility. In order to formalize this, assume again we are working in a neighborhood in $\mathcal{X} \times \mathcal{U}$ of the equilibrium point (2.4). With the initial point x^{o} and input sequence $u(k) = u^{o}$, for all $k \geq 0$, corresponds to constant output sequence $y(k) = y^{o} = h(x^{o},u^{o})$. The system (1.1) is said to be locally right-invertible at the equilibrium point x_0 if it is right-invertible at x_0 for all possible time functions $\xi(k)$, $k \geq 0$ in $\mathcal{Y}$ which are sufficiently close to y^{o}, i.e. $\|\xi(k)-y^{o}\| < \epsilon$ for some $\epsilon > 0$ and all $k \geq 0$. If one moreover requires that the corresponding state and input-sequence $(\bar{x}(k),\bar{u}(k))$ belongs to the given neighborhood in $\mathcal{X} \times \mathcal{U}$ it may happen that reproducibility property $y(k) = \xi(k)$ holds only on a finite time interval. We emphasize that each system which is (local) right-invertible at x_0 possesses a fixed <u>adjustment time</u> k_0 which represents an inherent delay before one is able to reproduce a desired output path. Based on the previously given references we give our main result on local right invertibility.

<u>Theorem 3.1</u> Consider the system (1.1) around the equilibrium point (2.4). Let $\sum_{k^*}$ be the precompensated system defined in section 2 around the corresponding equilibrium point and suppose this point is an interior point of the subset where its decoupling matrix has constant rank. Then the system (1.1) is locally (finite-time) right-invertible at x^{o} if and only if

$$r_* = m \tag{3.1}$$

<u>Proof</u> (outline) The main observation is that the properties for dynamic decoupling and local right invertibility are the same, see also Theorem 2.2. As (2.19) is the necessary and sufficient condition for the decoupling, it also is for invertibility. The rest of the proof is

analogous to the one given for continuous-time nonlinear systems in [21]. □

4. Linearization, decoupling and invertibility

Most often in practical control engineering one linearizes the given nonlinear plant and subsequently solves the design problem of interest. To what extent the herewith obtained linear solutions are useful in the controller design for the nonlinear model of course heavily depends on the ignored nonlinear characteristics. Also at a first instance it is not even clear whether or not the design problem for the nonlinear system is solvable, even if the problem is solvable for the linearized model. In this secton we show that generically the problems of local input-output decoupling and local right-invertibility for the nonlinear system (1.1) around the equilibrium (2.4) are solvable if and only if these questions are solvable for the linearized system.

So, consider again the system (1.1) around the equilibrium point (2.4) and let

$$\begin{cases} \bar{x}(k+1) = A\bar{x}(k) + B\bar{u}(k) \\ \bar{y}(k) = C\bar{x}(k) + D\bar{u}(k) \end{cases} \tag{4.1}$$

be the linearization of (1.1) around (x°, u°), i.e.

$$A = \frac{\partial f}{\partial x}(x^\circ, u^\circ),\ B = \frac{\partial f}{\partial u}(x^\circ, u^\circ),\ C = \frac{\partial h}{\partial x}(x^\circ, u^\circ),\ D = \frac{\partial h}{\partial u}(x^\circ, u^\circ), \tag{4.2}$$

To test if the linear system (4.1) is static state feedback input-output decouplable, we need to compute its decoupling matrix, cf. [7]. Therefore we first determine the characteristic numbers $\sigma_1, \ldots, \sigma_m$ of (4.1). That is σ_i is defined as the minimal integer ℓ such that $\bar{y}_i(k+\ell)$ explicitly depends upon $\bar{u}(k)$. It is straightforward to verify the relation with the characteristic numbers $\rho_1, \ldots, \rho_m$ of (1.1):

$$\rho_i \le \sigma_i, \qquad i = 1, \ldots, m \tag{4.3}$$

and equality in (4.3) occurs in case that

$$\frac{\partial}{\partial u} h_i^{\rho_i - 1}(f(x^\circ, u^\circ)) \neq 0 \tag{4.4}$$

which is generically the case because the analytic functions $h_i^{\rho_i - 1}(f(x,u))$ is nonvanishing on an open and dense submanifold of $\mathcal{X} \times \mathcal{U}$. Under the generic assumption that $\rho_i = \sigma_i$, $i = 1, \ldots, m$, we will show that the static state feedback decoupling problems for (1.1) and (4.1) are solvable under the same conditions.

Theorem 4.1 Consider the system (1.1) around the equilibrium point (2.4) and let (4.1) be the linearization of the system around this point. Assume that $\rho_i = \sigma_i$, $i = 1,\ldots,m$. Then the static state feedback decoupling problem for (1.1) is locally solvable around (x^o,u^o) if and only if the static state feedback decoupling problem is solvable for the linearized system (4.1).

Proof The basic observation for this result is that if $\rho_i = \sigma_i$, $i = 1,\ldots,m$, then the decoupling matrix of (4.1) equals the decoupling matrix $A(x,u)$ evaluated at (x^o,u^o). The rest of the proof follows as in the continuous-time case, cf. [9]. □

In a completely analogous way one obtains a similar result on dynamic state input-output decoupling. The (generically satisfied) extra requirement needed to do so is that the characteristic numbers of each of the precompensated systems $\sum_k$ (defined in section 2) and of their corresponding linearizations are the same. We refer to this as condition C.

Theorem 4.2 Consider the system (1.1) around the equilibrium point (2.4). Suppose condition C is satisfied. Then the dynamic state feedback decoupling problem for the system (1.1) is locally solvable if and only if the linearized system (4.1) is dynamic input-output decouplable.

Using the results of section 3 we immediately see that the above similarities between a system and its linearization as far as input-output decoupling concerns, extends to the problem of (local) right-invertibility.

5. An example

The methods and techniques employed in this paper will be illustrated on a nonlinear discrete time system stemming from economics. Consider the following model of a closed economy, cf. [25].

$$Y(k+1) = Y(k) + \alpha\{C(Y(k)) + I(Y(k),R(k),K(k)) + P(k)^{-1}G(k) - Y(k)\} \quad (5.1)$$

$$R(k+1) = R(k) + \beta\{L(Y(k),R(k)) - P(k)^{-1}M(k)\} \quad (5.2)$$

$$K(k+1) = K(k) + I(Y(k),R(k),K(k)) \quad (5.3)$$

$$Y(k) \quad = F(N(k),K(k)) \tag{5.4}$$

$$N(k) \quad = H(W(k),P(k)) \tag{5.5}$$

In this model the quantities have the following interpretation:

Y: real output

C: real private consumption

I: real private net investment

R: nominal interest rate

K: real capital stock

P: price level

G: nominal government spending

L: real money demand

M: nominal money stock

N: labour demand

W: nominal wage rate

α and β are positive constants.

Equation (5.1) is a dynamic *IS* equation and (5.2) is a dynamic *LM* equation. The capital accumulation is described via the dynamic Keynesian equation (5.3). Equation (5.4) is a macroeconomic production function and (5.5) defines the labour demand as a function of the real wage rate. In this model G and M are the instrument variables (controls), W is a known exogenous variable, which will be assumed to be constant, $W = \overline{W}$, and finally the real output Y and the price level P are the target variables (outputs). Input-output decoupling and right-invertibility (dynamic path controllability) are extremely important topics for economic systems, cf. [1,4,25]. Our discussion follows the one given in [19]. To study these questions for the system (5.1-5) we assume $(\overline{Y},\overline{R},\overline{K},\overline{W},\overline{G},\overline{M},\overline{N})$ is a particular steady state of interest. First we bring (5.1-5) into the usual state space form (1.1). Assuming that

$$\frac{\partial H}{\partial P}(\overline{W},\overline{P}) \neq 0, \qquad \frac{\partial F}{\partial N}(\overline{N},\overline{K}) \neq 0, \tag{5.6}$$

we may locally apply the Implicit Function Theorem yielding the equations

$$\begin{cases} P = \tilde{H}(W,N), & (\overline{P} = \tilde{H}(\overline{W},\overline{N})) \\ N = \tilde{F}(Y,K), & (\overline{N} = \tilde{F}(\overline{Y},\overline{K})) \end{cases} \tag{5.7}$$

and thus we obtain as the second target equation

$$P(k) = \tilde{H}(W(k),N(k)) = \tilde{H}(W(k),\tilde{F}(Y(k),K(k))). \tag{5.8}$$

Altogether we have obtained - locally - a model of the form (1.1)

$$\begin{cases} Y(k+1) = f_1(Y(k),R(k),K(k),\overline{W},G(k)) \\ R(k+1) = f_2(Y(k),R(k),K(k),\overline{W},M(k)) \\ K(k+1) = f_3(Y(k),R(k),K(k) \end{cases} \tag{5.9}$$

$$\begin{cases} Q_1(k) = Y(k) \\ Q_2(k) = P(k) = \tilde{H}\big(\overline{W},\tilde{F}(Y(k),K(k))\big) \end{cases} \tag{5.10}$$

where Q_1 and Q_2 denote the target variables and the functions f_1, f_2 and f_3 directly follow from (5.1-3) and (5.8). Note that both target variables Q_1 and Q_2 are <u>not</u> directly influenced by the instrument variables G and M. So the characteristic numbers $\rho_i > 0$, $i = 1,2$. From (5.10) we obtain

$$Q_1(k+1) = Y(k+1) = f_1(Y(k),R(k),K(k),\overline{W},G(k)) \tag{5.11}$$

$$\begin{aligned} Q_2(k+1) &= \tilde{H}\big(\overline{W},\tilde{F}(Y(k+1),K(k+1))\big) = \\ &= \tilde{H}\big(W(k+1),\tilde{F}\big(f_1(Y(k),R(k),K(k),\overline{W},G(k)),f_3(Y(k),R(k),K(k))\big)\big) \end{aligned} \tag{5.12}$$

From the defining equation for f_1, see (5.1), it follows that $\rho_1 = 1$ and from assumption (5.6) we conclude that $\rho_2 = 1$ also. Notice that both outputs $Q_1(k+1)$ and $Q_2(k+1)$ do not explicitly depend upon $M(k)$. Therefore the decoupling matrix (2.3) of the system has rank 1 and so the system is <u>not</u> input-output decouplable by static state feedback. Clearly for the linearization of (5.9,10) around $(\overline{Y},\overline{R},\overline{K},\overline{W},\overline{G},\overline{M},\overline{N})$ we find the characteristic numbers $\sigma_1 = \rho_1 = 1$ and $\sigma_2 = \rho_2 = 1$ and the same conclusion holds for this system. Next we turn attention to dynamic state feedback and we follow the algorithm of section 2. Let

$$\begin{cases} \tilde{G} = f_1(Y,R,K,\overline{W},G) \\ \tilde{M} = M \end{cases} \tag{5.13}$$

which by the Implicit Function Theorem, see (5.1), yields a control law around the equilibrium of the form

$$\begin{cases} G = \tilde{f}_1(Y,R,K,\overline{W},\tilde{G}) \\ M = \tilde{M} \end{cases} \tag{5.14}$$

Plugging (5.14) in (5.11) and (5.12) yields

$$Q_1(k+1) = \tilde{G}(k) \tag{5.15}$$

$$Q_2(k+1) = \tilde{H}\big(\overline{W},\tilde{F}\big(\tilde{G}(k),f_3(Y(k),R(k),K(k))\big)\big) \tag{5.16}$$

Next we introduce a 1-step delay for the instrument $\tilde{G}$. So define

$$\begin{cases} Z(k+1) = \hat{G}(k) \\ \tilde{G}(k) = Z(k) \end{cases} \tag{5.17}$$

The system (5.9) together with (5.17) becomes

$$\begin{cases} Y(k+1) = Z(k) \\ Z(k+1) = \hat{G}(k) \\ R(k+1) = f_2(Y(k),R(k),K(k),\overline{W},M(k)) \\ K(k+1) = f_3(Y(k),R(k),K(k)) \end{cases} \tag{5.18}$$

Note that in section 2 in the present case a 3 (= dimension of the state space) fold time delay is suggested, cf. (2.12a). However as we will see it suffices to take a 1-step delay.

Now compute $Q_1(k+2)$ and $Q_2(k+2)$

$$Q_1(k+2) = \hat{G}(k) \tag{5.19}$$

$$Q_2(k+2) = \tilde{H}\big(\overline{W},\tilde{F}\big(\tilde{G}(k+1),f_3(Y(k+1),R(k+1),K(k+1))\big)\big) = \tag{5.20}$$

$$= \tilde{H}\big(\overline{W},\tilde{F}\big(\hat{G}(k),f_3(Z(k)f_2(Y(k),R(k),K(k),\overline{W},M(k)),f_3(Y(k),R(k),K(k)))\big)\big)$$

Therefore the characteristic numbers of the precompensated system are $\rho_1^e = \rho_2^e = 2$. In order to determine the rank of the extended decoupling matrix we need to evaluate the dependency of the right-hand side of (5.20) on the instrument $M(k)$. Let

$$\frac{\partial \tilde{H}}{\partial N}(\overline{W},\overline{N}) \cdot \frac{\partial \tilde{F}}{\partial K}(\overline{Y},\overline{K}) \cdot \frac{\partial f_3}{\partial R}(\overline{Y},\overline{R},\overline{K}) \cdot \frac{\partial f_2}{\partial M}(\overline{Y},\overline{R},\overline{K},\overline{W},\overline{M}) \neq 0. \tag{5.21}$$

which by the particular structure of f_2, see (5.9) and (5.2) yields the equivalent condition

$$\frac{\partial \tilde{F}}{\partial K}(\overline{Y},\overline{K}) \cdot \frac{\partial f_3}{\partial R}(\overline{Y},\overline{R},\overline{K}) \neq 0. \tag{5.22}$$

Clearly if (5.22) and so (5.21) are satisfied, then, see (5.20), $Q_2(k+2)$ explicitly depends upon $M(k)$ and the extended decoupling matrix has rank 2. But this implies that provided (5.22) holds then the model is locally dynamic input-output decouplable and thus, see section 3, is locally right-invertible around the equilibrium point. On the other hand (5.22) also implies that the linearization of the precompensated system is input-output decouplable, and so the linearization of the system itself is dynamic state feedback decouplable and thus right-invertible. Note that the last conclusion about the linearized economic model was already obtained in [25]. Let us finally compute a decoupling control law. Define the "inverse" control law for the instrument M via

$$\hat{M} = \tilde{H}\big(W_1,\tilde{F}(\hat{G},f_3(Z,f_2(Y,R,K,W_2,M),f_2(Y,R,K))\big) \tag{5.23}$$

which by the Implicit Function Theorem is locally equivalent to the control policy

$$M = \xi(W_1, \hat{G}, Z, Y, R, K, W_2, \hat{M}). \tag{5.24}$$

Plugging (5.23) in into (5.19) and (5.20) yields the decoupled behaviour

$$\begin{cases} Q_1(k+2) = \hat{G}(k) \\ Q_2(k+2) = \hat{M}(k) \end{cases} \tag{5.25}$$

So indeed, in a local sense, from equation (5.25) it follows that the new instruments $\hat{G}$ and $\hat{M}$ respectively can be used to steer the real output Y and the price level target P respectively. In effect only the original G is needed to steer Y and the instrument M depending on the already chosen instrument G will be used in controlling P. In order to find the desired strategy yielding the prescribed targets $(\overline{Q}_1(k), \overline{Q}_2(k))$ use (5.25) and solve backwards using the equations (5.24), (5.17) and (5.14).

References

[1] Aoki, M., 1974, "Non-interacting control of macroeconomic variables, implications on policy mix considerations", Journal of Econometrics 2, 261-281.

[2] Aoki, M., 1975, "On a generalization of Tinbergen's condition in the theory of policy to dynamic models", Review of Economic Studies 42, 293-296.

[3] Aoki, M., 1976, Optimal control and system theory in dynamic economic analysis, (North-Holland, Amsterdam).

[4] Aoki, M. and M. Canzoneri, 1979, "Sufficient conditions for control of target variables and assignment of instruments in dynamic macroeconomic models", International Economic Review, 20, 605-615.

[5] Brockett, R.W. and M.D. Mesarovic, 1965, "The reproducibility of multivariable systems", Journal of Mathematical Analysis and Applications, 11, 548-563.

[6] Buiter, W.H., 1979, "Unemployment-inflation trade-offs with rational expectations in an open economy", Journal of Economic Dynamics and Control 1, 117-141.

[7] Falb, P.L. and W.A. Wolovich, 1967, "Decoupling in the design and synthesis of multivariable control systems", IEEE Trans Autom. Control 12, 651-659.

[8] Fliess, M., 1986, "Esquisses pour une theorie des systemes nonlineaires en temps descret", Rend. Sem. Mat. Univers. Politechnico Torino, 55-67.

[9] Gras, L.C.J.M. and H. Nijmeijer, 1989, "Decoupling in nonlinear systems: from linearity to nonlinearity". IEE Proceedings Part D (to appear).

[10] Grizzle, J.W., 1985, "Controlled invariance for discrete time nonlinear systems with an application to the disturbance decoupling problem", IEEE Trans. Autom. Control 30, 868-874.

[11] Grizzle, J.W., 1986, "Feedback linearization of discrete time systems", in Analysis and Optimization of Systems, LNCIS, 83, 273-281.

[12] Grizzle, J.W., 1986, "Local input-output decoupling of discrete time nonlinear systems", Intern. Journal of Control, 43, 1517,150.

[13] Grizzle, J.W., and H. Nijmeijer, 1986, "Zeros at infinity for nonlinear discrete time systems", Mathematical Systems Theory, 19, 79-93.

[14] Jakubczyk, B., 1987, "Feedback linearization of discrete time systems", Systems & Control Lett. 9, 411-416.

[15] Lee, H.G. and S.I. Marcus, 1986, "Approximate and local linearizability of nonlinear discrete time systems", Intern. Journal of Control, 44, 1103-1124.

[16] Lee, H.G. and S.I. Marcus, 1987, "On input-output linearizatio of discrete time nonlienar systems", Systems & Control Lett. 8, 249-260.

[17] Monaco, S. and D. Normand-Cyrot, 1984, "Sur la commande non interactive des systemes nonlineaires en temps discret", in Analysis and Optimization of Systems, LNCIS, 63, 364-377.

[18] Nijmeijer, H., 1987, "Local (dynamic) input-output decoupling of discrete time nonlinear systems", IMA Journ. Math. Control & Information, 4, 237-250.

[19] Nijmeijer, H., 1989, "On dynamic decoupling and dynamic path controllability in economic systems", Journ. Economic Dynamics and Control, 13, 21-39.

[20] Preston, A.J. and A.R. Pagan, 1982, The theory of economic policy (Cambridge University Press, Cambridge).

[21] Respondek, W. and H. Nijmeijer, 1988, "On local right-invertibility of nonlinear control systems", Control Theory & Advanced Technology, 4, 325-348.

[22] Sain, M.K. and J.L. Massey, 1969, "Invertibility of linear time-invariant dynamical systems", IEEE Trans. Autom. Control AC-14, 141-149.

[23] Wohltmann, H-W., 1984, "A note on Aoki's condition for path controllability of continuous-time dynamic economic systems", Review of Economic Studies, 51, 343-349.

[24] Wohltmann, H-W. and W. Krömer, 1983, "A note on Buiter's sufficient condition for perfect output controllability of a rational expectations model", Journal of Economic Dynamics and Control 6, 201-205.

[25] Wohltmann, H-W. and W. Krömer, 1984, "Sufficient conditions for dynamic path controllability of economic systems", Journal of Economic Dynamics and Control 7, 315-330.

[26] Wolovich, W.A., 1974, Linear Multivariable Systems (Springer Verlag, Berlin).

VARIATIONAL PROBLEMS ARISING IN STATISTICS

B. T. Polyak

1. Introduction.

Mathematical statistics is a nice source of nonstandard variational problems. As an example we can mention the famous Neyman–Pearson lemma on hypothesis testing: in optimization language it is a variational problem with integral type functional (not including derivatives) subject to specific constraints. In this paper we deal with variational problems of another kind arising in such areas of statistics as parameter estimation and nonparametric regression. Among them there are such nonstandard problems as minimization of a functional which is a ratio of two integrals, minimization of a matrix-valued criteria, finding a saddle point of a functional over some classes of functions etc. Some of these problems can be solved in explicit form by use of a technique which is untypical for the classical calculus of variations.

2. Parameter estimation (scalar case).

Let

$$y_i = \theta^* + \xi_i, \qquad i = 1, \ldots, n, \tag{1}$$

where θ^* is a true value of an unknown parameter, y_i are its measurements, ξ_i are random errors. The aim is to estimate θ^* under known values $y_1, \ldots, y_n$. A wide class of estimators (so called M-estimators) can be constructed as follows [1, 2]. Choose a function F: $R^1 \to [0, \infty)$, $F(0) = 0$, $F(x) > 0$ for $x \neq 0$, and find

$$\theta_n = \arg\min_{\theta} \sum_{i=1}^{n} F(y_i - \theta). \tag{2}$$

This one dimensional minimization problem can be solved explicitly for some functions F. For example, if $F(x) = x^2$, then $\theta_n = \sum_{i=1}^n y_i/n$ (arithmetical mean); if $F(x) = |x|$,then θ_n is the median of $y_1, \ldots, y_n$. Let us assume that ξ_i are independent identically distributed (i.i.d.) with density p, that the function F is twice differentiable and $\psi = f'$, and that $\int \psi(x)p(x)\,dx = 0$ (we do not give the most general conditions; the aim is just to illustrate the origin of a variational problem). Then under some additional technical assumptions it can be proved (see, e.g. [1, 2]) that $\sqrt{n}\,(\theta_n - \theta^*)$ has asymptotically normal distribution with mean 0 and variance σ^2; it is denoted as

$$\sqrt{n}\,(\theta_n - \theta^*) \sim N(0, \sigma^2). \tag{3}$$

The variance σ^2 is defined by

$$\sigma^2 = \sigma^2(\psi) = \frac{\int \psi^2 p\,dx}{(\int \psi' p\,dx)^2}. \tag{4}$$

Thus the accuracy of the estimator (2) can be measured by asymptotic variance σ^2. Now let us choose the function F (or the function ψ) to minimize σ^2. We arrive to the first variational problem

$$\min_{\psi} \frac{\int \psi^2 p\,dx}{\left(\int \psi' p\,dx\right)^2}. \tag{$P1$}$$

It has a nonstandard form (ratio of two integral functionals, infinite intervals of integration). However, its solution can be found immediately. Let us assume that p is absolutely continuous and $0 < I(p) = \int (p'/p)^2 p\,dx < \infty$ ($I(p)$ is called Fisher information). Then integrating by parts and applying Schwartz inequality we have

$$\left(\int \psi' p\,dx\right)^2 = \left(\int \psi(p'/p)p\,dx\right)^2 \leq \int \psi^2 p\,dx \int (p'/p)^2 p\,dx. \tag{5}$$

Thus $\sigma^2(\psi) \geq I(p)^{-1}$ and the equality is obtained iff $\psi = kp'/p$, k being a constant. The quantity $\sigma^2(\psi)$ does not depend on k, and one can take $k = -1$. Thus a solution of (P1) is

$$\psi^* = -p'/p = -(\log p)', \tag{6}$$

and corresponding F is $F^* = -\log p$. The estimator (2) transforms into

$$\theta_n = \arg\max_{\theta} \sum_{i=1}^{n} \log p(y_i - \theta)$$

and coincides with the maximum likelihood estimator. In particular, if $p(x) = (1/\sqrt{2\pi}\sigma)\exp(-x^2/2\sigma^2)$ (Gaussian distribution) then θ_n is the arithmetical mean, if $p(x) = (1/2a)\exp(-|x|/a)$ (Laplacian distribution), then θ_n is the median.

It is worth mentioning that the inequality $\sigma^2(\psi) \geq I(p)^{-1}$ can be taken as an extended definition of Fisher information $I(p)$. Let us define $I(p) = \sup_{\psi\in\mathcal{F}} \left(\int \psi' p\,dx\right)^2 / \int \psi^2 p\,dx$, $\mathcal{F}$ being a class of all continuously differentiable functions with compact support. Huber's theorem [1, 2] states that $I(p) < \infty$ iff p is absolutely continuous and $\int (p'/p)^2 p\,dx < \infty$.

3. Parameter estimation (vector case).

Let us consider the same problem (1) but with $\theta^* \in R^N$. We apply the estimator (2) with F: $R^N \to [0, \infty)$. Then under some natural assumptions

$$\sqrt{n}(\theta_n - \theta^*) \sim N(0, S), \tag{7}$$

where the covariance matrix S is defined by

$$S = S(\psi) = \Big(\int \psi \nabla^T p \, dx \Big)^{-1} \int \psi \psi^T p \, dx \Big(\int \nabla p \psi^T \, dx \Big)^{-1}, \tag{8}$$

where $\psi = \nabla F$. "The best" choice of ψ corresponds to "the least" value of S, but we have to define the precise sense of these words for matrices.

Let us write $A \geq B$ for two symmetric matrices A and B if the matrix $A - B$ is nonnegative definite. If $A \geq B$ then $(Ax, x) \geq (Bx, x)$ for all x, $\operatorname{Tr} A \geq \operatorname{Tr} B$ and so on, therefore if $S(\psi) \geq S(\psi^*)$ for some ψ^* and all ψ, then the covariance matrix $S(\psi^*)$ is "the least" in all natural statistical senses. Thus we have matrix valued optimization problem: find ψ^* such that

$$S(\psi) \geq S(\psi^*), \qquad \forall \psi. \tag{P2}$$

It is difficult to expect existence of a solution for the problem (P2), nevertheless it does exist. The following matrix analog of Schwartz inequality is valid [3].

Lemma. Let $A(\tau)$, $B(\tau)$ be matrices of dimension $n_1 \times N$ and $n_2 \times N$ respectively, depending on a parameter τ defined on a space with measure $d\mu(\tau)$. Then

$$\begin{aligned} &\int A(\tau) A^T(\tau) \, d\mu(\tau) \geq \\ &\geq \int A(\tau) B^T(\tau) \, d\mu(\tau) \Big(\int B(\tau) B^T(\tau) \, d\mu(\tau) \Big)^+ \int B(\tau) A^T(\tau) \, d\mu(\tau) \end{aligned} \tag{9}$$

where C^+ denotes Moore–Penrose pseudoinverse for a matrix C. Now let us assume that the density $p(X)$ is differentiable and

$$0 < I(p) = \int \nabla p \nabla^T p \, p^{-1} \, dx < \infty,$$

where $I(p)$ is called Fisher information matrix. If we choose

$$\psi^* = -\nabla p / p = -\nabla \log p \tag{10}$$

and apply the lemma (substituting $A(\tau)$ in place of $\psi(x)$, $B(\tau)$ in place of $\psi^*(x)$, $d\mu(\tau)$ in place of $p \, dx$) we arrive at the inequality (8). Thus the solution of matrix variational problem exists and is given by (10). The estimator (2) with $F^* = -\log p$ is the maximum likelihood estimator.

4. Robust estimation.

To construct the maximum likelihood estimator one should know the density p. In many real-life cases this is known just approximately, but a small deviation of the density from the preassumed one may lead to bad consequences, so optimal estimators are nonrobust. This situation is now well understood in statistics due to Huber's works [1, 2]. Huber's approach to robust estimation is based on introducing of some class of distributions $\mathcal{P}$ and on constructing the estimator which is optimal in a minimax sense for the class $\mathcal{P}$.

Let us suppose that (3) is valid for some function $F: R^1 \to [0,\infty)$ and all $p \in \mathcal{P}$. The main assumption that guarantees this is that F is even (or ψ odd) and all $p \in \mathcal{P}$ are symmetric: $p(x) = p(-x)$. Then the accuracy of the estimator (2) for a $p \in \mathcal{P}$ is given by (4); we shall denote σ^2 by $\sigma^2(\psi, p)$ to emphasize its dependance on p. The function ψ^* gives the best minimax estimator if

$$\sigma^2(\psi^*, p) \le \sigma^2(\psi^*, p^*) \le \sigma^2(\psi, p^*) \tag{P3}$$

for all $p \in \mathcal{P}$ and all ψ, where

$$\sigma^2(\psi, p) = \frac{\int \psi^2 p\, dx}{(\int \psi' p\, dx)^2}.$$

The solution of this saddle point problem is given by Huber [1, 2]:

$$p^* = \arg\min_{p\in\mathcal{P}} I(p), \quad I(p) = \int (p'/p)^2 p\, dx, \quad \psi^* = -(\log p^*)'. \tag{P4}$$

Thus one should find the least favourable distribution of the class $\mathcal{P}$ (i.e. the distribution with minimal Fisher information $I(p)$) and construct the maximum likelihood estimator for this distribution. So minimax problem (P3) is transformed to the variational problem (P4) under constraint $p \in \mathcal{P}$.

Existence and uniqueness of solutions (P4) can be proved for convex classes $\mathcal{P}$ [1, 2]. These results are based on convexity of the functional $I(p)$ over p. Let us examine various techniques for solving (P4) under some classes $\mathcal{P}$.

1. $\mathcal{P}_1 = \{p : p(x)$ is continuous at $x = 0$ and $p(0) \ge \varepsilon > 0\}$. This is the class of all nondegenerate distributions. Applying the inequality (5) for $\psi(x) = \operatorname{sign} x$ one has $I(p) \ge 4p^2(0)$ and the equality is attained for $p'/p = k \operatorname{sign} x$, that is $p^*(x) = \varepsilon \exp(-2\,|x|\,\varepsilon)$. Thus the least favourable distribution is Laplacian.

2. $\mathcal{P}_2 = \{p : \int x^2 p\, dx \le s^2\}$ — the class of distributions with bounded variances. The same approach with $\psi(x) = x$ gives $p^*(x) = (1/\sqrt{2\pi}\, s)\exp(-x^2/2s^2)$; the least favourable distribution is Gaussian.

3. $\mathcal{P}_3 = \{p = (1-\varepsilon)p_0 + \varepsilon p_1,\ p_0(x) = (1/\sqrt{2\pi})\exp(-x^2/2),\ p_1$ arbitrary, $\varepsilon > 0\}$. This is the class of contaminated normal distributions, ε being a parameter of contamination. The problem (P4) can be written as

$$\min \int (p'/p)^2 p\, dx$$

$$p(x) \ge (1-\varepsilon)p_0(x), \qquad -\infty < x < \infty \tag{11}$$

$$\int p\, dx = 1.$$

Let us look for a solution having the form

$$p^*(x) = (1-\varepsilon)p_0(x), \quad |x| \le \Delta, \quad p^*(x) > (1-\varepsilon)p_0(x), \quad |x| > \Delta.$$

Then as above we have $p^*(x) = c_1 \exp(-|x|\, c_2)$, $|x| > \Delta$. It is not difficult to find constraints c_1, c_2, Δ and to prove that such p^* is the unique solution of (11). The function $F^* = -\log p^*$ which corresponds to this distribution is the famous Huber's function

$$F^*(x) = \begin{cases} x^2, & |x| \le \Delta \\ c_2\,|x| + c_3, & |x| > \Delta \end{cases} \tag{12}$$

($F(x)$ is continuously differentiable). The robust Huber's estimator (2), (12) is an intermediate one between arithmetical mean and median.

4. $\mathcal{P}_4 = \left\{p\colon \int_{-a}^{a} p\,dx = 1-\varepsilon,\ 0 \le \varepsilon < 1\right\}$ — class of approximately finite distributions. The problem (P4) has the form

$$\begin{aligned} &\min \int (p'/p)^2\, p\,dx \\ &\int \chi p\,dx = 1-\varepsilon, \quad \chi(x) = 1, \quad |x| \le a, \quad \chi(x) = 0, \quad |x| > a, \\ &\int p\,dx = 1. \end{aligned} \tag{13}$$

The Lagrangian for this problem is

$$L(p, \lambda_1, \lambda_2) = \int \left((p'/p)^2 p + \lambda_1 p + \lambda_2 \chi p\right)\, dx$$

and Euler's equation is as follows

$$-(p'/p)^2 + \lambda_1 + \lambda_2 \chi - 2(p'/p)' = 0.$$

Denote $z = p'/p$ then

$$2z' + z^2 - \lambda_1 - \lambda_2 \chi = 0,$$

or for any interval $(-\infty, a)$, $(-a, a)$, (a, ∞)

$$2z' + z^2 - \lambda = 0 \tag{14}$$

where a constant λ depends on the interval. Partial solutions of (14) are $z = c$ and $z = k\,\mathrm{tg}(kx/2) + c$, corresponding p's are found from $p'/p = z$: $p = c_1 \exp(c_2 x)$, $p_3 = c_3 \cos^2(c_4 x + c_5)$. Combining them let us construct the function

$$p^*(x) = \begin{cases} c_1 \exp(c_2 x) & |x| > a \\ c_3 \cos^2(c_4 x) & |x| \le a \end{cases} \tag{15}$$

and choose c_i to make p^* continuously differentiable and satisfying constraints in (13). Such p^* satisfies necesary conditions for extremum (Euler's equation); employing the convexity of $I(p)$ and linearity of constraints it is not difficult to prove that p^* is the solution of (13).

Other examples of classes $\mathcal{P}$ can be found in [1–5]. Above results can be extended to the vector case. Let us consider (1) with $\theta^* \in R^N$ and apply the estimate (2), F: $R^N \to [0, \infty)$. Assume that $p \in \mathcal{P}$, $\mathcal{P}$ being a class of multidimensional densities. Vector analogue of Huber's theorem is as follows [4]. If there exist $p^* \in \mathcal{P}$ such that $0 < I(p^*) \leq I(p) < \infty$, $p \in \mathcal{P}$ (inequalities are understood in matrix sense, see above) then

$$S(\psi^*, p) \leq S(\psi^*, p^*) \leq S(\psi, p^*), \quad p \in \mathcal{P}, \forall \psi. \tag{P5}$$

Here $S(\psi, p)$(denoted earlier as $S(\psi)$) is defined by (8), $\psi^* = -(\log p^*)'$. Thus, constructing a robust vector estimator requires to solve the matrix variational problem: find a $p^* \in \mathcal{P}$, such that

$$I(p^*) \leq I(p), \qquad p \in \mathcal{P}. \tag{P6}$$

If $\mathcal{P}$ is the vector analogue of the class $\mathcal{P}_2$, the solution of (P6) exists and can be found in an explicit form [4]. Denote $\mathcal{P}_2 = \{\, p\text{: } \int xx^T p\, dx \leq S \,\}$, where $S > 0$ is a fixed matrix. Then p^* is the density of an N-dimensional normal distribution with the mean 0 and the covariance matrix S; this can be proved using the above formulated lemma.

If we assume that $\mathcal{P}$ includes radially symmetric densities only (i.e. $p(x) = p(y)$ for $|x| = |y|$, $p \in \mathcal{P}$) then matrix minimization of $I(p)$ is equivalent to scalar minimization of $\operatorname{Tr} I(p)$. Thus the following variational problem arises

$$\min \int \left(\|\nabla p\|^2 / p \right) dx, \qquad p \in \mathcal{P}. \tag{P7}$$

Substituting $p(x) = u^2(x)$ (it is correct since $p \geq 0$) we have

$$\begin{gathered} \min \int \|\nabla u\|^2 \, dx, \\ \int u^2 \, dx = 1, \quad u^2 \in \mathcal{P}. \end{gathered} \tag{P7}$$

Analogous variational problems are considered in mathematical physics; they are connected with finding eigenfunctions of the Laplace operator.

5. Nonparametric estimation.

The simplest problem of nonparametric regression is to reconstruct the function f^*: $[0,1] \to R^1$ under its measurements y_i at n points $x_i \in [0,1]$ disturbed by noise ξ_i.

$$y_i = f^*(x_i) + \xi_i, \qquad i = 1, \ldots, n. \tag{16}$$

The apriori information on f^* is that it belongs to some functional class $\mathcal{F}$. Nonparametric analogue of M-estimator has the form [6,7]:

$$f_n = \arg \min_{f \in \mathcal{F}} \sum_{i=1}^{n} F\left(y_i - f(x_i)\right). \tag{P8}$$

Let us suppose that the function $F\colon R^1 \to [0,\infty)$ is fixed. Then (P8) is a nonclassical optimization problem. Let us discuss its solution for various classes $\mathcal{F}$.

1. Classes of smooth functions. Denote

$$\mathcal{F}_q^l = \{\, f\colon \left\| f^{(l)} \right\|_q \le L \,\}. \tag{17}$$

Here $1 < q \le \infty$, $l \ge 1$, $\|a\|_q = \left(\int |a(x)|^q \, dx \right)^{1/q}$, $L > 0$. For $q = 1$ $\mathcal{F}_1^l$ denotes the class of functions such that $f^{(l-1)}$ has variation on $[0,1]$ bounded by L. Let us also introduce

$$\widetilde{\mathcal{F}}_q^l = \left\{ f \in \mathcal{F}_q^l\colon |f(x)| \le c,\ 0 \le x \le 1 \right\}.$$

Existence result for the problem (P8) is very simple [7]. If F is lower semicontinuous then a solution of (P8) for $\mathcal{F} = \widetilde{\mathcal{F}}_q^l$ exists. If in addition $F(x) \to \infty$ for $|x| \to \infty$ and $n \ge l$ (all points x_i being distinct), then a solution exists for $\mathcal{F} = \mathcal{F}_q^l$.

Optimality condition for (P8) with $\mathcal{F} = \mathcal{F}_q^l$, $q > 1$ and F convex is as follows [7]. The function $f_n \in \mathcal{F}$ is the solution of (P8) if and only if there exist $\lambda_0 \ge 0$, $\lambda_1, \ldots, \lambda_n$ such that

$$\begin{aligned}
&\lambda_i \in \partial F(y_i - f_n(x_i)), \quad i = 1, \ldots, n \\
&\sum_{i=1}^{n} \lambda_i x_i^r = 0, \quad r = 0 \ldots, l-1 \\
&\lambda_0 f_n^{(l)}(x) = |u(x)|^{\frac{1}{q-1}} \operatorname{sign} u(x),\ 0 \le x \le 1,\ \ \lambda_0 \left(L - \left\| f_n^{(l)} \right\|_q \right) = 0,\ q < \infty \\
&f_n^{(l)}(x) \in L \operatorname{sign} u(x), \quad 0 \le x \le 1, \quad q = \infty,
\end{aligned} \tag{18}$$

where ∂F is a subgradient of F, $u(x) = \sum_{i=1}^{n} \lambda_i (x_i - x)_+^{l-1}$, $a_+ = \max\{0, a\}$, $\operatorname{sign} 0 = [-1, 1]$. The case $q = 2$, $F(x) = x^2$ has been intensively studied in approximation theory (see, eg. [8]).

This optimality condition gives full description of the solution. Let us describe it for two cases. Assume $q = 2$, then the solution is a polynomial spline of degree $2l - 1$ with nodes at the points x_i. If $q = \infty$, then the solution is a polynomial spline of degree l (e.g., for $l = 1$ it is a piecewise linear function with x_i as nodes).

Iterative methods for solving (P8), $\mathcal{F} = \mathcal{F}_q^l$, can be constructed [7]; we do not discuss them here.

2. *Class of monotone functions.*

$$\mathcal{F}_{\text{mon}} = \{\, f\colon f(x) \ge f(y),\ 0 \le y \le x \le 1\, ,\}. \tag{19}$$

In this case solution of (P8) is equivalent to solution of the following finite dimensional mathematical programming problem

$$\min \sum_{i=1}^{n} F(y_i - t_i)$$

$$t_{i+1} \geq t_i, \quad i = 1, \dots n-1, \tag{P9}$$

if we set $f(x_i) = t_i$, where $f(x)$ is an arbitrary monotone continuation of these values for $x \neq x_i$. Let F be convex and $F(x) \to \infty$ for $|x| \to \infty$. Then a solution $t_1^*, \dots, t_n^*$ exists [7] if and only if

$$t_i^* = \max_{k \leq i} \min_{l \geq i} \arg\min_t \sum_{s=k}^{l} F(y_s - t). \tag{20}$$

For instance if $F(x) = x^2$ we have

$$t_i^* = \max_{k \leq i} \min_{l \geq i} (y_k + \dots + y_l)/(l-k) \tag{21}$$

and if $F(x) = |x|$ then

$$t_i^* = \max_{k \leq i} \min_{l \geq i} \operatorname{med}(y_k, \dots, y_l), \tag{22}$$

where $\operatorname{med}(y_k, \dots, y_l)$ denotes the median of $y_k, \dots, y_l$.

It is not difficult to construct optimization methods based on these criteria.

6. Conclusions.

We have not surveyed all variational problems arising in estimation theory. In addition we can mention variational problems connected with parameter estimation of autoregression model [9], with kernel nonparametric estimators [10] and kernel-type methods in stochastic approximation [11]. But the above examples are nice illustrations of the diversity and complexity of optimization problems in statistics.

References

[1] P. J. Huber, Robust estimation of a location parameter, *Ann. Math. Stat.*, **35** (1964), pp. 13–101.

[2] P. J. Huber, *Robust statistics*, Wiley, New York, 1981.

[3] B. T. Polyak, Ya. Z. Tsypkin, Optimal pseudogradient adaptation algorithms, *Autom. and Remote Contr.*, **41** (1981), pp. 1101–1110.

[4] B. T. Polyak, Ya. Z. Tsypkin, Robust pseudogradient adaptation algorithms, *Autom. and Remote Contr.*, **41** (1981), pp. 1404–1409.

[5] B. T. Polyak, Ya. Z. Tsypkin, Robust identification, *Automatica*, **16** (1980), pp. 53–69.

[6] A. S. Nemirovskii, B. T. Polyak, A. B. Tsybakov, Estimators of maximum likelihood type for nonparametric regression, *Soviet Math. Dokl.*, *28* (1983), pp. 788–792.

[7] A. S. Nemirovskii, B. T. Polyak, A. B. Tsybakov, Signal processing by the nonparametric maximum likelihood method, *Probl. Inform. Transmiss.* **20** (1984), pp. 177–191.

[8] C. H. Reinsch, Smoothing by spline functions I, II, *Numer. Math.* **10** (1967), pp. 177–183 and **16** (1971), pp. 451–454.

[9] B. T. Polyak, Ya. Z. Tsypkin,Optimal and robust estimation of autoregression coefficients, *Engrg. Cybern.*, **21** (1983), No. 1.

[10] L. Devroye, L. Gyorgi, *Nonparametric density estimation: L_1-view*, Wiley, New York, 1985.

B. T. Polyak
Institute of Control Problems
Profsojuznaja 65
117–342 Moscow, B 279, USSR

NECESSARY CONDITIONS FOR AN EXTREMUM, PENALTY FUNCTIONS AND REGULARITY

Boris N. Pshenichnyj

Abstract

The aim of this paper is to show that the three concepts mentioned in the title are closely related when considering the most general optimization problems. Each of these concepts has been studied in numerous works where different relations among them have been established. Here we give the most general statements and we show that in the presence of regularity the necessary conditions for an extremum in the form of the Lagrange multipliers rule are always fulfilled. It should be emphasized that for the most complete description of the point of extremum in problems with non-smooth constraints the optimality conditions must be formulated not in the form of a single rule of multipliers but in the form of the whole family of such rules. Only this makes it possible to avoid a situation when necessary conditions are fulfilled at points which trivially cannot be optimal.

1. Preliminaries.

Let X be a Banach space whose conjugate space is denoted by X^*. For $x \in X$, $x^* \in X^*$, we denote by $\langle x, x^* \rangle$ the value of a linear continuous functional x^* on the element x. Lest the details should encumber the main contents of the paper all functions considered below will be assumed to satisfy the local Lipschitz condition.

If $A, B \subseteq X$ then

$$\varrho_A(x) = \inf_y \{\|y - x\| : y \in A\},$$

$$\varrho(B, A) = \sup_x \{\varrho_A(x) : x \in B\},$$

$$\Delta(A, B) = \max\{\varrho(B, A), \ \varrho(A, B)\}.$$

For a given function f we set $f^+ = f$, $f^- = -f$ and

$$\mathcal{D}f(x, \overline{x}) = \varliminf_{\overline{y} \to \overline{x}} \sup_{\lambda \downarrow 0} \frac{f(x + \lambda \overline{y}) - f(x)}{\lambda},$$

$$\mathcal{D}^0 f(x, \overline{x}) = \lim_{y \to x} \sup_{\lambda \downarrow 0} \frac{f(y + \lambda \overline{x}) - f(y)}{\lambda}.$$

The function $\mathcal{D}^0 f$ is known to possess the following properties: it is convex, positively homogeneous with respect to $\overline{x}$ and

$$\mathcal{D}^0 f(x, \overline{x}) \leq L \|\overline{x}\|,$$

where L is a local Lipschitz constant.

If $M \subseteq X$ and $x \in M$, then we set

$$T_M^0(x) = \{\overline{x}\colon \mathcal{D}^0 \varrho_M(x, \overline{x}) \leq 0\}.$$

By virtue of the property of operation $\mathcal{D}^0$, $T_M^0(x)$ is a convex closed cone. The given properties of operation $\mathcal{D}^0$ were studied in many works. It was firstly introduced into general use by F. H. Clarke in [1].

We call the direction $\overline{x}$ tangent to M at a point x if

$$\lim_{\lambda \downarrow 0} \frac{\varrho_M(x + \lambda \overline{x})}{\lambda} = 0.$$

It is obvious that the cone $T_M^0(x)$ consists of tangent directions. We call a convex cone $T_M(x)$ a tangent cone to M at point x if it consists of tangent directions.

2. Upper convex approximations (u.c.a.).

Definition. A function $h(\overline{x})$ is called an upper convex approximation of f at the point x if it is positively homogeneous, convex, closed and

$$h(\overline{x}) \geq \mathcal{D}f(x, \overline{x}), \qquad \forall \overline{x}.$$

The set

$$\partial f(x) = \{\, x^* \in X^*\colon \langle \overline{x},\, x^* \rangle \leq h(\overline{x}) \,\}$$

is called a subdifferential of f at the point x.

From the convex analysis [2] it is well known that, under the given assumptions, $\partial f(x) \neq \emptyset$ and

$$h(\overline{x}) = \max_{x^*}\{\, \langle \overline{x},\, x^* \rangle\colon\ x^* \in \partial f(x) \,\}.$$

It should be emphasized that u.c.a. is not defined uniquely and therefore it is desirable to construct possibly most complete families of u.c.a. for function f at a given point x. This makes it possible to formulate the most complete necessary conditions for an extremum.

Since it is obvious that

$$\mathcal{D}^0 f(x, \overline{x}) \geq \mathcal{D}f(x, \overline{x}),$$

then $\mathcal{D}^0 f(x, \overline{x})$ is always an u.c.a. Its corresponding subdifferential is denoted by $\partial^0 f(x)$. Here it is elementary to check that

$$\mathcal{D}^0 f^-(x, \overline{x}) = \mathcal{D}^0 f(x, -\overline{x})$$

and therefore $\partial^0 f^-(x) = -\partial^0 f(x)$. For arbitrary subdifferentials this relation is not valid.

The class of functions which admit upper convex approximations is very broad. Methods of calculation of u.c.a. are given in considerable detail in [2]. Note also that quasidifferential functions studied in [3] also have u.c.a. which are easily calculated by their sub- and superdifferentials [3]. For illustration we give the following theorem which can be easily proved.

Theorem 1. Let f be a continuous convex function. Then

$$\mathcal{D}f(x, \overline{x}) = f'(x, \overline{x}) \equiv \lim_{\lambda \downarrow 0} \frac{f(x + \lambda\overline{x}) - f(x)}{\lambda}$$

is the u.c.a., and the corresponding subdifferential ∂f coincides with the subdifferential $\partial^0 f$ and with the usual subdifferential of the convex function.

At the same time

$$\mathcal{D}f^-(x, \overline{x}) = -f'(x, \overline{x}) \leq \langle \overline{x}, -x^* \rangle$$

for any element $x^* \in \partial f(x)$, and therefore $-x^*$ is the subdifferential of $f^- = -f$ for any choice $x^* \in \partial f(x)$. This theorem shows a distinguishing feature. The Clarke subdifferential $\partial^0 f^-$ is defined uniquely and equals $-\partial^0 f(x)$. At the same time the use of the u.c.a. gives the whole family of subdifferentials.

3. Penalty functions.

Let now $f_0, f_1, \ldots, f_n$ be locally Lipschitz functions, and M be an arbitrary set. Consider the following problem $P(0)$: find

$$V(0) = \inf_x \{f_0(x)\colon\ f_i(x) \leq 0,\ i = 1, \ldots, m,\ x \in M\}.$$

The fact that we consider only constraints of the inequality type does not mean any loss of generality since the constraint

$$f(x) = 0$$

is equivalent to two inequalities

$$f(x) \leq 0, \qquad -f(x) \leq 0,$$

and we shall use this in future.

Let

$$V(y) = \inf_x \{ f_0(x)\colon\ f_i(x) \leq y_i,\ i = 1, \ldots, m,\ x \in M \},$$
$$F(x) = \max\{ 0,\ f_1(x), \ldots,\ f_m(x) \},$$
$$\Phi_N(x) = f_0(x) + NF(x).$$

The following chain of inequalities is evident:

$$
\begin{aligned}
\inf_{x\in M} \Phi_N(x) &= \inf_{x,\lambda}\{ f_0(x) + N_\lambda\colon\ 0 \le \lambda,\ f_1(x) \le \lambda, \dots,\ f_m(x) \le \lambda,\ x \in M \} \\
&= \inf_{\lambda \ge 0} [V(\mathbf{1}\lambda) + N\lambda], \\
\mathbf{1} &= \begin{pmatrix} 1 \\ \vdots \\ 1 \end{pmatrix} \in R^m.
\end{aligned} \tag{1}
$$

From this it follows that if the lower bound on the right-hand side of (1) is attained only at $\lambda = 0$, i.e.

$$V(\mathbf{1}\lambda) - V(0) > -N\lambda, \qquad \lambda > 0, \tag{2}$$

then

$$V(0) = \inf_x \{ \Phi_N(x)\colon\ x \in M \}. \tag{3}$$

Thus, if (2) is satisfied then the lower bound in problem $P(0)$ coincides with the lower bound of $\Phi_N(x)$ for $x \in M$.

We shall show that there is valid a stronger result, namely the points at which the minimum is attained in problem $P(0)$ and the points of the minimum $\{ \Phi_N(x),\ x \in M \}$ coincide.

Theorem 2. Let

$$\inf_{\lambda > 0} \frac{V(\mathbf{1}\lambda) - V(0)}{\lambda} = -L > -\infty, \tag{4}$$

and $N > L$. Then the points of minimum of the problem $P(0)$ and of the problem $\inf\{ \Phi_N(x),\ x \in M \}$ coincide.

Remark. Since $V(\mathbf{1}\lambda)$ is evidently a decreasing function of λ, then $L \ge 0$.

Proof. From (4) it follows that $V(\mathbf{1}\lambda) + N\lambda > V(0)$ for $\lambda > 0$. Let $x_0 \in M$ and $\Phi_N(x_0) = \inf\{ \Phi_N(x),\ x \in M \}$. Then,

$$\inf_x \{ f_0(x) + NF(x)\colon\ x \in M \} \le \inf_x \{ f_0(x)\colon\ f_i(x) \le 0,\ i = 1, \dots, m.\ x \in M \},$$

i.e.,

$$V(0) \ge \inf_x \{ \Phi_N(x)\colon\ x \in M \}. \tag{5}$$

We show that $F(x_0) = 0$. Assume the oposite, i.e., that $\lambda_0 = F(x_0) > 0$. We have

$$
\begin{aligned}
f_0(x_0) + NF(x_0) &= \min_x \{ f_0(x) + NF(x)\colon\ x \in M \} \\
&\le \min_x \{ f_0(x) + NF(x)\colon\ x \in M,\ F(x) \le \lambda_0 \} \\
&\le \min_x \{ f_0(x) + N\lambda_0\colon\ x \in M,\ F(x) \le \lambda_0 \} \\
&= V(\mathbf{1}\lambda_0) + N\lambda_0.
\end{aligned}
$$

But, on the other hand,

$$\begin{aligned} f_0(x_0) + NF(x_0) =& f_0(x_0) + N\lambda_0 \\ &\geq \min_x \{\, f_0(x) + N\lambda_0\colon\ x \in M,\ F(x) \leq \lambda_0 \,\} \\ &\geq V(\mathbf{1}\lambda_0) + N\lambda_0. \end{aligned}$$

Thus,

$$f_0(x_0) + NF(x_0) = V(\mathbf{1}\lambda_0) + N\lambda_0.$$

From this equality together with (4), (5) and the fact that $N > L$ we obtain

$$V(0) \geq f_0(x_0) + NF(x_0) = V(\mathbf{1}\lambda_0) + N\lambda_0 > V(0),$$

since $\lambda_0 = F(x_0) > 0$. Thus, $F(x_0) = 0$, from which it follows that the point $x_0 \in M$ satisfies all constraints of the problem $P(0)$. Therefore $f_0(x_0) \geq V(0)$. But, according to (5),

$$V(0) \geq f_0(x_0) + NF(x_0) = f_0(x_0).$$

Thus, $f_0(x_0) = V(0)$, i.e., x_0 is the solution of the problem $P(0)$.

Conversely, let x_0 be the solution of the problem $P(0)$. Then

$$\begin{aligned} f_0(x_0) + NF(x_0) &= f_0(x_0) = V(0) \\ &\leq \inf_{\lambda \geq 0} \left[V(\mathbf{1}\lambda) + N\lambda\right] = \inf_x \{\, \Phi_N(x)\colon\ x \in M \,\}, \end{aligned}$$

i.e., x_0 minimizes $\Phi_N(x)$ on M. ■

The condition (4) has a global character since, for large λ, it takes into account the behavior of the functions f_i at infinity. At the same time the derivation of necessary conditions for an extremum is based on a local consideration. Therefore it is necessary to introduce some additional notions.

Let further x_0 be a point of a local minimum in the problem $P(0)$, and Ω be some neighborhood of x_0.

Definition. A point x_0 of a local minimum in the problem $P(0)$ is regular if there exists a neighborhood Ω of x_0 such that

$$\liminf_{\lambda \downarrow 0} \frac{V_\Omega(\mathbf{1}\lambda) - V_\Omega(0)}{\lambda} > -\infty, \tag{6}$$

$$V_\Omega(0) = V(0), \tag{7}$$

where

$$V_\Omega(y) = \inf_x \{\, f_0(x)\colon\ f_i(x) \leq y_i,\ i = 1, \ldots, m,\ x \in M \cap \Omega \,\}.$$

It should be recalled that, by the definition of a local minimum, the neighborhood Ω satisfying (7) always exists.

Theorem 3. If the point x_0 is regular, then for a sufficiently large number $N > 0$, it is a point of minimum of the function $\Phi_N(x)$ on the set $M \cap \Omega$.

Proof. By virtue of condition (6) there exists such a number $\lambda_0 > 0$ that

$$\frac{V_\Omega(\mathbf{1}\lambda) - V_\Omega(0)}{\lambda} \geq -L, \qquad 0 < \lambda \leq \lambda_0.$$

Further, let

$$\alpha = \inf_x \{ f_0(x) \colon x \in \Omega \}.$$

The number α is finite since f_0 satisfies the local Lipschitz condition. Then

$$\frac{V_\Omega(\mathbf{1}\lambda) - V_\Omega(0)}{\lambda} \geq \frac{\alpha - V(0)}{\lambda_0}, \qquad \lambda \geq \lambda_0.$$

Now take

$$N > \max\left\{ L,\ \frac{V(0) - \alpha}{\lambda_0} \right\}.$$

We have

$$\frac{V_\Omega(\mathbf{1}\lambda) - V_\Omega(0)}{\lambda} > -N$$

for all $\lambda > 0$, and Theorem 3 follows from Theorem 2. ■

The importance of Theorem 3 consists in the fact that under the condition of regularity it reduces the construction of necessary conditions for a local minimum in problem $P(0)$ to the derivation of the necessary conditions for a local minimum of function $\Phi_N(x)$ on the set $M \cap \Omega$.

4. Regularity.

Let X, Y be Banach spaces, and let $M \subseteq X$, $x \in M$. Consider an operator $\mathcal{F}$ which maps some neighborhood of x into Y, and let $\mathcal{F}(x) = 0$.

Definition. A point x is called $(\mathcal{F}, M)$-regular if there exist a neighborhood Ω of x and a number N such that

$$\varrho_{\mathcal{D}_0}(y) \leq N \, \|\mathcal{F}(y)\|, \qquad \forall y \in M \cap \Omega,$$

$$\mathcal{D}_0 = \{ y \in M \cap \Omega \colon \mathcal{F}(y) = 0 \}.$$

The notion of regularity is closely related to costruction of tangent manifolds and to implicit function theorems. A review of these results can be found in [4]. We present two results for illustration. The first is the Ljusternik theorem [4], the second is obtained in [5].

Theorem 4. a) Let the operator $\mathcal{F}$ be continuously Fréchet differentiable in a neighborhood of a point x and $\mathcal{F}'(x)X = Y$. Then the point x is $(\mathcal{F}, X)$-regular.

b) Let Y be finite-dimensional, $\mathcal{F}$ be continuously differentiable in a neighborhoood of the point x and

$$\mathcal{F}'(x)T_M^0(x) = Y.$$

Then the point x is $(\mathcal{F},\, M)$-regular.

The study of $(\mathcal{F},\, M)$-regularity in the general case is a difficult problem, important in many areas of mathematics.

Let us associate now the notion of regularity of solution x_0 to problem $P(0)$ with the introduced notion of the $(\mathcal{F},\, M)$-regularity.

Theorem 5. If the point x_0 is $(F,\, M)$-regular, where the mapping $F\colon\, X \to R^1$ is given by the formula

$$F(x) = \max\{\, 0,\, f_1(x), \ldots,\, f_m(x)\,\}$$

then x_0 is a regular solution of problem $P(0)$.

Proof. The proof is by contradiction. Assume that x_0 is not a regular solution of problem $P(0)$, i.e. for an arbitrarily small neighborhood Ω of the point x_0

$$\liminf_{\lambda \downarrow 0} \frac{V_\Omega(\mathbf{1}\lambda) - V_\Omega(0)}{\lambda} = -\infty.$$

If $N > 0$ is fixed then there exists a small $\lambda > 0$ such that

$$V_\Omega(\mathbf{1}\lambda) - V_\Omega(0) \le -N\lambda.$$

We choose $y \in M \cap \Omega$, $F(y) \le \lambda$, such that

$$f_0(y) - V_\Omega(\mathbf{1}\lambda) \le \lambda^2.$$

Then

$$f_0(y) - \lambda^2 - V_\Omega(0) \le -N\lambda \tag{8}$$

and

$$f_0(x) \ge V_\Omega(0) \ge f_0(y) - \lambda^2 + N\lambda$$

for any point x of the set

$$\mathcal{D}_0 = \{\, x \in M \cap \Omega\colon\; F(x) = 0\,\}.$$

Therefore

$$L_0\, \|y - x\| \ge f_0(x) - f_0(y) \ge \lambda N - \lambda^2,$$

where L_0 is a Lipschitz constant for f_0 in the neighborhood Ω. Hence we obtain

$$\varrho_{\mathcal{D}_0}(y) \ge \left(\frac{N}{L_0} - \frac{\lambda}{L_0}\right)\lambda.$$

If λ is chosen sufficiently small so that $N - \lambda \geq 0.5N$, then we finally obtain

$$\varrho_{\mathcal{D}_0}(y) \geq 0.5NL_0^{-1}\lambda \geq 0.5NL_0^{-1}F(y), \tag{9}$$

since y is chosen such that $y \in M \cap \Omega$, $F(y) \leq \lambda$. Then, by virtue of (8)

$$f_0(y) \leq V_\Omega(0) - \lambda(N - \lambda) < V_\Omega(0).$$

Hence y cannot belong to $\mathcal{D}_0$ and therefore $F(y) > 0$. Since Ω can be arbitrary small and N can be chosen arbitrary, (9) contradicts the definition of the (F, M) -regularity of the point x_0. The obtained contradiction proves the theorem. ■

Corollary. If the point x_0 of a local minimum in problem $P(0)$ is (F, M)-regular, then there exists a neighborhood Ω such that for all sufficiently large N, x_0 is a point of minimium of $\Phi_N(x)$ on the set $M \cap \Omega$.

5. Necessary conditions for a minimum.

Let, as before, x_0 be a solution of problem $P(0)$.

Theorem 6. Let $h_i(\overline{x})$, $i = 0, 1 \ldots, m$, be an upper convex approximations of f_i at the point x_0, and $T_M(x_0)$ be a convex tangent cone to M at this point. Suppose that $\operatorname{dom} h_i = X$ and the point x_0 is (F, M)-regular. Then there exist numbers $\lambda_i \geq 0$, not all of them equal to zero, such that $\lambda_0 = 1$ and

$$\begin{aligned} \sum_{i=0}^{m} \lambda_i h_i(\overline{x}) \geq 0, \qquad & \overline{x} \in T_M(x_0), \\ \lambda_i f_i(x_0) = 0, \qquad & i = 1, \ldots, m, \end{aligned} \tag{10}$$

$$\Big(\sum_{i=0}^{m} \lambda_i \partial f_i(x_0)\Big) \cap T_M^*(x_0) \neq \emptyset, \tag{11}$$

where $\partial f_i(x_0)$ are subdifferentials corresponding to h_i and $T_M^*(x_0)$ is the cone conjugate to $T_M(x_0)$.

Proof. Relation (11) follows directly from (10) by virtue of the known results of convex analysis. Therefore we focus our attention on inequality (10). Since the point x_0 is (F, M)-regular, then there exists a neighborhood Ω of x_0 such that x_0 is a point of minimum of $\Phi_N(x)$ on $M \cap \Omega$. Since x_0 is an internal point of Ω, then $T_M(x_0)$ is simultaneously a tangent cone to $M \cap \Omega$.

Denote

$$\begin{aligned} &\varphi_0(x) = f_0(x) \\ &\varphi_i(x) = f_0(x) + N f_i(x), \qquad i = 1, \ldots, m. \end{aligned}$$

Then from the fact that x_0 minimizes $\Phi_N(x)$ on $M \cap \Omega$ it follows that the point x_0, $\xi_0 = f_0(x_0)$ is a solution to the problem

$$\min_{x,\xi}\{\, \xi\colon\ \varphi_i(x) - \xi \leq 0,\ i = 0, 1, \ldots, m,\ x \in M \cap \Omega \,\}. \tag{12}$$

It is obvious that the functions $h_0(\overline{x}) - \overline{\xi}$, $h_0(\overline{x}) + Nh_i(\overline{x}) - \overline{\xi}$ are upper convex approximations of $\varphi_0(x) - \xi$ and $\varphi_i(\overline{x}) - \xi$, $i = 1, \ldots, m$, at the point x_0.

Now we use the known necessary conditions of a minimum for problem (12) (see Theorem 4.4 in [6], and [7]) to conclude that there exist numbers $\lambda_i \geq 0$ and $\gamma \geq 0$, not all of them equal to zero, such that

$$\gamma\overline{\xi} + \lambda_0\big(h_0(\overline{x}) - \overline{\xi}\big) + \sum_{i=1}^{m} \lambda_i\big(h_0(\overline{x}) + Nh_i(\overline{x}) - \overline{\xi}\big) \geq 0,$$

$$\overline{x} \in T_M(x_0), \qquad \overline{\xi} \in R^1, \tag{13}$$

$$\lambda_i\big(\varphi_i(x_0) - \xi_0\big) = 0, \qquad i = 0, 1, \ldots, m.$$

Since $\xi_0 = f_0(x_0)$, then from the second relation it directly follows that

$$\lambda_i N f_i(x_0) = 0, \qquad i = 1, \ldots, m. \tag{14}$$

We rewrite the first of these relations in the form

$$\Big(\gamma - \sum_{i=0}^{m} \lambda_i\Big)\overline{\xi} + \Big(\sum_{i=0}^{m} \lambda_i\Big) h_0(\overline{x}) + \sum_{i=1}^{m} \lambda_i N h_i(\overline{x}) \geq 0,$$
$$\overline{x} \in T_M(x_0), \qquad \xi \in R^1.$$

Since $\overline{\xi}$ is arbitrary, then the inequality is satisfied if and only if

$$\sum_{i=0}^{m} \lambda_i = \gamma.$$

Here $\gamma > 0$, since if $\gamma = 0$ then all $\lambda_i = 0$ by virtue of their non-negativity. Without loss of generality, we may assume that $\gamma = 1$. Then we finally obtain

$$h_0(\overline{x}) + \sum_{i=1}^{m} \lambda_i N h_i(\overline{x}) \geq 0, \qquad \overline{x} \in T_M(x_0).$$

If we redenote $\lambda_i N$ again by λ_i, then taking into account (14) we obtain (10). ∎

The proved theorem shows that the assumption of regularity eliminates all problems arising in the derivation of the necessary conditions for a minimum. In particular, without additional efforts, we can include an arbitrary number of additional equality-type constraints if each of them is replaced by two inequalities as it was stated above. We give the corresponding result.

Theorem 7. Let x_0 be a solution of the problem

$$\min\{\, f_0(x)\colon\ f_i(x) \leq 0,\ i = 1, \ldots, m,\ f_i(x) = 0,\ i = m+1, \ldots, k,\ x \in M \,\}.$$

Denote

$$F_0(x) = \max\{\, f_1(x), \ldots, f_m(x),\ |f_{m+1}(x)|\,, \ldots, |f_k(x)|\,\}.$$

If the point x_0 is $(F_0,\, M)$-regular, then there exist numbers

$$\lambda_i \geq 0, \quad i = 0, 1, \ldots, m, \quad \lambda_i^+, \lambda_i^- \geq 0, \quad i = m+1, \ldots, k, \quad \lambda_0 = 1,$$

such that

$$\sum_{i=0}^{m} \lambda_i h_i(\overline{x}) + \sum_{i=m+1}^{k} (\lambda_i^+ h_i^+(\overline{x}) + \lambda_i^- h_i^-(\overline{x})) \geq 0, \qquad \overline{x} \in T_M(x_0),$$

$$\lambda_i f_i(x_0) = 0, \qquad i = 1, \ldots, m.$$

Here h_i, h_i^+, h_i^- are u.c.a. for f_i, f_i^+, f_i^- at the point x_0 and it is assumed that $\operatorname{dom} h_i$, $\operatorname{dom} h_i^+$, $\operatorname{dom} h_i^-$ coincide with X.

Besides,

$$\Big[\sum_{i=0}^{m} \lambda_i^+ \partial f_i(x_0) + \sum_{i=m+1}^{k} (\lambda_i^+ \partial f_i^+(x_0) + \lambda_i^- \partial f_i^-(x_0))\Big] \cap T_M^*(x_0) \neq \emptyset.$$

Let us illustrate applications of this theorem by the following example. Let $f_0(x)$ and $f(x)$ be a smooth and a continuous convex function respectively. Consider the problem

$$\min\{\, f_0(x)\colon\ f(x) = 0\,\},$$

and let x_0 be its solution. If the point x_0 is $(|f|\,, X)$ regular, then Theorem 7 is applicable. Further, f_0 is a smooth function and therefore $\partial f_0(x_0) = \{\, f_0'(x_0)\,\}$. Since f is convex, then $\partial f^+(x_0) = \partial^0 f(x_0)$ is a usual subdifferential of the convex function f. At the same time, according to Theorem 1, $-x^*$ is a subdifferential of $f^- = -f$ for any choice of x^* from $\partial^0 f(x_0)$. Since the use of Theorem 7 makes it possible to choose any subdifferentials and for $M = X$, $T_M(x_0) = X$, $T_M^*(x_0) = \{0\}$, then we obtain the following result: for any $x^* \in \partial^0 f(x_0)$ there exist numbers $\lambda^+ \geq 0$, $\lambda^- \geq 0$ such that

$$0 \in f_0'(x_0) + \lambda^+ \partial^0 f(x_0) - \lambda^- x^*.$$

If we denote

$$\operatorname{con} M = \{\, \lambda x\colon\ \lambda \geq 0,\ x \in M\,\},$$

then finally we can formulate the result in the following form:

for any $x^* \in \partial^0 f(x_0)$

$$(\operatorname{con} x^*) \cap \big(f_0'(x_0) + \operatorname{con} \partial^0 f(x_0)\big) \neq \emptyset. \tag{15}$$

Comparing formula (15) with results of [1], we find that there exist $x^* \in \partial^0 f(x_0)$ and numbers $\lambda_0 \geq 0$, $\lambda \in R^1$, which are not at the same time equal to zero, such that

$$\lambda_0 f_0'(x_0) + \lambda x^* = 0.$$

It is clear that these results are comparable only if x_0 is a point of smoothness of function f. In other cases the condition (15) is considerably stronger.

6. Conclusion.

The obtained results show that the regularity assumption of the problem allows to construct the first-order necessary conditions for an extremum, which are significantly more comprehensive in the case of non-smooth data of the problem, than it is possible without the regularity assumptions.

Naturally checking the regularity conditions is not a simple problem, as it is shown by Theorem 4, and a further study of this problem is necessary. At the same time, the necessary conditions for a minimum obtained without the requirement of the regularity conditions (especially for non-smooth problems) are often either too weak or are fulfilled at points which obviously are not the points of minimum. Such examples can be found in abundance. Therefore it is of great interest to investigate a reasonable compromise between required test of regularity conditions and possibility of constructing sufficiently complete necessary conditions of an extremum without a complete preliminary test of regularity.

References

1. Clarke F.H. A new approach to Lagrange multipliers, Math. of Operations Research. **1** (1976), 165–174.
2. Pshenichnyj B. N. Convex analysis and extremal problems, Nauka, Moscow 1980 (in Russian).
3. Dem'yanov V. F., Rubinov A. M. Quasidifferential Calculus, Optimization Software, New York 1986.
4. Dmitr'yuk A. W., Milyutin A. A., Osmolovskii, N. P. Lusternik's theorem and extremum theory, Uspiekhy Matematichieskih Nauk 35 (1980), 11–46 (in Russian).
5. Aubin J.-P., Frankowska H. On inverse function theorems for self-valued maps. IIASA, WP–84–68, 1984, 1–21.
6. Pshenichnyj B. N. Necessary conditions of extremum, M.N. 1982 (in Russian).
7. Neustadt L. Optimization: A Theory of Necessary Conditions. — Princeton University Press. Princeton, N.J., 1976.

Boris N. Pshenichnyj
V.M.Glushkov Institute of Cybernetics
Academy of Sciences of the Ukrainian SSR
252207 Kiev 207, USRR

TRANSFORMATIONS AND REPRESENTATIONS OF NONLINEAR SYSTEMS

A.J. van der Schaft

Abstract. This paper deals with nonlinear systems described by sets of smooth algebraic and (higher-order) differential equations in the external variables (inputs and outputs), as well as auxiliary variables (states, driving variables). A general theorem for transforming such a system into a locally equivalent one, assuming constant rank conditions, is formulated. With the aid of this theorem it is shown how one can eliminate the auxiliary variables in the system description. Conversely, necessary and sufficient conditions for representing a system only involving inputs and outputs as a state space system are derived. Finally the techniques are applied to the construction of inverse systems.

1. INTRODUCTION

One of the central themes in system theory is the problem of *representing* a system in a form which is convenient for the particular purpose one has in mind, and of *transforming* one representation into another. Especially for linear systems it is well-known that the various representations which are being employed (such as state space representations, transfer matrix and impulse response matrix representations, matrix polynomial descriptions) all have their own advantages for control, modelling, identification, theoretical purposes, etc., and that it is very useful to have ways (cq. algorithms) of readily transforming one representation into another. For nonlinear systems the representations which have been dominant in the literature so far, are the state space description

$$(1.1) \qquad \begin{aligned} \dot{x} &= f(x,u) \qquad x \in \mathbb{R}^n, \qquad u \in \Omega \subset \mathbb{R}^m, \qquad y \in \mathbb{R}^p \\ y &= h(x,u) \end{aligned}$$

where x are local coordinates for the n-dimensional state space manifold, Ω is some subset of $\mathbb{R}^m$, and f and h are smooth (i.e. C^∞) mappings, and various input-output map descriptions (Volterra series, generating power series, general non-anticipating maps), and the relations between these representations are by now fairly well-understood (see [8] and the survey [10] and the references quoted there). However it has become more and more apparent that also the representation as an *input-output differential system*

$$(1.2) \qquad P_i(y,\dot{y},..,y^{(k)},u,\dot{u},..,u^{(k)}) = 0 \quad i = 1,..,\ell, \quad y \in \mathbb{R}^p, \ u \in \mathbb{R}^m$$

with $y^{(j)},u^{(j)}$, $j = 0,1,..,k$, denoting the j-th time-derivatives of the time-functions $y(t),u(t)$, up to some arbitrary order k, and P_i smooth functions, may arise naturally in modelling [24,25,17,22], and are very important in understanding many properties of the system (see e.g. Fliess [4,5,6]).

In this paper we want to focus on the relations between the state space representation (1.1), and the input-output differential representation (1.2), continuing and partly surveying our earlier work [17,18,19,20, 21]. (We remark that the connections between input-output differential representations and input-output maps are largely unexplored up to now, see however [23]). The mathematical tools that will be employed here are mainly the implicit function theorem (thereby restricting basically to a *local* analysis *avoiding singularities*), and some recent notions from geometric nonlinear control theory. As such our approach will be quite elementary, and there is certainly a need for more sophisticated analytical and differential algebraic (and geometric) methods, see already [3,6,7,13,15].

2. EQUIVALENCE TRANSFORMATIONS

The principal aim of this section is to state a basic theorem on the local transformation of a set of higher-order differential equations into an *equivalent* but *reduced* set of higher-order differential equations, provided some constant rank assumptions are satisfied. This theorem will be instrumental to all the further developments in this paper.

Let us first recall some terminology, see e.g. [18,21]. Consider a single higher-order differential equation

$$(2.1) \qquad P(z,\dot{z},..,z^{(k)},\xi,\dot{\xi},..,\xi^{(k)}) = 0, \qquad z \in \mathbb{R}^q, \ \xi \in \mathbb{R}^s.$$

Primarily we look at (2.1) as an *algebraic* equation in the indeterminates

$$z,\dot{z},..,z^{(k)},\xi,\dot{\xi},..,\xi^{(k)} \in \mathbb{R}^{(k+1)(q+s)}$$

Indeed, let $(\bar{z}(t),\bar{\xi}(t))$, $t \in (-\epsilon,\epsilon)$, $\epsilon > 0$, be a smooth solution of the differential equation (2.1) for all $t \in (-\epsilon,\epsilon)$. We denote the point in the k-jet space $\mathbb{R}^{(k+1)(q+s)}$ defined by taking the time-derivatives of $(\bar{z}(t),\bar{\xi}(t))$ up to order k in $t = 0$ by $(\bar{z},\bar{\xi})$. Clearly $(\bar{z},\bar{\xi})$ is a solution point of (2.1) regarded as an algebraic equation.
The *order* σ of (2.1) with respect to the variables ξ in a solution point $(\bar{z},\bar{\xi})$ is defined as the largest integer such that

$$(2.2) \qquad \frac{\partial P}{\partial \xi^{(\sigma)}}(z,\dot{z},..,z^{(k)}, \xi,..,\xi^{(k)}) \neq 0 \qquad \text{in } (\bar{z},\bar{\xi}).$$

If σ is not defined (i.e. if P does not depend on ξ), then we set $\sigma = 0$. The first-order *prolongation* of (2.1) is defined as

$$(2.3) \qquad P^{(1)}(z,\dot{z},..,z^{(k+1)}, \xi,\dot{\xi},..,\xi^{(k+1)}) :=$$

$$\frac{\partial P}{\partial z}\dot{z} + .. + \frac{\partial P}{\partial z^{(k)}} z^{(k+1)} + \frac{\partial P}{\partial \xi}\dot{\xi} + .. + \frac{\partial P}{\partial \xi^{(k)}} \xi^{(k+1)} = 0,$$

(using obvious vector notation), and inductively we define the j-th order prolongation by setting $P^{(j)} = (P^{(j-1)})^{(1)}$, $j = 1,2,..$ (with $P^{(0)} = P$). It immediately follows that if $(\bar{z}(t),\bar{\xi}(t))$, $t \in (-\epsilon,\epsilon)$, is a solution of (2.1), then it is also a solution of (2.3) and of all prolonged equations $P^{(j)} = 0$, $j = 1,2,..$. It thus follows that we may add to (2.1) all prolonged equations $P^{(j)} = 0$, $j = 1,2,..$, without changing the solution set of time-functions $(z(t),\xi(t))$. (Note that in principle we are working in a smooth category, i.e. the time-functions $(z(t),\xi(t))$ are assumed to be arbitrarily often differentiable in $t = 0$.) Motivated by this we call two *sets* of higher-order differential equations in the variables $z \in \mathbb{R}^q$, $\xi \in \mathbb{R}^s$

$$(2.4) \qquad P_i(z,\dot{z},..,z^{(k)},\xi,\dot{\xi},..,\xi^{(k)}) = 0, \quad i = 1,...,\ell$$

and

$$(2.5) \qquad R_i(z,\dot{z},..,z^{(\bar{k})},\xi,\dot{\xi},..,\xi^{(\bar{k})}) = 0, \quad i = 1,...,\bar{\ell}$$

equivalent if the following holds. Equations (2.4) define some subset of $\mathbb{R}^{(k+1)(q+s)}$, while equations (2.5) define some subset of $\mathbb{R}^{(\bar{k}+1)(q+s)}$. Now (2.4) and (2.5) are called *equivalent* if there exists some $\tilde{k}$, with $\tilde{k} \geq k$ and $\tilde{k} \geq \bar{k}$, such that the subset of $\mathbb{R}^{(\tilde{k}+1)(q+s)}$ defined by equations (2.4) together with some suitable prolonged equations

$P_i^{(j)} = 0$, equals the subset of $\mathbb{R}^{(\tilde{k}+1)(q+s)}$ defined by equations (2.5) together with some suitable prolonged equations $R_i^{(j)} = 0$. The systems (2.4) and (2.5) are called *locally equivalent* around a common solution point $(\bar{z},\bar{\xi})$ if *locally* around $(\bar{z},\bar{\xi})$ they define, as above, the same subset of $\mathbb{R}^{(\tilde{k}+1)(q+s)}$.

Theorem 2.1. [21]. A set of higher-order differential equations (2.4) having a solution point $(\bar{z},\bar{\xi})$, is, under constant rank assumptions, locally equivalent around $(\bar{z},\bar{\xi})$ to a set of higher-order differential equations of the form

$$\text{(2.6a)} \qquad \tilde{P}_i(z,\dot{z},..,z^{(\tilde{k})},\xi,\dot{\xi},..,\xi^{(\tilde{k})}) = 0, \qquad i = 1,...,\tilde{\ell}.$$

$$\text{(2.6b)} \qquad \tilde{P}_i(z,\dot{z},..,z^{(\tilde{k})}) = 0, \qquad i = \tilde{\ell} + 1,...,\ell,$$

satisfying (with σ_i the order of $\tilde{P}_i$, $i = 1,..,\tilde{\ell}$)

$$\text{(2.7)} \qquad \left(\frac{\partial \tilde{P}_i}{\partial \xi_j^{(\sigma_i)}} \right)_{\substack{i=1,..,\tilde{\ell} \\ j=1,..,s}} = \tilde{\ell}, \qquad \text{around } (\bar{z},\bar{\xi}).$$

Furthermore, possibly after permutation of the variables $\xi_1,..,\xi_s$ and the equations $\tilde{P}_1,..,\tilde{P}_{\tilde{\ell}}$, we can ensure that

$$\text{(2.8)} \qquad \sigma_1 \le \sigma_2 \le \dots \le \sigma_{\tilde{\ell}}$$

and for $i = 1,...,\tilde{\ell}$,

$$\text{(2.9a)} \qquad \frac{\partial \tilde{P}_i}{\partial \xi_j^{(\tau)}} = 0, \qquad \sigma_j \le \tau \le \sigma_i, \quad j < i,$$

around $(\bar{z},\bar{\xi})$

$$\text{(2.9b)} \qquad \frac{\partial \tilde{P}_i}{\partial \xi_i^{(\sigma_i)}} \ne 0.$$

For a precise statement of the constant rank assumptions involved in Theorem 2.1 we refer to [21]. The proof of the theorem consists of a constructive algorithm for converting (2.4) into (2.6). Basically at every step of the (finite) algorithm one equation is replaced, with the aid of the implicit function theorem, by another equation (depending on the original equations and (prolongations of the) other equations) which has lower order than the original equation.

We will now give some immediate applications of Theorem 2.1. First, in recent papers [24,25] Willems has proposed the general system representation (also covering the Rosenbrock description [16])

(2.10) $\quad P_i(w,\dot{w},..,w^{(k)},\xi,\dot{\xi},..,\xi^{(k)}) = 0, \qquad i = 1,...,\bar{\ell}.$

where $w \in \mathbb{R}^q$ are the *external* variables (inputs and outputs taken together, and not necessarily distinguished from each other), and $\xi \in \mathbb{R}^s$ are the auxiliary or *latent* variables. Application of Theorem 2.1 (with $z = w$) gives that locally around a solution point $(\bar{w},\bar{\xi})$ (2.10) is equivalent to a system

(2.11a) $\quad \tilde{P}_i(w,\dot{w},..,w^{(\tilde{k})},\xi,\dot{\xi},..,\xi^{(\tilde{k})}) = 0, \qquad i = 1,..,\tilde{\ell}$

(2.11b) $\quad \tilde{P}_i(w,\dot{w},..,w^{(\tilde{k})}) = 0, \qquad i = \tilde{\ell}+1,..,\ell$

where furthermore (2.11a) satisfies (2.7),(2.8),(2.9) (with z replaced by w). It follows from these last properties (see [22]) that for every $w(t)$ close to $\bar{w}(t)$, $t \in (-\epsilon,\epsilon)$, there exists a $\xi(t)$ close to $\bar{\xi}(t)$ such that equations (2.11a) are satisfied around $(\bar{w},\bar{\xi})$. Hence the behavior of the external variables w around $(\bar{w},\bar{\xi})$ is completely described by the equations (2.11b), which only involve the external variables. Thus we have been able to *eliminate* the latent variables ξ in our system representation. (An alternative algorithm for eliminating ξ in (2.10), applying to P_i's which are *polynomials*, has been recently proposed in [3], using tools from differential algebra and avoiding the constant rank assumptions of our approach.)

A very special case of (2.10) is formed by the state space system (1.1), written in implicit form as

$$
(2.12) \qquad \begin{aligned} \dot{x}_i - f_i(x,u) &= 0, \qquad i = 1,..,n, \\ y_j - h_j(x,u) &= 0, \qquad j = 1,..,p, \end{aligned}
$$

where we take $(y,u) \in \mathbb{R}^{p+m}$ to be the external variables and $x \in \mathbb{R}^n$ as the latent variables. In this case the orders of the equations (2.12) with respect to x are 1 or 0, and moreover $\dot{x}$ is written as an explicit function of x and u. Due to this special structure the algorithm underlying Theorem 2.1 can be cast into a much more simple and explicit form, see [20], and under constant rank assumptions the system (2.12) can be shown to be locally equivalent to a system of the form [20]

(2.13a) $\dot{x}^1 - f^1(x^1,x^2,u) = 0, \qquad x = (x^1,x^2),$

(2.13b) $x^2 = \psi(y,\dot{y},..,y^{(\tilde{k}-1)},u,\dot{u},..,u^{(\tilde{k}-1)}) = 0, \qquad \tilde{k} \le n,$

(2.13c) $\tilde{P}_i(y,\dot{y},..,y^{(\tilde{k})},u,\dot{u},..,u^{(\tilde{k})}) = 0, \qquad i = 1,..,p,$

where the new coordinate functions $x = (x^1,x^2)$ are such that $h_j(x,u)$ only depends on x^2, $j \in \underline{p}$. It follows [20] that (2.13a) describes the dynamics of the *unobservable* part x^1 of the system. Furthermore, (2.13b) expresses the *observable* part x^2 of the state as a function of y and u and their time-derivatives, and (2.13c) describes the input-output differential system corresponding to (2.12).
Alternative approaches to convert a state space system (1.1) into an input-output differential system can be found in [1,7,15].
As remarked before, Theorem 2.1 and its specialization to state space systems heavily relies on constant rank assumptions (contrary to the differential algebraic approach as initiated in [4,6,14,3,7], see also [13,15]), which is a major drawback of our approach.

Example 2.2. Consider the state space system

(2.14) $\dot{x} - u = 0, \qquad y - x^2 = 0$

This system is equivalent to [20]

(2.15) $\dot{x} - u = 0, \qquad y - x^2 = 0, \qquad \dot{y} - 2xu = 0$

If $x > 0$ then the constant rank assumptions of [20] are satisfied, and the system is equivalent to

(2.16) $x = \sqrt{y}, \qquad \dot{y} - 2\sqrt{y}\,u = 0$

(and similarly for $x < 0$). Around $x = 0$, however, we have to consider further prolongations; we refer to similar considerations in [13,15].

3. REALIZATIONS

In the preceeding section we showed how under constant rank assumptions a state space system (1.1) can be transformed into an input-output differential system

(3.1) $P_i(y,\dot{y},..,y^{(k)},u,\dot{u},..,u^{(k)}) = 0, \; i = 1,...,\ell.$

In this section we want to consider the converse problem of *realizing* an input-output differential system (3.1) as a state space system (1.1). This is a much more complex problem, already because we have to "invent" state variables x instead of eliminating them.

First we have to go a little in detail about the role of the inputs u and outputs y in (3.1). Clearly (3.1) is the nonlinear generalization of a linear system

$$(3.2) \qquad D(\frac{d}{dt})y(t) = N(\frac{d}{dt})u(t)$$

with D(s) and N(s) polynomial matrices of appropriate dimensions.

Usually it is assumed that (3.2) corresponds to a *proper* transfer matrix from u to y. This means that D(s) is a square matrix with det D(s) not identically zero, such that $G(s) := D^{-1}(s)N(s)$ is a proper (i.e. $\lim_{s\to\infty} G(s)$ exists) transfer matrix. This generalizes to the nonlinear case as follows. Consider (3.1); by an application of Theorem 2.1 (with $\xi = (y,u)$, and z void) we can transform (under constant rank assumptions) (3.1) around a solution point $(\bar{y},\bar{u})$ into a locally equivalent input-output differential system

$$(3.3) \qquad \bar{P}_i(y,\dot{y},..,y^{(\bar{k})},u,\dot{u},..,u^{(\bar{k})}) = 0, \qquad i = 1,..,\bar{\ell}(\leq \ell)$$

with orders σ_i, $i \in \underline{\ell}$, satisfying

$$(3.4) \qquad \text{rank} \left(\frac{\partial \bar{P}_i}{\partial w_j^{(\sigma_i)}}\right)_{\substack{i=1,..,\bar{\ell} \\ j=1,..,p+m}} = \bar{\ell}, \qquad w = (y,u)$$

together with $\ell - \bar{\ell}$ trivial equations 0 = 0 (which have order zero).

Defintion 3.1. An input-output differential system (3.3) satisfying (3.4) is called *proper* if $\bar{\ell} = p$ and

$$(3.5) \qquad \text{rank} \left(\frac{\partial \bar{P}_i}{\partial y_j^{(\sigma_j)}}\right)_{\substack{i=1,..,p \\ j=1,..,p}} = p$$

Remark 1. For a linear system (3.2) Definition 3.1 is easily seen ([12,p.385]) to be equivalent to requiring that $D^{-1}(s)N(s)$ is proper.

Remark 2. A non-proper system (3.3) can be always made proper by a different choice of input and output variables, see [21].

Now let us restrict attention to proper input-output differential systems. It follows that without loss of generality we may assume that the system is given as

$$(3.6) \qquad P_i(y,\dot{y},..,y^{(k)},u,\dot{u},..,u^{(k)}) = 0, \qquad i = 1,...,p.$$

satisfying rank $\left(\dfrac{\partial P_i}{\partial y_j^{(\sigma_j)}}\right)_{\substack{i=1,..,p\\ j=1,..,p}} = p$. Furthermore by an application

of Theorem 2.1 (with $\xi = y$ and $z = u$) we may assume without los of generality that

$$\sigma_1 \leq \sigma_2 \leq \ldots \leq \sigma_p$$

$$(3.7) \qquad \frac{\partial P_i}{\partial y_j^{(\tau)}} = 0, \qquad \sigma_j \leq \tau \leq \sigma_i, \ j < i,$$

$$\frac{\partial P_i}{\partial y_i^{(\sigma_i)}} \neq 0, \quad i = 1,..,p$$

(In particular we may set $k = \sigma_p$ in (3.6).) Let d be the largest integer such that $\sigma_d = 0$. Then the first d equations of (3.6) are purely *algebraic* equations, from which by (3.7) and the implicit function theorem we may locally solve for $y_1,..,y_d$, i.e.

$$(3.8) \qquad y_i = \varphi_i(y_{d+1},..,y_p,u_1,..,u_m), \qquad j = 1,..,d.$$

These static equations will not play any role in the following analysis, and thus for clarity of exposition we will throughout assume that $d = 0$ (so that $\sigma_i \geq 1$, $i \in \underline{p}$). Then by (3.7) and the implicit function theorem we may locally solve from (3.6) for $y^{(\sigma_1)} = \frac{d}{dt} y^{(\sigma_1 -1)}$, $...,y^{(\sigma_p)} = \frac{d}{dt} y^{(\sigma_p -1)}$, see [21], so as to obtain

$$(3.9a) \qquad \begin{aligned} \frac{d}{dt} y^{(\sigma_1 -1)} &= k_1(s,u,\dot{u},..,u^{(\sigma_1)}) \\ &\vdots \\ \frac{d}{dt} y^{(\sigma_p -1)} &= k_p(s,u,\dot{u},..,u^{(\sigma_p)}) \end{aligned}$$

for certain smooth functions $k_1,..,k_p$ depending on

$$(3.10) \qquad s:=(y_1,\dot{y}_1,..,y_1^{(\sigma_1 -1)},y_2,..,y_2^{(\sigma_2 -1)},...,y_p,..,y_p^{(\sigma_p -1)})$$

and $u = (u_1,..,u_m)$ and its derivatives.

Let us rewrite (3.9) as a *driven state space system* (see [18,19]).

First we add to (3.9a) the equations

$$\text{(3.9b)} \qquad \frac{d}{dt} y_i^{(r_i-1)} = y_i^{(r_i)}, \qquad r_i = 1,..,\sigma_i - 1, \qquad i = 1,..,p,$$

thus obtaining together with (3.9a) a $(\sigma_1 + \ldots + \sigma_p)$-dimensional set of first-order differential equations in the variables s (defined by (3.10)), parametrized by u and its derivatives. Then we add to (3.9a,b) m parallel (σ_{p+1})-fold strings of integrators

$$\text{(3.9c)} \qquad \begin{aligned} &\frac{d}{dt} u = \dot{u} \\ &\frac{d}{dt} \dot{u} = \ddot{u} \\ &\quad \vdots \\ &\frac{d}{dt} u^{(\sigma_p)} = v \end{aligned}$$

where the m-vector v (called the vector of *driving variables*) consists of arbitrary functions of time t. (This expresses the fact that no constraints are imposed on the time-evolution of $u^{(\sigma_p)}$.) Equations (3.9a,b,c) define a driven state space system, concisely written as

$$\text{(3.11)} \qquad \begin{aligned} &\dot{z} = g_0(z) + \sum_{j=1}^{m} g_j(z) v_j \\ &w_i = H_i(z), \qquad i = 1,..,p+m, \qquad w = (y,u), \end{aligned}$$

with state $z = (s,u,\dot{u},..,u^{(\sigma_p)})$. (Notice that (3.11) is of the form (2.10) with $\xi = (z,v)$.) Pictorially (3.9) has the form

(3.12)

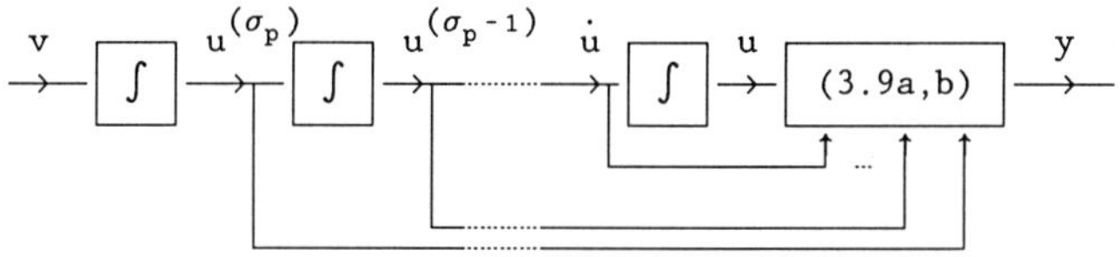

The realization problem is now to find an input-state-output system (1.1) which is equivalent to (3.9) in the sense that the *projections* of the solutions sets of equations (1.1) and (3.9) (and their prolongations) on some $\tilde{k}$-jet space of *only* the output and input variables are *equal*.

The key tool to this problem will be the algorithm for computing the *minimal conditionally invariant distribution* for the driven state space system (3.9) (see [2,7] for related approaches).

Indeed define the nested sequence of distributions [9]

$$(3.13)\qquad \begin{aligned} S_1 &= \text{span}\,\{g_1(z),..,g_m(z)\} \\ S_k &= \bar{S}_{k-1} + \Big[g_0, \bar{S}_1 \cap \Big(\bigcap_{i=1}^{p+m} \ker dH_i\Big)\Big], \qquad k = 2,3,.. \end{aligned}$$

where $\bar{S}$ denotes the involutive closure of a distribution S. Due to the special structure of (3.9) we have

Lemma 3.2. Suppose the distributions $S_1, S_2, .., S_{\sigma_p+1}$ are all *involutive*, then

$$(3.14a)\qquad S_k \subset \bigcap_{i=1}^{p+m} \ker dH_i, \qquad k = 1,..,\sigma_p$$

$$(3.14b)\qquad S_{\sigma_p+1} \cap \Big(\bigcap_{i=p+1}^{p+m} \ker dH_i\Big) = S_{\sigma_p+1} \cap \Big(\bigcap_{i=1}^{p+m} \ker dH_i\Big) = S_{\sigma_p}$$

$$(3.14c)\qquad S_{\sigma_p+j} = S_{\sigma_p+1}, \qquad j = 1,2,...$$

(3.14d) $S_1,..,S_{\sigma_p+1}$ all have constant dimension and $\dim S_k = k \times m$.

Proof. Clearly $S_1 = \text{span}\left\{\dfrac{\partial}{\partial u_j^{(\sigma_p)}}\right\}$ and

$$S_2 = S_1 + \text{span}\left\{\frac{\partial}{\partial u_j^{(\sigma_p-1)}} + \sum_i \frac{\partial k_i}{\partial u_j^{(\sigma_p)}} \frac{\partial}{\partial y_i^{(\sigma_p-1)}}\right\}.$$ Therefore

$$S_3 = S_2 + \text{span}\left\{\frac{\partial}{\partial u_j^{(\sigma_p-2)}} + \sum_i \alpha_{ij} \frac{\partial}{\partial y_i^{(\sigma_p-2)}} + \sum_i \beta_{ij} \frac{\partial}{\partial y_i^{(\sigma_p-1)}}\right\}$$

for certain functions α_{ij}, β_{ij}. In general for $k \le \sigma_p + 1$

$$(3.15)\qquad S_k = S_{k-1} + \text{span}\left\{\frac{\partial}{\partial u_j^{(\sigma_p-k+1)}} + \sum_{r=1}^{k-1} \sum_i \gamma_{ij}^r \frac{\partial}{\partial y_i^{(\sigma_p-r)}}\right\}$$

from which (3.14) immediately follows. □

The following basic theorem was already proved in [S2]. Based on Lemma 3.2 we obtain a much simpler proof.

Theorem 3.3. The proper input-output differential system (3.6) can be locally realized by a state space system (1.1) if the distributions $S_1,..,S_{\sigma_p+1}$ are all involutive.

Proof. The proof is very similar to the proof of the feedback linearization theorem [11]. By involutivity of $S_1,..,S_{\sigma_p+1}$ and (3.14d), an extension of Frobenius' theorem, see [11], yields local coordinates $z = (z^1,..,z^{\sigma_p+1},\tilde{z})$, $\dim z^i = m$, $i \in \underline{\sigma_p+1}$, on the space $(s,u,\dot{u},..,u^{(\sigma_p)})$ such that $S_i = \text{span}\left\{\frac{\partial}{\partial z^1},...,\frac{\partial}{\partial z^i}\right\}$. It is easily seen that we can take $z^1 = u^{(\sigma_p)}$. Furthermore by (3.14a,b) it follows that we can take $z^{\sigma_p+1} = u$. By (3.13) and (3.14a) we have $S_k = S_{k-1} + [g_0, S_{k-1}]$, $k = 2,..,\sigma_p + 1$ and thus ([11]) the equations (3.9) in the above coordinates take the form

$$\text{(3.16a)}\quad \begin{aligned} \dot{z}^1 &= v \\ \dot{z}^2 &= g_0^2(z^1,..,z^{\sigma_p+1},\tilde{z}) \\ &\vdots \\ \dot{z}^{\sigma_p+1} &= g_0^{\sigma_p+1}(z^{\sigma_p},z^{\sigma_p+1},\tilde{z}) \end{aligned}$$

$$\text{(3.16b)}\quad \dot{\tilde{z}} = \tilde{g}_0(z^{\sigma_p+1},\tilde{z})$$

with

$$\text{(3.17)}\qquad \text{rank}\,\frac{\partial g_0^i}{\partial z^{i-1}} = m, \qquad i = 2,..,\sigma_p + 1.$$

Inductively we transform $z^{\sigma_p},z^{\sigma_p-1},..,z^1$ into $x^{\sigma_p},..,x^1$ as follows. First set $x^{\sigma_p} := g_0^{\sigma_p+1}(z^{\sigma_p},z^{\sigma_p+1},\tilde{z})$. By (3.17) replacement of z^{σ_p} by x^{σ_p} defines a coordinate transformation, and it is easily checked that in the new coordinates (3.9) remains of the form (3.16). Then set $x^{\sigma_p-1} := g_0^{\sigma_p}(z^{\sigma_p-1},x^{\sigma_p},z^{\sigma_p+1},\tilde{z})$ and replace z^{σ_p-1} by x^{σ_p-1}. Continuing in this way (3.16a) transforms into the strings of integrators

$$\text{(3.18)}\quad \begin{aligned} \dot{x}^1 &= \tilde{v} \\ \dot{x}^2 &= x^1 \\ &\vdots \\ \dot{x}^{\sigma_p} &= x^{\sigma_p-1} \\ \dot{x}^{\sigma_p+1} &= x^{\sigma_p} \end{aligned}$$

Furthermore from (3.14b) it follows that

$$\text{(3.19)}\qquad dy_j \in d\tilde{z} + \text{span}\,\{du_1,..,du_m\},\ j \in \underline{p},$$

and thus [18] there exist locally smooth functions h_j such that

$$\text{(3.16c)}\qquad y_j = h_j(\tilde{z},u) \qquad j \in \underline{p}$$

Recalling that $z^{\sigma_p+1} = u$, it immediately follows that (3.16b) together with (3.16c) is a state space realization of (3.6). □

Remark 1. Note that the proof of Theorem 3.3 is based on transforming the signal-flow diagram (3.12) into

(3.20)

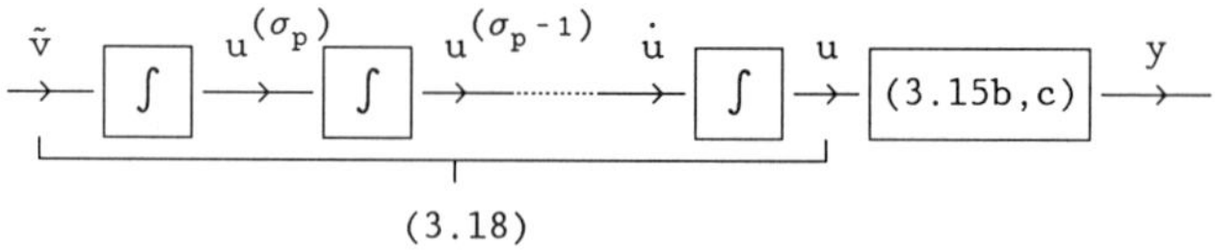

It can be immediately checked that the orders $\sigma_1,\ldots,\sigma_p$ equal the *observability indices* of the input-state-output system (3.16b,c). Furthermore, (3.16b,c) is *locally observable* in the sense defined in [20]. (Indeed the state $\tilde{z}$ can be expressed as function of y and u and their time-derivatives.)
Note that (3.16b,c) *cannot* be expected to be controllable in some reasonable sense, since also systems without inputs are included in the system representation (3.6).

We will now show that the conditions of involutivity of $S_1,\ldots,S_{\sigma_p+1}$ are also *necessary* for (3.6) to be realizable as an input-state-output system (1.1), at least provided some constant rank assumptions are satisfied. First consider the algorithm (3.13). Since the sequence S_k, $k = 1,2,\ldots$, is increasing there will exist under general constant rank assumptions some r such that $S_{r+j} = S_r$ for all $j = 0,1,\ldots$. Denote $S_k = S^*$ then it immediately follows [9] that S^* is involutive and

(3.21) $$[g_0, S^* \cap (\bigcap_{i=1}^{p+m} \ker dH_i)] \subset S^*$$

(S^* is called *conditionally invariant*.) Furthermore S^* is the *minimal* distribution containing S_1 with this property (see [9]). Now assume that S^* has constant dimension. Then define P^* as the annihilating codistribution, i.e.

(3.22) $$S^*(x) = \ker P^*(x) := \operatorname{span}\{X(x) \mid X \text{ vectorfield such that } \alpha(X) = 0 \text{ for all one-forms } \alpha \text{ in } P^*\}$$

It follows from (3.21) (see [9,18]) that P^* is the *maximal* codistribution P with the property that

$$(3.23) \qquad L_{g_0} P \subset P + \mathrm{span}\{dy_1,..,dy_p,du_1,..,du_m\}$$
$$g_j \in \ker P, \qquad j = 1,\dots,m$$

By Frobenius' theorem we can find local coordinates $z = (\bar{z},\tilde{z})$ such that $P^* = \mathrm{span}\ \{d\tilde{z}\}$ (or equivalently $S^* = \mathrm{span}\ \{\frac{\partial}{\partial \bar{z}}\}$). Then (3.21) implies that

$$(3.24) \qquad \dot{\tilde{z}} = \tilde{g}_0(\tilde{z},y_1,..,y_p,u_1,..,u_m)$$

for some function $\tilde{g}_0$. Hence the coordinate functions $\tilde{z}$ can be viewed as the *maximal* part of the state $(s,u,\dot{u},..,u^{(\sigma_p)})$ of (3.9) for which the time-evolution depends only on itself and the external variables $y_1,..,y_p,u_1,..,u_m$. We are now ready to formulate the converse of Theorem 3.3.

Theorem 3.4. Consider the proper input-output differential system (3.6). Assume that S^* exists and has constant dimension for the driven state space system (3.9). If at least one of the distributions S_i, $i \in \underline{\sigma_p + 1}$, is *not* involutive, then (3.6) does *not* admit a local state space representation (1.1) that can be transformed into (2.13).

Proof. Let $s \in \underline{\sigma_p + 1}$ be the first integer such that S_s is *not* involutive. Then it follows from the form of the distributions S_i, $i = 1,\dots,s$ (cf.(3.15)) that there exists a vector in S^*

$$\sum_{r=1}^{\sigma_p} \sum_{i=1}^{p} \delta_i^r \frac{\partial}{\partial y_i^{(\sigma_p - r)}}$$

with $\delta_i^{\sigma_p} \neq 0$ for some i. Since $S^* \supset S_{\sigma_p+1}$ this implies by (3.14b) that

$$(3.25) \qquad \dim \left[S^* \cap \left(\bigcap_{i=p+1}^{p+m} \ker dH_i \right) \right] > \dim \left[S^* \cap \left(\bigcap_{i=1}^{p+m} \ker dH_i \right) \right]$$

Equivalently, $\dim \left(P^* + \mathrm{span}\ \{du_1,..,du_m\}\right) <$ $\dim \left(P^* + \mathrm{span}\ \{dy_1,..,dy_p,du_1,..,du_m\}\right)$, and thus not all outputs $y_1,..,y_p$ are expressible as function of $\tilde{z}$ (with $P^* = \mathrm{span}\ \{d\tilde{z}\}$) and $u_1,..,u_m$. Now let (1.1) be a state space realization of (3.6) that can be transformed into (2.13). It follows that x^2 defined by (2.13b) consists of functions of $(s,u,\dot{u},..,u^{(\sigma_p)})$ whose time-evolution depends only on x^2 and $u_1,\dots,u_m$. By maximality of P^* with regard to (3.23) it follows that $dx^2 \in P^*$. However since $y_j = h_j(x,u) = h_j(x^2,u)$, $j \in \underline{p}$ this means that all outputs are expressible as functions of $\tilde{z}$ and $u_1,..,u_m$, which is a contradiction. □

Remark. Let ρ be the largest integer such that $S_{\rho+1}$ is involutive and $S_\rho \subset \bigcap_{i=1}^{m} \ker du_i$. If $\rho < \sigma_p$ then as shown in Theorem 3.4, the system can not be realized by an input-state-output system (1.1). However we can still factor out the driven state space system (3.9) by the distribution $S_{\rho+1}$, resulting in a system of the form

$$
\begin{aligned}
\dot{x} &= f(x,u,\dot{u},..,u^{(\sigma_p-\rho)}) \\
y &= h(x,u)
\end{aligned}
\tag{3.25}
$$

4. **INVERSE SYSTEMS**

Let us once again consider an input-output differential system (1.2) (not necessarily proper). By applying Theorem 2.1, but now with $z = y$ and $\xi = u$, around a solution point $(\bar{y},\bar{u})$ we obtain, under constant rank assumptions, a locally equivalent system

$$
\tilde{P}_i(y,\dot{y},..,y^{(\tilde{k})},u,\dot{u},..,u^{\tilde{k}}) = 0, \qquad i = 1,..,\tilde{\ell}, \tag{4.1a}
$$

$$
\tilde{P}_i(y,\dot{y},..,y^{(\tilde{k})}) = 0 \qquad i = \tilde{\ell}+1,..,\ell, \tag{4.1b}
$$

with (4.1a) satisfying

$$
\operatorname{rank}\left(\frac{\partial \tilde{P}_i}{\partial u_j^{(\tau_i)}}\right)_{\substack{i=1,..,\tilde{\ell}\\ j=1,..,m}} = \tilde{\ell}, \qquad \text{around } (\bar{y},\bar{u}), \tag{4.2}
$$

where τ_i are the orders of $\tilde{P}_i$, with respect to u, $i = 1,...,\tilde{\ell}$. This immediately suggests the following definitions, very much in the spirit of definitions given by Fliess [4,5].

Definition 4.1. Consider an input-output differential system (1.2), which can be transformed into the form (4.1). The system is called *right-invertible*, if $\tilde{\ell} = \ell$ (i.e. if (4.1b) is void), and *left-invertible* if $\tilde{\ell} = m$.

Remark. For a *proper* input-output differential system (Definition 3.1) we have $\ell = p$ (cf.(3.6)). In this case we have $\tilde{\ell} \leq \min(p,m)$ and the system is right-(left-)invertible if $\tilde{\ell} = p(= m)$. The integer $\tilde{\ell}$ coincides in this case with the *differential output rank* as introduced by Fliess [4,5] for systems defined by polynomials P_i. (Of course, in our approach we had to assume constant rank conditions.)

Let us now assume *left-invertibility* of (4.1), and let us investigate the existence of an inverse system. First of all by Theorem 2.1 (with $z = y$, $\xi = u$) we may assume without loss of generality that (setting $\tilde{\ell} = m$)

$$(4.3) \qquad \begin{gathered} \tau_1 \le \tau_2 \le \dots \le \tau_m \\ \frac{\partial \tilde{P}_i}{\partial u_j^{(\tau)}} = 0, \ \tau_j \le \tau \le \tau_i, \ \ j < i, \qquad \frac{\tilde{\partial} P_i}{\partial u_i^{(\tau_i)}} \ne 0, \ i = 1,..,m \end{gathered}$$

Let d be the largest integer such that $\tau_d = 0$. Then by the implicit function theorem we can locally solve for $u_1,..,u_d$ from the first d equations of (4.1a)

$$(4.4) \qquad u_j = \psi_j(u_{d+1},..,u_m,y_1,..,y_p), \qquad j = 1,..,d.$$

Furthermore by left-invertibility we may locally for $u_{d+1}^{(\tau_d+1)},..,u_m^{(\tau_m)}$ from the last m-d equations of (4.1a) yielding a system of the form

$$(4.5a) \qquad \begin{aligned} \frac{d}{dt}u_{d+1}^{(\tau_{d+1}-1)} &= k_{d+1}(s,y,\dot{y},..,y^{(\tilde{k})}) \\ &\vdots \\ \frac{d}{dt}u_m^{(\tau_m-1)} &= k_m(s,y,\dot{y},..,y^{(\tilde{k})}) \end{aligned}$$

with $s := (u_{d+1}, \dot{u}_{d+1},..,u_{d+1}^{(\tau_{d+1}-1)},...,u_m,..,u_m^{(\tau_m-1)})$. Completing these equations with the integrators

$$(4.5b) \qquad \frac{d}{dt}\, u_i^{(r_i-1)} = u_i^{(r_i)}, \quad r_i = 1,..,\tau_i - 1, \quad i = d+1,..,m,$$

we obtain the $(\tau_{d+1} + \dots + \tau_m)$-dimensional system (4.5) with state s, driven by the "inputs" $y,\dot{y},..,y^{(\tilde{k})}$, and yielding the "outputs" $u_{d+1},..,u_m$ together with $u_1,..,u_d$ given by (4.4). Thus in principle (4.5) together with (4.4) constitutes an *inverse system*. (It reconstructs the inputs from the outputs.)

However the order of time-derivatives of y appearing in (4.5) may be unnecessarily large. Indeed we may apply to (4.5) the same procedure as used in Theorem 3.3. By adding to (4.5) p parallel strings of integrators

$$(4.6) \qquad \begin{aligned} \frac{d}{dt}\, y &= \dot{y} \\ \frac{d}{dt}\, \dot{y} &= \ddot{y} \\ &\vdots \\ \frac{d}{dt}\, y^{(\tilde{k})} &= v \end{aligned}$$

with v driving variables, we obtain a driven state space system. We can thus set up the minimal conditionally invariant distribution algorithm (3.13) for this driven state space system. Let now ρ be the largest integer such that $S_{\rho+1}$ is involutive and $S_\rho \subset \bigcap_{i=1}^{m} \ker du_i$. Then as in Theorem 3.3 (see the Remark after Theorem 3.4) we may factor out the driven state space system (4.5) together with (4.6) by $S_{\rho+1}$, so as to obtain a system of the same form but now involving y and its time-derivatives only up to order $\tilde{k}-\rho$.

Example 4.2. Consider the linear system

$$(4.7) \qquad \ddot{y} = \ddot{u}$$

An inverse system is given by

$$(4.8) \qquad \frac{d}{dt} u = \dot{u}, \ \frac{d}{dt} \dot{u} = \ddot{y}.$$

However, adding the string of integrators

$$(4.9) \qquad \frac{d}{dt} y = \dot{y}, \ \frac{d}{dt} \dot{y} = \ddot{y}, \ \frac{d}{dt} \ddot{y} = v$$

we obtain a driven state space system, for which

$$S_1 = \text{span} \{\frac{\partial}{\partial \ddot{y}}\}, \ S_2 = S_1 + \text{span} \{\frac{\partial}{\partial \dot{u}} + \frac{\partial}{\partial \dot{y}}\}, \ S_3 = S_2 + \text{span} \{\frac{\partial}{\partial u} + \frac{\partial}{\partial y}\},$$

so that $\rho = 2$. The leaves of the foliation defined by S_3 are the level sets of the functions $x_1 = u-y$, $x_2 = \dot{u}-\dot{y}$. Factoring out (4.8)-(4.9) by S_3 thus results in the inverse system

$$(4.10) \qquad \dot{x}_1 = x_2, \quad \dot{x}_2 = 0, \qquad u = x_1 + y.$$

5. CONCLUSIONS

Based on the local equivalence Theorem 2.1, we have shown how one can eliminate the latent variables (including state space variables) in differential system descriptions. After giving a simple proof of the sufficiency of the conditions of Theorem 3.3 for state space realizability we have shown the necessity (modulo constant rank assumptions) of these conditions as well (Theorem 3.4). Finally we have applied the same techniques to the problem of constructing inverse systems involving a minimal order of differentiations. Other applications of the same methodology are currently under investigation.

REFERENCES

[1] G. Conte, C.M. Moog, A. Perdon, Un théorème sur la représentation entrée-sortie d'un système non linéaire, *C.R. Acad.Sci. Paris*, t.307, Serie I (1988), 363-366.

[2] P.E. Crouch, F. Lamnabhi-Lagarrigue, State space realizations of nonlinear systems defined by input-output-differential equations, *Analysis and Optimization of Systems* (eds. A. Bensoussan, J.L. Lions), Lect.Notes.Contr.Inf.Sci., 111, Springer, Berlin (1988), 138-149.

[3] S. Diop, A state elimination procedure for nonlinear systems, to appear in Proc. *Colloque International Automatique Nonlinéaire*, Nantes, June 1988.

[4] M. Fliess, A note on the invertibility of nonlinear input-output systems, *Syst.Contr.Lett.*, 8 (1986), 147-151.

[5] M. Fliess, Nonlinear control theory and differential algebra, *Modelling and Adaptive Control* (eds. C.I. Byrnes, A. Kurzhanski), Lect.Notes Contr.Inf.Sc., 105, Springer, Berlin (1988), 134-145.

[6] M. Fliess, Automatique et corps différentiels, *Forum Math.*, 1 (1989), to appear.

[7] S.T. Glad, Nonlinear state space and input output descriptions using differential polynomials, to appear in Proc. *Colloque International Automatique Nonlinéaire*, Nantes, June 1988.

[8] A. Isidori, *Nonlinear Control Systems: an Introduction*, Lect. Notes Contr.Inf.Sc., 72, Springer, Berlin (1985).

[9] A. Isidori, A.J. Krener, C. Gori-Giorgi, S. Monaco, Nonlinear decoupling via feedback: a differential geometric approach, *IEEE Trans.Aut.Contr.*, 26 (1981), 331-345.

[10] B. Jakubczyk, Realizations of nonlinear systems: three approaches, *Algebraic and Geometric Methods in Non-linear Control Theory* (eds. M. Fliess, M. Hazewinkel), Reidel, Dordrecht (1986).

[11] B. Jakubczyk, W. Respondek, On linearization of control systems, *Bull.Acad.Polon.Sci.* Sér.Sci.Math., 28 (1980), 517-522.

[12] T. Kailath, *Linear Systems*, Prentice-Hall, Englewood Cliffs (1980).

[13] F. Lamnabhi-Lagarrigue, P.E. Crouch, I. Ighneiwa, Tracking through singularities, to appear in Proc.*Colloque International Automatique Non-linéaire*, Nantes, June 1988.

[14] J.F. Pommaret, Géométrie différentielle algébrique et théorie du Contrôle, *C.R. Acad.Sci.Paris*, t.302, Serie I (1986), 547-550.

[15] W. Respondek, Transforming a nonlinear control system into a differential equation in the inputs and outputs, preprint Polish Academy of Sciences (1989).

[16] H.H. Rosenbrock, *State-space and Multivariable Theory*, Nelson, London (1970).

[17] A.J. van der Schaft, *System Theoretic Descriptions of Physical Systems*, CWI Tract 3, Centrum voor Wiskunde en Informatica, Amsterdam (1984).

[18] A.J. van der Schaft, On realization of nonlinear systems described by higher-order differential equations. *Math. Syst. Theory*, 19 (1987), 239-275, Correction *Math. Syst. Theory*, 20 (1987), 305-306.

[19] A.J. van der Schaft, A realization procedure for systems of non-linear higher-order differential equations, Preprints 10th *IFAC World Congress*, Munich, 1987, 97-102.

[20] A.J. van der Schaft, Representing a nonlinear state space system as a set of higher-order differential equations in the inputs and outputs, *Syst. Contr. Lett.*, 12 (1989), 151-160.

[21] A.J. van der Schaft, Transformations of nonlinear systems under external equivalence, Memorandum 700, University of Twente (1988), to appear in Proc. *Colloque International Automatique Non-linéaire*, Nantes, June 1988.

[22] J.M. Schumacher, Transformations of linear systems under external equivalence, *Lin.Alg.and its Appl.*, 102 (1988), 1-33.

[23] E.D. Sontag., Bilinear realizability is equivalent to existence of a singular affine I/O equation, *Syst. Contr. Lett.*, 11 (1988), 181-187.

[24] J.C. Willems, System theoretic models for the analysis of physical systems, *Ricerche di Automatica*, 10 (1979), 71-106.

[25] J.C. Willems, From time-series to linear system. Part I: Finite dimensional linear time invariant systems, *Automatica*, 22 (1986), 561-580.

DEPARTMENT OF APPLIED MATHEMATICS

UNIVERSITY OF TWENTE, P.O. BOX 217,

7500 AE ENSCHEDE, THE NETHERLANDS.

WHY REAL ANALYTICITY IS IMPORTANT IN CONTROL THEORY

H. J. Sussmann

1. Introduction .

In the recent development of nonlinear control theory, properties of real analytic functions and maps, and their associated classes of sets —i.e. analytic, semianalytic and subanalytic sets— have come to play an increasingly important role. The purpose of this survey is to explain why this is so, by outlining some results where analyticity matters, and by giving general reasons why it ought to matter.

The paper is organized as follows. After introducing some basic notations and definitions, we describe some facts about integral manifolds, reachable sets, and nonlinear realization theory, which make use of analyticity in an "elementary way," i.e. through the analytic continuation principle. We then turn to deeper properties, such as the stratification theory of semianalytic and subanalytic sets, and present results in control theory where methods based on these results play a role. Finally, we turn to the even deeper theory of resolution of singularities, and some of its applications in control. In all cases, we try to explain why analyticity is crucial, but we do not give details of proofs.

2. Basic notations .

Throughout this paper, "smooth" means "of class C^∞," and "analytic" means "real analytic." *Manifolds* are supposed to be Hausdorff and countable at infinity, but *not* of pure dimension, so that a manifold M consists of a finite or countable number of connected components, each of which is paracompact, and these components are allowed to be of different dimensions. It is *not* required that there be an upper bound for these dimensions.

If M is a smooth manifold and $x \in M$, then T_xM denotes the tangent space to M at x. If X is a smooth vector field on M, and $x \in M$, then xX will mean the same as $X(x)$, i.e. the value of X at x. Similarly, if $\Phi : M \to N$ is a (possibly partially defined) smooth map, then $x\Phi$ will be an alternative notation for $\Phi(x)$.

We use exponential notation for the *flow* of a vector field, so that $t \to xe^{tX}$ is the integral curve of X that goes through x at time $t = 0$. If $v \in T_xM$, then $v\Phi$ denotes the value at Φ of the differential of Φ at x, i.e. the vector usually denoted by $\Phi_*(v)$ or $d\Phi(x)v$.

Since X is a first-order differential operator acting on smooth functions on M, the result of this action on a particular function f —sometimes denoted $L_X f$ and called the "Lie derivative" of f with respect to X— will simply be denoted Xf.

The *Lie bracket* $[X, Y]$ of two smooth vector fields X and Y is the differential operator $[X,Y] \stackrel{\text{def}}{=} XY - YX$. The function $t \to xe^{tX}Ye^{-tX}$ takes values in the linear space T_xM. It turns out that this function is smooth, and its derivative is given by

$$\frac{d}{dt}\left(xe^{tX}Ye^{-tX}\right) = xe^{tX}[X,Y]e^{-tX} \ . \tag{2.1}$$

A *polysystem* is a differential equation

$$\dot{x} = f(x,u) \ , \ x \in M \ , \ u \in U \ , \tag{2.2}$$

where M is a smooth manifold, U is a set, and $f(x,u) \in T_xM$ for each $(x,u) \in M \times U$. To such a system we associate the collection $V_f = \{f_u : u \in U\}$ of vector fields f_u, where $f_u(x) \stackrel{\text{def}}{=} f(x,u)$. The polysystem (2.2) is *smooth* (resp. *analytic*) if all the vector fields f_u are smooth (resp. analytic). If (2.2) is smooth, then the *controllability Lie algebra of* f —denoted by Λ_f— is the Lie algebra of vector fields generated by V_f.

A *control* for a polysystem (2.2) is a U-valued function η defined on some compact interval $[a,b] \subseteq \mathbb{R}$. Two controls that coincide almost everywhere are identified. A *trajectory* for a control $\eta : [a,b] \to U$ is an absolutely continuous curve $\gamma : [a,b] \to M$ such that $\dot{\gamma}(t) = f(\gamma(t), \eta(t))$ for almost all $t \in [a,b]$. A *control system* consists of the specification of a polysystem (2.2) together with a collection $\mathcal{U}$ of controls. We call the members of $\mathcal{U}$ *admissible controls*, and require that the class $\mathcal{U}$ be invariant under time translations (i.e. if $\eta : [a,b] \to U$ is in $\mathcal{U}$, $\tau \in \mathbb{R}$, and $\eta_\tau : [a+\tau, b+\tau] \to U$ is defined by $\eta_\tau(t) = \eta(t-\tau)$, then $\eta_\tau \in \mathcal{U}$) and concatenations (i.e. if $\eta : [a,b] \to U$ and $c \in]a,b[$ are such that the restrictions of η to $[a,c[$ and $]c,b]$ are in $\mathcal{U}$, then $\eta \in \mathcal{U}$), and we also demand that $\mathcal{U}$ contain all constant controls. It then follows that $\mathcal{U}$ contains all piecewise constant controls as well. A trajectory that corresponds to an admissible control is an *admissible trajectory*. If φ is any function with domain $[a,b]$ (e.g. a control or a trajectory), then we write $T_-(\varphi)$ for a, $T_+(\varphi)$ for b. When φ is a trajectory, we write $x_-(\varphi)$, $x_+(\varphi)$ for $\varphi(a)$, $\varphi(b)$, respectively.

If Σ is a control system, (2.2) the associated polysystem, and $\mathcal{U}$ the class of admissible controls, we write f_Σ, V_Σ, Λ_Σ, M_Σ, U_Σ, $\mathcal{U}_\Sigma$, for f, V_f, Λ_f, M, U, $\mathcal{U}$, respectively.

We use $\mathcal{A}_\Sigma$ to denote the set of all *admissible pairs* of Σ, i.e. of all pairs $\Gamma = (\gamma, \eta)$ such that $\eta \in \mathcal{U}$ and γ is a trajectory for η. (We then use $T_\pm(\Gamma)$, $x_\pm(\Gamma)$, to denote $T_\pm(\gamma)$ and $x_\pm(\gamma)$, respectively.) The system Σ has the *piecewise constant approximation property* (PCAP) if, whenever γ is an admissible trajectory, and $c \in [a,b]$, then there exists a sequence $\{\gamma_j\}$ of trajectories corresponding to piecewise constant controls, such that $T_-(\gamma_j) = a$, $T_+(\gamma_j) = b$, $\gamma_j(c) = \gamma(c)$, and $\gamma_j \to \gamma$ uniformly on $[a,b]$ as $j \to \infty$.

If Σ is a control system, $x \in M_\Sigma$ and $T \geq 0$, we use $\mathcal{R}_\Sigma(x, \leq T)$ to denote the set of all points of the form $\gamma(t)$, for all admissible trajectories γ and all t such that $x(\gamma) = x$, $T_-(\gamma) = 0$, $T_+(\gamma) = t \leq T$.

3. Analytic continuation, integral manifolds and reachable sets .

Undoubtedly, the simplest property of analytic maps is the *analytic continuation principle* (ACP) for functions of one variable, according to which, if f is a real- or vector-valued analytic function on an open interval $I \subseteq \mathbb{R}$, and the set of zeros of f has an accumulation point in I, then $f \equiv 0$. This result already has an important control-theoretic implication that makes analytic control systems fundamentally different from C^∞ ones, as we now show.

Recall that, if Λ is any set of smooth vector fields on a smooth manifold M, an *integral manifold* of Λ is a smooth connected embedded submanifold S of M such that T_xS is equal to the linear span of $\Lambda(x)$ for each $x \in S$. (Naturally, $\Lambda(x) \stackrel{\text{def}}{=} \{X(x) : X \in \Lambda\}$.) An integral manifold S is *maximal* if, whenever S' is another integral manifold such that $S' \cap S \neq \emptyset$, it follows that S' is an open subanifold of S. Then the ACP implies:

Theorem 3.1. *Let Λ be a Lie algebra of analytic vector fields. Then through every point $x \in M$ there passes a unique maximal integral manifold of Λ.*

We will use $I(\Lambda, x)$ to denote the unique maximal integral manifold whose existence is assured by Theorem 3.1. The above result is due to Hermann [12] and Nagano [20]. We now sketch one method of proof based on the results of [26], in order to make it clear that the only fact about analytic functions being used is the ACP.

It is shown in [26] that, if Λ is any set of smooth vector fields, then the *orbits* of Λ are smooth submanifolds. By definition, the orbit $\mathcal{O}(\Lambda, x)$ of Λ through a point $x \in M$ is the set of all $y \in M$ of the form

$$y = xe^{t_1X_1}e^{t_2X_2}\dots e^{t_mX_m}, \tag{3.1}$$

for all choices of m, of the X_i in Λ and the $t_i \in \mathbb{R}$. Moreover, if we define $\Phi^x_{\mathbf{X}}(\mathbf{t})$ to be the right-hand side of (3.1) (with $\mathbf{X} = (X_1, X_2, \dots, X_m)$, $\mathbf{t} = (t_1, t_2, \dots, t_m)$), then the tangent space $T_y\mathcal{O}(\Lambda, x)$ of $\mathcal{O}(\Lambda, x)$ at y is spanned by the images of the differentials $d\Phi^x_{\mathbf{Z}}(\boldsymbol{\tau})$, ranging over all choices of $\mathbf{Z}$ and $\boldsymbol{\tau}$ such that $\Phi^x_{\mathbf{Z}}(\boldsymbol{\tau}) = y$. In addition, it is obvious that $\Lambda(y) \subseteq T_y\mathcal{O}(\Lambda, x)$. It is easy to show that every integral manifold of Λ is contained in an orbit. Using all this, it is clear that Theorem 3.1 will follow if we show that, if M is analytic and Λ is a Lie algebra of analytic vector fields, then $\operatorname{Im} d\Phi^x_{\mathbf{X}}(\mathbf{t}) \subseteq \Lambda(\Phi^x_{\mathbf{X}}(\mathbf{t}))$ for each $x \in M$ and each choice of $\mathbf{X}$, $\mathbf{t}$. But $\operatorname{Im} d\Phi^x_{\mathbf{X}}(\mathbf{t})$ is spanned by the vectors $v_k = y_kX_ke^{t_{k+1}X_{k+1}}\dots e^{t_mX_m}$ for $k = 1, \dots, m$, where $y_k = xe^{t_1X_1}\dots e^{t_kX_k}$. So all we need is to show that, if $X \in \Lambda$, then $xXe^{t_1X_1}\dots e^{t_mX_m} \in \Lambda(xe^{t_1X_1}\dots e^{t_mX_m})$ for all choices of x, m, the X_i and the t_i. This is clearly equivalent to the statement that $v = ze^{-t_mX_m}\dots e^{-t_1X_1}Xe^{t_1X_1}\dots e^{t_mX_m} \in \Lambda(z)$ for all z, m, X_i, t_i. Clearly, $v \in T_zM$ and, if we fix z, m and the X_i, then v is an analytic function of the t_i whose domain is a connected open subset of $\mathbb{R}^m$. Moreover, it follows from (2.1) that

$$\left.\frac{\partial^{\nu_1+\dots+\nu_m}v}{\partial t_1^{\nu_1}\dots t_m^{\nu_m}}\right|_{t_1=\dots=t_m=0} = \mathrm{ad}^{\nu_1}_{X_1}\dots\mathrm{ad}^{\nu_m}_{X_m}(X) \tag{3.2}$$

for all $\nu_1, \dots, \nu_m$, where ad_Y is the operator $Z \to [Y, Z]$. So all the partial derivatives of v with respect to the t_i, of all orders, are in $\Lambda(z)$. Because of the ACP, it follows that $v \in \Lambda(z)$, and the proof is complete. ∎

Theorem 3.1 provides some interesting information about properties of reachable sets. To see this, we first quote two results that are valid for smooth systems as well. The first one is the "positive form of Chow's Theorem" (cf. [14] for a proof). Recall that the system Σ is said to satisfy the *Lie algebra full rank condition* (LAFRC) at a point x if $\Lambda_\Sigma(x) = T_x M_\Sigma$. Then we have

Theorem 3.2. *Assume that Σ satisfies the LAFRC at x. Then for every $T > 0$ the interior of the reachable set $\mathcal{R}_\Sigma(x, \leq T)$ is nonvoid.* ∎

The second result is elementary, so we omit the proof.

Theorem 3.3. *If a smooth control system Σ satisfies the PCAP, then every trajectory of Σ is entirely contained in an orbit of V_Σ.* ∎

Remark 3.1. The PACP plays a crucial role in Theorem 3.3. To see this, consider the system $\dot{x} = ux$, $x \in \mathbb{R}$, $u \in \mathbb{R}$, with $\mathcal{U}$ the class of all measurable $\mathbb{R}$-valued functions. Then $x = 1$ is reachable from $x = 0$ using the control $\eta(t) = \frac{1}{2t}$ and corresponding trajectory $\gamma(t) = t^2$, but no trajectory that corresponds to a piecewise constant —or Lebesgue integrable— control and goes through $x = 0$ can ever leave $x = 0$. Therefore Σ does not have the *PCAP*. It is clear that $\{0\}$ is an orbit, but γ is not entirely contained in it. ∎

Combining these results, and using $I(\Sigma, x)$ to denote the integral manifold $I(\Lambda_\Sigma, x)$, and $I(\Sigma, x)$-closure (resp. $I(\Sigma, x)$-interior) to denote relative closure (resp. relative interior) with respect to $I(\Sigma, x)$, one gets immediately the following theorem (cf. [25]):

Theorem 3.4. *For an analytic system Σ, if $x \in M_\Sigma$ and $T > 0$, then the reachable set $\mathcal{R}_\Sigma(x, \leq T)$ is a subset of $I(\Sigma, x)$ and is actually contained in the $I(\Sigma, x)$-closure of its $I(\Sigma, x)$-interior.*

The proof of Theorem 3.4 uses analyticity only once, to conclude that $I(\Sigma, x)$ exists. So Theorem 3.4 is "elementary," in the sense that it only depends on the ACP. However, the theorem already gives interesting control-theoretic information. For instance, it implies in particular that the reachable sets $\mathcal{R}_\Sigma(x, \leq T)$ have integer dimension (using any reasonable dimension theory, e.g. Hausdorff). Also, it shows that the reachable sets do not have "smaller dimensional tails," (i.e. nonvoid relatively open subsets of smaller dimension). The latter conclusion fails for smooth systems, as can be seen by considering the system $\dot{x} = 1$, $\dot{y} = \varphi(x)u$, where $u \in \mathbb{R}$ and φ is a smooth function which vanishes for $x \leq 0$ and is > 0 for $x > 0$. (The reachable set from $(-1,0)$ in time ≤ 2 is the union of the segment $\{(x,0) : -1 \leq x \leq 0\}$ and the strip $\{(x,y) : 0 < x \leq 1\}$.) It is not known whether the former conclusion —on the integrality of the dimension— holds for smooth systems as well.

4. Controllability and minimal realizations .

The theorem on the existence of integral manifolds makes it possible to develop a theory of minimal realizations for nonlinear analytic systems which retains some of the features of linear realization theory. Recall that in the linear theory the key concepts are *controllability* and *observability.* A linear system $\dot{x} = Ax + Bu$, $x \in \mathbb{R}^n$, with output $y = Cx$, is defined to be *minimal* if it is both controllable and observable. One of the fundamental results of the theory is that every linear input-output map arising from a linear system as above can also be obtained from a minimal one, and this *minimal realization* is unique up to an obvious diffeomorphism. To extend the theory to the nonlinear case one has to look for the appropriate extensions to the nonlinear framework of the concepts of controllability and observability, and then seek to prove results on minimal realizations.

For linear systems, controllability can be characterized in several equivalent ways. For instance, one could use the algebraic criterion provided by the Kalman condition (i.e. $\operatorname{rank}(B, AB, A^2B, \ldots, A^{n-1}B) = n$), or the property of *controllability from the origin* (i.e. that the reachable set from the origin is the whole space), or *complete controllability* (i.e. the condition that from every point of the state space it is possible to reach every point), or the *accessibility property* (i.e. the condition that the reachable set from the every point has a nonvoid interior). These four properties are not equivalent in the nonlinear case. When we seek a nonlinear generalization, it is clear that the Kalman condition, as formulated in terms of matrices, is too tied to linearity to even make sense in the nonlinear setting (although, as will be pointed out below, it does have a good nonlinear generalization if properly reformulated in terms of Lie brackets). The second and third properties have an obvious drawback: suppose we start with a very simple nonlinear system such as $\dot{x} = y^2$, $\dot{y} = u$, with an output (say $z = x$). The system is not "controllable" in the sense of one of these conditions. By analogy with linear theory we would like to make it "controllable" by making the state space smaller. This means that we would have to find a subset S of $\mathbb{R}^2$ such that the restriction of our system to S is "controllable," and then replace the original system by its restriction to this subset. We can certainly make the system controllable from zero by taking S to be the reachable set from zero, but this set is no longer a manifold. To pursue the theory along these lines would force us to consider systems on very general state spaces, including at least all reachable sets. This is quite unpleasant, and in addition would make the next step — quotienting out the unobservable part to achieve observability— impossible. If we seek to make our system completely controllable rather than just controllable from zero, then the situation is even worse: there is *no subset* of M to which our system can be restricted and become completely controllable. It turns out, however, that the system can easily be made "controllable" in the sense of the fourth definition, i.e. the accessibility property. In fact, our system has this property, as can be verified by computing Lie brackets and showing that the controllability Lie algebra has dimension two at every point, or directly by studying the reachable sets.

It turns out that, in general, if one wishes to develop a good theory of nonlinear minimal realizations, then *the "right" class of nonlinear systems is that of complete analytic systems, and the accessibility property is the "right" concept of controllability for such systems.* (A system Σ is *complete* if all the vector fields in V_Σ are complete. The first one to single out the accessibility property as a crucial element in the theory was D. L. Elliott; cf., e.g. [9].) We will discuss the first assertion below but first let us present the evidence for the second one. To begin with, it is always true that, if we are given an *initialized analytic system with outputs* (i.e. an analytic system Σ together with a point $\bar{x} \in M_\Sigma$ and an analytic "output" function $\varphi : M_\Sigma \to N$, where N is some other analytic manifold), then one can restrict Σ to a submanifold S of M_Σ so that the new system will have the accessibility property and give rise to the same input-output map as the original one. (Indeed, this can be achieved by just taking S to be the maximal integral manifold of Λ_Σ through $\bar{x}$.) Moreover, once this is done, then it is possible to "quotient out the unobservable part" and achieve observability as well. (Remarkably, the difficulties we encountered when trying to define controllability in the nonlinear setting do not arise for observability. The concept of observability has a very natural nonlinear generalization: we call (Σ, φ) *observable* if, whenever x_1, x_2 are two different initial states, it follows that they are distinguishable. Here we call x_1, x_2 *distinguishable* if there are admissible pairs $\Gamma_i = (\gamma_i, \eta_i)$, $i = 1, 2$, such that $x_-(\gamma_i) = x_i$, $\eta_1 \equiv \eta_2$, but $\varphi(\gamma_1(t)) \neq \varphi(\gamma_2(t))$ for some t.) In addition, the accessibility property, as opposed to the other candidates for a nonlinear analogue of controllability, lends itself to a simple algebraic characterization, because it holds if and only if the LAFRC is satisfied at every point. (Not surprisingly, for linear systems the LAFRC turns out to be exactly the Kalman controllability criterion.)

We now turn to the assertion that analyticity, rather than smoothness, is the natural property of nonlinear systems that is required in order to have a good theory. Having explained that a good theory exists in the analytic case, we must now show why such a theory cannot exist for C^∞ systems. In our outline of the real analytic theory, the first point where analyticity comes in is in the existence of integral manifolds of Λ_Σ. This might appear not to be a serious obstacle, because in the C^∞ case something quite similar to the integral manifolds still exists, namely, the orbits. Notice, however, that it is not true that, if we restrict a smooth system to one of its orbits, then the resulting restriction has the accessibility property. This objection might be countered by trying a new definition of controllability. Define a system Σ to be *orbit-minimal* if the state space M_Σ is itself an orbit of Σ. (This is equivalent to requiring complete controllability with piecewise constant controls of the *symmetrized system* Σ^{sym}, i.e. the system arising from the polysystem $\{f_u : u \in U_\Sigma\} \cup \{-f_u : u \in U_\Sigma\}$.) We could then remark that the properties of accesibility and orbit-minimality are equivalent for analytic systems, and seek to extend the theory to smooth systems by using the latter property to be our nonlinear analogue of controllability. If we do this, then we have a way of restrciting an arbitrary smooth intialized system $(\Sigma, \bar{x})$ to a submanifold of M_Σ to make it "controllable," but *it becomes impossible to carry out the other two steps.*

To see why this is so, consider first the step of "quotienting out the unobservable part." The precise meaning of this is that we define an equivalence relation $\sim$ by declaring x_1 and x_2 to be equivalent if they are not distinguishable, i.e. if for every input both states give rise to the same output. One then needs to know that $\sim$ is "regular," in the sense that the quotient $M_\Sigma / \sim$ is still a manifold. As stated before, this is true in the analytic case. However, in the C^∞ case $\sim$ *may fail to be regular.* For instance, if $\varphi : \mathbb{R} \to \mathbb{R}$ is smooth, $\varphi(x) = 0$ for $x \geq 0$, and $\varphi'(x) < 0$ for $x < 0$, and we consider the system $\dot{x} = 1$ with output φ, then $x_1 \sim x_2$ iff $x_1 = x_2$ or x_1 and x_2 are both ≥ 0.

In the analytic case, $\sim$ is regular because of the main theorem of [27], according to which an equivalence relation $\sim$ on a manifold M is regular if (a) $\sim$ is closed as a subset of $M \times M$, and (b) $\sim$ admits a sufficiently large class V of "symmetry vector fields." We recall that a *symmetry vector field* for $\sim$ is a vector field X such that, if $x \sim \tilde{x}$, then it follows that $xe^{tX} \sim \tilde{x}e^{tX}$ for all t for which both sides are defined. "Sufficiently large" means that $V(x) = T_xM$ for all x. Notice that X is a symmetry vector field if and only if, whenever $p = (x, \tilde{x}) \in \sim$ —where $\sim$ is regarded as a subset of $M \times M$— then it follows that the integral trajectory of $X \oplus X$ through p is entirely contained in $\sim$. This shows that $\sim$ is a union of orbits of the set $\mathbf{S}$ of all $X \oplus X$ such that X is a symmetry vector field. It then follows that, if L_{sym} is the Lie algebra of vector fields generated by the symmetry vector fields, then every member of L_{sym} is a symmetry vector field as well. If Σ is analytic and has the accessibility property, the existence of a sufficiently large class of symmetry vector fields will follow if we prove that every f_u is a symmetry vector field. Now, it is easy to see from the definition of indistinguishablity that, if $x \sim \tilde{x}$, then $xe^{tf_u} \sim \tilde{x}e^{tf_u}$ for all $t \geq 0$. Using analyticity one can then conclude that the same is true for $t < 0$, and therefore f_u is indeed a symmetry vector field.

Finally, we claim that the third step, i.e. the equivalence of minimal realizations, also depends on analyticity. It is easy to show by means of examples that equivalence can fail in the smooth case. To understand why it works in the analytic case, we simply observe that the input-output map only depends on the control system and the output function *on the reachable set from the initial state.* Hence it will be impossible to show that two systems are equivalent using only the equality of the input-output maps, unless one can somehow infer, when the systems coincide on the reachable sets from the initial states, that they coincide everywhere. This can be done under the hypothesis of analyticity, because in that case the reachable sets have novoid interior —due to accessibility— and then the vector fields f_u and the output map φ are determined everywhere —because of analytic continuation— once they are known on the reachable sets.

For a detailed account of the existence and uniqueness results, the reader is referred to [28].

5. The Weierstrass Preparation Theorem and its consequences .

Real analytic functions of one variable "behave locally like polynomials." This can be made precise in several ways, e.g by observing that, if φ is analytic, then for y near a point x, $\varphi(y)$ is equal to $p(y)u(y)$, where $p(y)$ is a polynomial and u is an analytic function such that $u(x) \neq 0$. The higher dimensional generalization of this fact is the *Weierstrass Preparation Theorem* (WPT): if φ is analytic real-valued on an open set $\Omega \subseteq \mathbb{R}^n$, and

$\bar{x} \in \Omega$, then, after making a linear change of coordinates, one can write $\varphi(x) = p(x)u(x)$ on some neigborhood of $\bar{x}$, where u is a analytic function such that $u(\bar{x}) \neq 0$ and p is a *distinguished polynomial*, i.e. $p(x) = (x_1 - \bar{x}_1)^m + \sum_{i=1}^m p_i(x^*)(x_1 - \bar{x}_1)^{m-i}$, where $x^* \stackrel{\text{def}}{=} (x_2, \ldots, x_m)$, and the p_j are analytic functions near $\bar{x}^* = (\bar{x}_2, \ldots, \bar{x}_m)$ that vanish at $\bar{x}^*$.

An important consequence of the WPT is the *Noetherian property* of the ring of germs of analytic functions. Precisely, let $\mathcal{O}_x$ denote, for $x \in \mathbb{R}^n$, the set of germs at x of analytic functions. (By definition, a member of $\mathcal{O}_x$ is an equivalence class of real-valued analytic functions φ defined on some neighborhood N_φ of x, the equivalence relation being the one that identifies two such functions φ, ψ, if there some neighborhood N of x such that $N \subseteq N_\varphi \cap N_\psi$ and $\varphi \equiv \psi$ on N.) Then it can be proved, using the WPT, that *the ring $\mathcal{O}_x$ is Noetherian* (i.e. every ideal is finitely generated).

The Noetherian property of the rings of germs implies theorems about the number of zeros of certain functions, which in turn have control theoretic consequences. We illustrate this by briefly describing a bang-bang theorem proved in [31]. (Several other theorems giving bounds on the number of switchings have been proved, also making heavy use of analyticity, cf. e.g. [34], [36], [37], [42], [43], [44].)

Consider a system $\dot{x} = f(x) + ug(x)$, $-1 \leq u \leq 1$, with f and g analytic. A natural condition one can impose in order to guarantee that all time-optimal trajectories will be bang-bang is the following (cf. [31] for an explanation of why the condition is natural):

(C) for every x, k there is a neighborhood N of x and analytic functions α_{ik} for $i = 0, \ldots, k+1$, such that $[g, \mathrm{ad}_f^k(g)] = \sum_{i=0}^{k+1} \alpha_{ik}\, \mathrm{ad}_f^i(g)$ on N, and $|\alpha_{k,k+1}(x)| < 1$.

If (C) holds, then it was proved in [31] that, whenever there is a control η that steers a point x time-optimally to a point y, then there is an η' that steers x to y time-optimally and is bang-bang with at most ν switchings, where ν can be chosen to be a fixed number as long as $T = T_+(\eta) - T_-(\eta)$ stays bounded and the trajectory γ that corresponds to η and starts at x stays in a fixed compact set. The proof of this result is rather technical, and we shall limit ourselves here to describing the role of analyticity. If (γ, η) is time-optimal, let $t \to \lambda(t)$ be an adjoint vector, as specified by the Pontryagin Maximum Principle(cf. [21]). The switching function $\varphi(t)$ is then given by $\varphi(t) =$

$\langle\lambda(t), g(\gamma(t))\rangle$. Differentiation yields $\dot{\varphi}(t) = \langle\lambda(t), [f,g](\gamma(t))\rangle$. Differentiating again we get $\ddot{\varphi}(t) = \langle\lambda(t), [f,[f,g]](\gamma(t))\rangle + \eta(t)\langle\lambda(t), [g,[f,g]](\gamma(t))\rangle$. If we are in a neighborhood N of some point z where the conclusion of Condition (C) holds, then we can write $[g,[f,g]] = \alpha_{01}g + \alpha_{11}[f,g] + \alpha_{21}[f,[f,g]]$, and express $\ddot{\varphi}(t)$ as a linear combination of $\langle\lambda(t), g(\gamma(t))\rangle$, $\langle\lambda(t), [f,g](\gamma(t))\rangle$ and $\langle\lambda(t), [g,[f,g]](\gamma(t))\rangle$, the coefficients being $\eta(t)\alpha_{01}(t)$, $\eta(t)\alpha_{11}(t)$ and $1 + \eta(t)\alpha_{21}(t)$. If N is chosen small enough so that $|\alpha_{21}| \leq c_2 < 1$ throughout N, then the last coefficient is positive and bounded away from zero. If we let $\varphi_k(t) = \langle\lambda(t), \mathrm{ad}_f^{k-1}(g)(\gamma(t))\rangle$ (so that $\varphi_1 \equiv \varphi$), then we have shown that $\dot{\varphi}_1 = \varphi_2$ and $\dot{\varphi}_2 = \theta_{21}\varphi_1 + \theta_{22}\varphi_2 + \theta_{23}\varphi_3$, where the functions θ_{2i} are measurable and bounded by a fixed constant which does not depend on η, and θ_{23} is also positive and bounded below by a fixed constant. One can differentiate further and write $\dot{\varphi}_k = \sum_{i=1}^{k+1} \theta_{ki}\varphi_i$, with the θ_{ki} measurable and bounded, and $\theta_{k,k+1}$ bounded below. Notice that, in principle, each equation involves a new φ_j, and it appears that the system of differential equations might never close. However, due to the Noetherian property, the system does indeed close: the space of germs at z of analytic vector fields is a finitely generated module over the Noetherian ring $\mathcal{O}_z$. Hence every submodule is finitely generated. If we apply this to the submodule generated by the germs of the vector fields $\mathrm{ad}_f^k(g)$, we see that eventually one of these germs will be a linear combination of the preceding ones. This implies that one of the φ_k's will be a linear combination of the φ_j for $j < k$. So the system of differential equations described above does close after all. Therefore, for some k, we can write $\dot{\Phi}_k(t) = M(t)\Phi_k(t)$, where Φ_k is the vector $(\varphi_1, \ldots, \varphi_k)$, and M is a matrix whose entries m_{ij} vanish whenever $j > i+1$. Moreover, all the m_{ij} are measurable and bounded by a fixed constant, and the $m_{i,i+1}$ are also positive and bounded below by a fixed constant. Notice that one particular case of this situation would be $m_{i,i+1} \equiv 1$ and $m_{ij} \equiv 0$ if $i < k$ and $j \neq i+1$. In that case, if we let $\beta_i = \theta_{ki}$, we see that φ_1 satisfies the k-th order differential equation $\varphi_1^{(k)} = \beta_1\varphi_1 + \beta_2\dot{\varphi}_1 + \ldots + \beta_k\varphi_1^{(k-1)}$, and the functions β_i are measurable and bounded by a fixed constant. In this particular case we can therefore conclude that there is a fixed bound on the number of zeros of φ_1, using well known results on zeros of solutions of linear differential equations. It turns out that the general case is not fundamentally different: we can allow the entries above the main diagonal —i.e. the $m_{i,i+1}$— to be measurable functions not necessarily equal to 1, as long as they are positive, bounded, and bounded below, and we can allow the entries in the main diagonal and below —i.e. the m_{ij} with $j \leq i$— to be nonzero, as long as they stay bounded. (The crucial fact is that the m_{ij} for $j > i+1$ have to vanish.) Under these conditions, we proved in [31] that the number of zeros of φ_1 has to stay bounded, and this of course gives the bound on the number of switchings.

6. Analytic sets and stratifications .

Many finer properties of analytic maps are *finiteness results*. The simplest such result for functions φ of one variable is the fact that, if φ is analytic on an open interval I, then the set $Z(\varphi)$ of zeros of φ is discrete, unless $Z(\varphi) = I$. This follows, for instance, from the observation that φ can be written locally as a product of a polynomial and an invertible function. In higher dimensions, the analogue of this observation is the WPT, and this also has deep consequences about sets of zeros. Roughly speaking, the set of zeros $Z(\varphi)$ of an analytic function must be very special. More precisely, it can be shown that $Z(\varphi)$ has, locally, a finite number of connected components. One can actually prove results that are much stronger, in that they apply to larger classes of sets, and that they give stronger conclusions. To make this precise, define a *semianalytic subset* of an analytic manifold M to be a set $S \subseteq M$ with the property that, if x is any point of M, then there exists a neighborhood N of x and a finite collection $\mathcal{F}$ of real-valued analytic functions on N, such that $S \cap U$ belongs to the Boolean algebra of subsets of N generated by the sets $\{y : f(y) = 0\}$, $\{y : f(y) > 0\}$, for all $f \in \mathcal{F}$. It follows easily from the definition that the class $\mathrm{Sem}\, M$ of semianalytic subsets of M is a Boolean algebra (i.e. is closed under the operations of complementation, finite union, and finite intersection).

It turns out that semianalytic sets have a "very special structure," in the sense that they admit *stratifications*, as we now explain. Define a *stratum* in a smooth manifold M to be a smooth connected embedded submanifold of M. The *frontier* $\mathrm{Fron}\, S$ of a subset S of M is the set $\mathrm{Clos}\, S \cap (M - S)$. A *stratification* in M is a set $\mathcal{S}$ of pairwise disjoint strata, such that (i) $\mathcal{S}$ is locally finite (i.e. for every compact subset K of M, the set $\mathcal{S}_K = \{S : S \in \mathcal{S}\,,\, S \cap K \neq \emptyset\}$ is finite), and (ii) for every $S \in \mathcal{S}$, the frontier of S is a union of members of $\mathcal{S}$ of dimension strictly less than $\dim S$. The *support* $|\mathcal{S}|$ of $\mathcal{S}$ is the union of all members of $\mathcal{S}$. If $|\mathcal{S}| = A$ then we call $\mathcal{S}$ a *stratification of* A. We call $\mathcal{S}$ *analytic* if every stratum is a analytic submanifold of M, and *semianalytic* if every stratum is a semianalytic subset of M.

A stratification $\mathcal{S}$ is *compatible* with a subset A of M if A is a union of members of $\mathcal{S}$, i.e. if $A = \bigcup\{S : S \in \mathcal{S}_A\}$. With this terminology, one of the main theorems on semianalytic sets can be stated simply as follows: if $A \subseteq M$ is semianalytic, then there exists an analytic, semianalytic stratification $\mathcal{S}$ of M which is compatible with A. Actually, a much stronger result is still true: first of all, if one has a finite, or even locally finite collection $\mathcal{A}$ of semianalytic sets, then one can use the same stratification $\mathcal{S}$ for all of them. Second, $\mathcal{S}$ can be chosen to have a number of extra properties, such as the Whitney conditions, and the fact that all the strata are actually "nice." (For instance, one can take $\mathcal{S}$ such that all the strata are diffeomorphic to balls, and one can even construct an $\mathcal{S}$ which is a *triangulation.*) We shall not pursue these details, which can be found, e.g., in the work of Lojasiewicz (cf. [17]), and we limit ourselves to stating the following relatively weaker result:

Theorem 6.1. *If M is an analytic manifold and $\mathcal{A}$ is a locally finite set of semianalytic subsets of M, then there exists an analytic, semianalytic stratification $\mathcal{S}$ of M which is compatible with all the members of $\mathcal{A}$.* ∎

7. Real analyticity and trajectory regularity .

It was stated in §6 that the set $Z(\varphi)$ of zeros of an analytic function is very special. Theorem 6.1 enables us to make this precise in a number of ways. Indeed, the set $Z(\varphi)$ is clearly semianalytic, and so Theorem 6.1 tell us, for instance, that its intersection with every compact semianalytic set has finitely many connected components.

This observation has at least one important consequence, on the conceptual level, pertaining to one of the most fundamental issues in Control Theory, namely, the problem of the *regularity properties of optimal trajectories.* Consider an optimal control problem obtained by the specification of a control system Σ together with a "Lagrangian" L, i.e. a function $L : M_\Sigma \times U \to \mathbb{R}$. Asssume that, for every admissible pair $\Gamma = (\gamma, \eta) \in \mathcal{A}_\Sigma$, the Lebesgue integral $J_{\Sigma,L}(\Gamma) = \int_{T_-(\gamma)}^{T_+(\gamma)} L(\gamma(t), \eta(t))\, dt$ is defined (i.e., if $h(t) = L(\gamma(t), \eta(t))$, then h is measurable and at least one of the functions $h_+ = \max(h, 0)$, $h_- = \max(-h, 0)$ is integrable). Then we call a particular Γ *optimal* if $J_{\Sigma,L}(\Gamma) \leq J_{\Sigma,L}(\Gamma')$ for every admissible Γ' such that $x_-(\Gamma') = x_-(\Gamma)$ and $x_+(\Gamma') = x_+(\Gamma)$. It is known that, under suitable technical conditions, optimal trajectories must satisfy certain special properties, known as *necessary conditions for optimality,* of which the Pontryagin Maximum Principle (cf. [21] or [16]) is the best known example. One may then ask whether these conditions actually impose some nontrivial *a priori* restriction on optimal controls. For instance, one might begin by asking whether, for problems in which U is a reasonably smooth or piecewise smoooth set (e.g. a closed cube in $\mathbb{R}^m$), and f and L are smooth in x and u, it follows that, if $\Gamma = (\gamma, \eta)$ is optimal, then η is necessarily piecewise smooth. The answer to this question is "no," as shown, e.g., the well known example of "Fuller's problem" (cf. [19], [15]).

But one may then ask whether the answer might be "yes" if less regularity is required. For instance, does optimality imply that the set of points of discontinuity of η is necessarily countable? Or perhaps it may be uncountable but it has to be of measure zero?

It is clear that the answer to the above questions, as formulated, is "no" for a very trivial reason, namely, that there are "degenerate" systems for which *every* admissible pair is optimal. (This happens, e.g., if $L \equiv 0$, or if we are dealing with a minimum time problem —i.e. $L \equiv 1$— but one of the coordinates —say x_1— satisfies an equation $\dot{x}_1 = 1$.) In order to rule out such situations, we reformulate the question in terms of the possibility of *choosing* a special η. Given a class $\mathcal{P}$ of optimal control problems (Σ, L), all of them having the same control space U and the same class $\mathcal{U}$ of admissible controls, we call a subset $\mathcal{U}'$ of $\mathcal{U}$ *sufficient* for $\mathcal{P}$ if, whenever $(\Sigma, L) \in \mathcal{P}$ and $\Gamma = (\gamma, \eta)$ is optimal for (Σ, L), then there exists an admissible $\Gamma' = (\gamma', \eta')$ such that $x_-(\Gamma') = x_-(\Gamma)$, $x_+(\Gamma') = x_+(\Gamma)$, $J_{\Sigma,L}(\Gamma') = J_{\Sigma,L}(\Gamma)$, and $\eta' \in \mathcal{U}'$.

We can, for instance, take U to be the unit cube in $\mathbb{R}^m$, and $\mathcal{U}$ to be the class of all measurable U-valued functions defined on some compact subinterval of $\mathbb{R}$, and define $\mathcal{P}^\infty$ (resp. $\mathcal{P}^\omega$) to be the class of all (Σ, L) such that M_Σ is a smooth (resp. analytic) manifold and f, L are smooth (resp. analytic) as functions of x and u. We then ask whether there exists *some* class $\mathcal{U}'$ which is a *proper subset* of $\mathcal{U}$ and is sufficient for $\mathcal{P}^\infty$ or for $\mathcal{P}^\omega$.

It turns out that *the answer is "no" for $\mathcal{P}^\infty$ but "yes" for $\mathcal{P}^\omega$.* We will prove the first assertion, because the proof is quite simple and shows very clearly why the realm of C^∞ systems allows pathology that is definitely excluded in the analytic case. To establish the first assertion, it suffices to show that, if an arbitrary measurable function $\eta : [a, b] \to U$ is given, then we can construct a $(\Sigma, L) \in \mathcal{P}^\infty$, and states $x_\pm \in M_\Sigma$ such that η optimally steers x_- to x_+ but no other control does. We will take Σ to be of the special form $\dot{x}_0 = 1$, $\dot{x}_1 = u_1, \ldots, \dot{x}_m = u_m$, $\dot{x}_{m+1} = \varphi(x_0, x_1, \ldots, x_m)$, where φ is a smooth function to be chosen in a special way. We will select φ so that there are states $x_\pm$ such that η steers x_- to x_+ and no other control does. One can then choose L arbitrarily (e.g. $L \equiv 0$, or $L \equiv 1$), and the desired conclusion will follow.

In order to choose φ, we first define $\zeta : [a,b] \to \mathbb{R}^m$ by $\zeta(t) = \int_a^t \eta(s)\,ds$, and we let $K = \{(x_0, \zeta(x_0)) : a \le x_0 \le b\}$. Then ζ is a continuous function on $[a,b]$, and K is its graph, so K is a compact subset of $\mathbb{R}^{m+1}$. We then use the fact that, given any closed subset C of a smooth manifold M, then there exists a smooth function on M whose set of zeros is exactly C. Clearly, this function can be taken to be > 0 outside C. We apply this with $C = K$, $M = \mathbb{R}^{m+1}$, thereby obtaining our desired φ. We then take $x_- = (a, 0, \ldots, 0)$, $x_+ = (b, \zeta(b), 0)$. It is then easily verified that a control η' steers x_- to x_+ iff $\eta' = \eta$ a.e.

The preceding construction works *because every closed subset of a Euclidean space is the zero set of some smooth function.* As explained above, a similar conclusion is not true for analytic functions, since the zero sets of such functions, besides being closed, are very special. Naturally, this does not yet prove the existence of nontrivial special properties of optimal controls for analytic problems (i.e. the existence of nontrivial classes $\mathcal{U}'$ that are sufficient for $\mathcal{P}^\omega$) but it strongly supports the conjecture that such properties might exist. This suggestion has actually turned out to be a true theorem, as discussed below.

8. Subanalytic sets, reachable sets and synthesis .

The theory of semianalytic sets and their associated stratifications turns out to lack one important ingredient. This is most easily seen by considering the analogous theory of *semialgebraic sets.* Recall that a semialgebraic subset of $\mathbb{R}^n$ is a set which belongs to the Boolean algebra generated by the sets $\{x : p(x) = 0\}$ and $\{x : p(x) > 0\}$, where $p : \mathbb{R}^n \to \mathbb{R}$ is an arbitrary polynomial function. It is then more or less clear that the theories of semialgebraic and semianalytic sets are somewhat similar, the latter being a kind of "local" analogue of the former, in which polynomials are replaced by analytic germs. Since analytic germs are in some sense like polynomials —e.g. thanks to the WPT— it is natural to expect that the two theories will be similar, except possibly for the fact that theorems that hold globally for semialgebraic sets will hold locally for semianalytic sets. This is indeed true to some extent. For instance, there is a *stratification theorem*, according to which, if $\mathcal{A}$ is a finite family of semialgebraic subsets of $\mathbb{R}^n$, then there exists a *finite* stratification $\mathcal{S}$ of $\mathbb{R}^n$ by semialgebraic subsets which are analytic submanifolds, such that $\mathcal{S}$ is compatible with every member of $\mathcal{A}$. It is clear that Theorem 6.1 is, in a very precise sense, a local analogue of this result.

There is, however, one very important property of semialgebraic sets whose semianalytic analogue is false. The *Tarski-Seidenberg Theorem* says that, if $P : \mathbb{R}^n \to \mathbb{R}^m$ is a polynomial map, and $S \subseteq \mathbb{R}^n$ is semialgebraic, then $P(S)$ is semialgebraic in $\mathbb{R}^m$. The "local semianalytic" analogue of this would be the statement that, if $F : M \to N$ is a proper analytic map between analytic manifolds, then the image under F of a semianalytic subset S of M is semianalytic in N. (The restriction to proper maps is a natural one, because one is seeking a local theory. Moreover, if we do not impose such a restriction the results fails for trivial reasons, as seen, e.g., by taking $F : \mathbb{R}^2 \to \mathbb{R}$ to be the projection onto the first coordinate, and $S = \{(1/n, n) : n = 1, 2, \ldots\}$.) Unfortunately, this statement is false. However, it can be made true by enlarging the class of semianalytic sets and defining a new type of sets, called *subanalytic* (cf. [10], [11], [13], [49]). By definition, a subanalytic subset of a analytic manifold M is a subset S such that there exist a analytic manifold N, a proper analytic map $F : N \to M$, and a semianalytic subset S' of N, such that $F(S') = S$. It is then obvious from this definition that the image of a subanalytic set under a proper analytic map is subanalytic, and that every semianalytic set is subanalytic. It is slightly less obvious, but true and easy to prove, that the class of subanalytic sets is closed under finite unions, finite intersections, and inverse images by analytic maps. It is a lot less obvious, but still true, that the complement of a subanalytic set is subanalytic. Finally, there is a stratification theorem for subanalytic sets which is identical to Theorem 6.1, except only that the word "semianalytic" has to be replaced throughout by "subanalytic."

This suggests that the class of subanalytic sets perhaps qualifies better than that of semianalytic sets as the analytic analogue of the class of polynomials. Moreover, we claim that *the property of being closed under images is precisely what is needed in many applications in control theory*, and therefore the class of subanalytic sets is bound to play a fundamental role in control.

To see why this should be so, let us first observe that *subanalytic sets are, by definition, images, whereas reachable sets are also images.* (Assuming global existence and uniqueness of trajectories, the reachable set in time let $\leq T$ from a point x is the image of the set $\mathcal{U}^T$ of all admissible controls whose domain is $[0, t]$ for some $t \in [0, T]$ under the "terminal state map" X_+^x which assigns to each control η the terminal point $x_+(\gamma^{\eta,x})$, where $\gamma^{\eta,x}$ is the trajectory for η such that $\gamma(0) = x$.) This suggests that reachable sets might be subanalytic, in which case we could apply the stratification theorems to derive nontrivial conclusions about their structure. Perhaps more importantly, one ought to be able to get conclusions about the regularity of the value function for optimal control problems. Indeed, given an optimal control problem (Σ, L), we can consider the *augmented control system* Σ^L obtained from Σ by adding a new scalar variable y and the equation $\dot{y} = -L(x,u)$. It is clear that (x,y) can be driven to $(x',0)$ by means of a trajectory of Σ^L iff x can be driven to x' by a trajectory of Σ with cost y. In particular, if we use $\mathcal{C}_\Sigma(x)$ to denote, for a control system Σ and a point $x \in M_\Sigma$, the set of all points that can be driven to x, then the graph G of the value function for (Σ, L) with target point x is the "lower boundary" $\partial_-\mathcal{C}_{\Sigma_L}(x,0) = \{(z,y) : y = \inf\{y' : (z,y') \in \mathcal{C}_{\Sigma_L}(x,0)\}\}$. Since $\mathcal{C}_{\Sigma_L}(x,0)$ is a reachable set (the reachable set from $(x,0)$ for the "backward" system whose trajectories are those of Σ^L run backwards), we see that subanalyticity theorems for reachable sets are likely to imply results on the structure of value functions.

Unfortunately, the observation that both subanalytic sets and reachable sets are images is no more than a very vague analogy, from which we have no right to conclude anything. While it may be true that both kinds of sets are images, they are images of very different things. In one case they are images of semianalytic sets, i.e. in particular of finite-dimensional sets, whereas in the other case they are images of infinite-dimensional function spaces. So there appears to be no serious reason to expect that the theory of subanalytic sets will apply to the study of reachable sets.

This is a very valid objection, but it can be countered with the following argument, which is still far from rigorous but represents a definite improvement over our initial observation. If we wish to understand the structure of reachable sets, then what matters is their boundary. And the trajectories that go from a point x to the boundary of the reachable set from x have to satisfy a necessary condition, as specified by the Prontryagin Maximum Principle. This condition says that there exists an adjoint vector with certain properties. The important thing for us is that the system of equations for the state vector x and the adjoint vector λ is a system of ordinary differential equations of the usual kind, with existence and uniqueness of solutions, except for the fact that the right-hand side involves the control u, which is to be chosen as a function of x and λ so as to minimize the Hamiltonian. The minimizing value of u need not be unique and, even when it is unique, the vector field in (x, λ) space that results from plugging it in need not be locally Lipschitz, or even continuous, so existence and uniqueness of solutions is not guaranteed. If we ignore this last problem for the time being, we see that there are serious reasons to suspect that the boundary of the reachable set might be, after all, an image of a finite-dimensional set, namely, the set of all pairs $(\lambda(0), t)$.

If we want to turn the above remarks into a rigorous theory, it is now clear that what needs to be done is to study the question of *finite-dimensional analytic reduction* (FDAR) of a control system, i.e. the question whether we can find a finite dimensional analytic manifold A_0, a subanalytic subset A of A_0, and a family $\{\gamma_\alpha : \alpha \in A\}$ of trajectories, such that (a) whenever it is possible to go from a point x to a point x' by *some* trajectory γ of Σ, then this can actually be done by means of one of the γ_α, and (b) the maps $X_- : \alpha \to x_-(\gamma_\alpha)$ and $X_+ : \alpha \to x_+(\gamma_\alpha)$ are analytic, i.e. —to be completely precise— have analytic extensions to some open set containing the closure of A_0. (If in addition the γ_α of (a) can be chosen so that $T_+(\gamma_\alpha) - T_-(\gamma_\alpha) = T_+(\gamma) - T_-(\gamma)$, then we will say that we have an *equal-time* FDAR.)

If such a family can be found, then it is clear that the reachable set from any point x is subanalytic, provided only that an extra properness condition holds. Indeed, the reachable set from x is just the image under X_+ of the set $\{\alpha \in A_0 : X_-(\alpha) = x\}$, which is clearly subanalytic in A. Therefore the reachable set will be subanalytic if X_+ is proper, or even under the weaker condition that the map $\alpha \to (X_-(\alpha), X_+(\alpha))$ is proper. Similarly, if we just consider the reachable set up to time T, then this set will be subanalytic if our FDAR is equal-time and the time function $\tau : \alpha \to T_+(\gamma_\alpha) - T_-(\gamma_\alpha)$ is analytic, provided —as before— that a properness condition holds. Indeed, the set $\mathcal{R}_\Sigma(x, \leq T)$ is just the image under X_+ of $\{\alpha : X_-(\alpha) = x \, , \, \tau(\alpha) \leq T\}$, so $\mathcal{R}_\Sigma(x, \leq T)$ is subanalytic provided that the map $\alpha \to (X_-(\alpha), X_+(\alpha), \tau(\alpha))$ is proper.

To see how this can actually be applied, we now outline, using our terminology, some of the results of Brunovsky [7], who was the first to notice that the theory of subanalytic sets was bound to play an important role in control theory. We consider the simple example of a linear control system $\dot{x} = Ax + Bu$, $x \in \mathbb{R}^n$, with a control constraint $u \in K$, where K is a *polyhedron*, i.e. the convex hull of a finite set $\mathcal{K}$ of points. It is then easy to prove a *bang-bang theorem with a bound on the number of switchings*, according to which there exist $c_1 > 0$, $c_2 > 0$ with the property that, whenever it is possible to go from a point x to another point y in time T, then it is also possible to go from x to y in time T by means of a bang-bang trajectory with at most $c_1 + c_2 T$ switchings. Using this, we take our manifold A to have connected components A^σ, where σ ranges over the set of all *bang-bang control strategies*, i.e. all finite sequences $\sigma = (u_1, \ldots, u_r)$ (of arbitrary length $r \stackrel{\text{def}}{=} |\sigma|$) of points of $\mathcal{K}$. For each σ, A^σ is just a copy of $\mathbb{R}^n \times \mathbb{R}^{|\sigma|}$. (The resulting manifold is not of pure dimension, and has connected components of arbitrarily high dimension but, as explained in §2, this is permitted.) The set A_0 is the union of sets $A_0^\sigma \subseteq A^\sigma$, where $A_0^\sigma = \{\alpha = (x, t_1, \ldots, t_{|\sigma|}) : x \in \mathbb{R}^n \,,\, t_1 \geq 0 \,, \ldots,\,,\, t_{|\sigma|} \geq 0 \,,\, 1 + c_1 + c_2 T \geq |\sigma|\}$. For each $\sigma = (u_1, \ldots, u_r)$, $\alpha = (x, t_1, \ldots, t_r) \in A_0^\sigma$ we let γ_α be the bang-bang trajectory that starts at x at time 0, then is a trajectory corresponding to the constant control u_1 during time t_1, then a trajectory for the control u_2 during time t_2, and so on. This is indeed an equal-time FDAR of our control system, due to the bound on the number of switchings. The properness condition follows because, if x and T stay bounded —say $x \in J$, with J compact, and $0 \leq T \leq \bar{T}$— then $\{\alpha \in A_0^\sigma : X_-(\alpha) \in J \,,\, \tau(\alpha) \leq \bar{T}\}$ is obviously compact for each σ *and in addition is void for all but finitely many* σ (precisely, for all σ such that $|\sigma| > 1 + c_1 + c_2 \bar{T}$). Notice that, if we had not included the requirement that $1 + c_1 + c_2 T \geq |\sigma|$ in the definition of A_0^σ, we would still have obtained an equal time FDAR, but the properness would have failed. This shows that the role of the bound on the number of switchings is to give the necessary properness.

The above construction shows that, for a linear system with a polyhedral control constraint as above, the reachable sets $\mathcal{R}_\Sigma(x, \leq T)$ are compact and subanalytic. The stratification theorem then implies that these sets are finite unions of pairwise disjoint connected embedded analytic submanifolds.

The time-optimal control problem for linear systems with a polyhedral control constraint can be studied by a similar method. This results in the proof of the existence, for such systems, of a "regular synthesis." (The precise definition of "regular synthesis" needed here is a modification of the one given by Boltyanski in [1], and can be found, e.g. in Brunovsky's papers [7], [8], or in [38] or [39]. For other results on synthesis, also making heavy use of real analyticity, cf. [30], [33], [36], [37], [42], [43], [44].)

The previous result for linear systems is of course very special, but it is clear that it only depends on some features of the linear problem, notably the bang-bang theorem with bounds on the number of switchings. Moreover, the fact that the controls used are bang-bang is not really very important: the same result will clearly follow if singular arcs are also allowed, provided that these arcs are analytically parametrized by a finite-dimensional parameter. This has led to renewed interest in a question which, in any case, ought to have been regarded from the very beginning as one of the most basic problems in optimal control, namely, that of the structure of optimal trajectories, and in particular under what conditions such trajectories can be proved to be finite concatenations of bang-bang and singular arcs, with bounds on the number of switchings. (For recent results on the structure of trajectories, cf. [2], [3], [4], [5], [6], [22], [23], [45], [47], [51], [52].)

9. Some results on optimal trajectories .

Motivated by the relationship with the problem of the structure of reachable sets and the properties of the value functions, we now return to the question of the structure of optimal trajectories, which was briefly touched upon in §5. We showed in §5 that, for general smooth systems, one cannot derive nontrivial regularity properties for optimal controls. The situation is quite different for analytic systems, as shown by the following theorem:

Theorem 9.1. *Consider a system* $\dot{x} = f(x,u)$, $x \in M$, $u \in U$, *where* M *is an analytic manifold,* U *is a compact subanalytic subset of an analytic manifold* N, *and* f *is analytic as function of* x *and* u. *Assume that* x, y *are points of* M *such that there is a trajectory* γ *corresponding to some measurable control* $\eta : [a,b] \to U$, *that goes from* x *to* y. *Then there is a trajectory from* x *to* y *that corresponds to a control* η' *that is analytic on an open dense subset of the interval* $[T_-(\eta'), T_+(\eta')]$.

This result is proved in [46] for systems $\dot{x} = f(x) + ug(x)$, $-1 \le u \le 1$, and in [48] for the general case. (Cf. also [50] for a stronger result for bang-bang trajectories.) As is to be expected —since the result is *not* true for smooth systems— the proof makes heavy use of properties of analytic maps and associated sets. The proof for the case considered in [46] uses stratifications, and the proof of the general case requires even deeper tools, namely *desingularization* (cf. below).

It is a very interesting open problem to decide to what extent the very weak regularity result of Theorem 9.1 can be improved. For instance, is it possible to improve it by having "open dense subset of full measure" instead of just "open dense subset"? Can one do even better and obtain an open subset whose complement is *countable*? Ideally, one would want to obtain controls with finitely many switchings, but this is known to be impossible, as shown by Fuller's example.

10. Desingularization and its applications .

H. Hironaka's *resolution of singularities theorem* (cf. [13], and [49] for an elementary account) says that, if M is a connected analytic manifold and $\varphi : M \to \mathbb{R}$ is an analytic function which does not vanish identically, then φ admits a *desingularization*, i.e. there exists an analytic manifold N and a proper, surjective, analytic map $\Phi : N \to M$ which is an analytic diffeomorphism on an open dense subset Ω of N, and is such that the composite function $\varphi\Phi$ is *locally monomial*, i.e. for every point $z \in N$ there is an analytic coordinate chart $(x_1, \ldots, x_n)$ on a neighborhood Z of z such that, on Z, $\varphi\Phi$ is given by $\varphi\Phi(x) = H(x)x_1^{a_1}x_2^{a_2}\ldots x_n^{a_n}$, where the a_i are nonnegative integers and $H(x)$ nowhere vanishes.

One of the consequences of the resolution of singularities theorem is the *desingularization of subanalytic sets.* A subanalytic subset of M is a locally finite union of sets $\psi_j(C_j)$, where the C_j are open cubes in Euclidean spaces $\mathbb{R}^{n_j}$, and each ψ_j is an analytic map on a neigborhood Z_j of the closure of C_j. In particular, it follows easily that a compact subanalytic set is the image of a compact manifold under an analytic map.

The desingularization theorems have had at least three applications to significant control theory problems. First, desingularization plays a crucial role in the proof of Theorem 9.1. Second, the theory has been used to prove a regularity theorem for linear time-optimal control problems with a non-polyhedral control constraint. The results of [18] showed that, even for a very simple control set, such as the set $\{(u_1,u_2) : u_2^2 - 1 \leq u_1 \leq 1 - u_2^2\}$, it is possible for the reachable sets $\mathcal{R}_\Sigma(x, \leq T)$ to fail to be subanalytic, and for the value function for the minimum time problem not to be subanalytic either. However, in [40] it was shown that, if the control set U is compact, convex, and subanalytic, then the value function is analytic on the complement of a finite or countable union of connected analytic submanifolds of positive codimension.

Finally, the third application is to the theory of synthesis. It is well known that, if a family of Pontryagin extremals $\{(\gamma_x, \eta_x) : x \in \Omega\}$, with $x(\gamma_x) = x$, is "sufficiently smooth" in some sense, and satisfies a normality condition, then the gradient at a point $\bar{x}$ of the cost function $V : x \to J(\gamma_x, \eta_x)$ is equal to the value at $T_-(\gamma_x)$ of the adjoint variable $t \to \lambda_x(t)$ whose existence is asserted by the Pontryagin Maximum Principle. This fact plays a crucial role in the proof of sufficiency theorems for optimality, as shown, e.g., in [1]. It is an important question to find good conditions under which the existence of a family of optimal trajectories with the appropriate smoothness can be guaranteed. It turns out that, using desingularization, one can prove the existence of a family that has the desired properties on a reasonably large set, provided the optimal control problem admits a FDAR which satisfies some extra technical conditions. (For a more detailed account, cf. [53]).

11. Other applications of analyticity .

We conclude by briefly listing some other developments where analyticity has played an important role:

1. the existence of universal inputs was proved in [32] using stratification theory,
2. the convergencew of the Chen series for analytic systems has applications in local controllability theory (cf., e.g., [35], [41]),
3. the existence of piecewise analytic stabilizing feedback controls for completely controllable systems was established in [29] using subanalytic sets,
4. a complete analysis of the structure of time-optimal trajectories for analytic systems $\dot{x} = f(x) + ug(x)$, $|u| \leq 1$ in the plane was carried out in [42], [43], [44].

REFERENCES

[1] Boltyansky, V.G., "Sufficient conditions for optimality and the justification of the Dynamic Programming Principle," *S.I.A.M. J. Control* **4** (1966), pp. 326-361.

[2] Bressan, A., "A high-order test for optimality of bang-bang controls," *S.I.A.M. J. Control Opt.* **23** (1985), pp. 38-48.

[3] Bressan, A., "The generic local time-optimal stabilizing controls in dimension 3," *S.I.A.M. J. Control Opt.* **24** (1986), pp. 177-190.

[4] Bressan, A., "Local asymptotic approximation of nonlinear control systems," *Int. J. Control*, 41: 1331-1336 (1985).

[5] Bressan, A., "Directional convexity and finite optimality conditions," *J. Math. Anal. Appl.*, 80: 102-129 (1981).

[6] Bressan, A., "Dual variational methods in optimal control theory," to appear in *Finite Dimensional Controllability and Optimal Control*, H.J. Sussmann ed., to be published by M. Dekker, Inc.

[7] Brunovsky, P., "Every normal linear system has a regular synthesis," *Mathematica Slovaca*, **28** (1978), pp. 81-100.

[8] Brunovsky, P., "Existence of regular synthesis for general problems," *J. Diff. Equations*, **38** (1980), pp. 317-343.

[9] Elliott, D. L., "A consequence of controllability," J. Differential Equations **10** (1971), pp. 364-370.

[10] Gabrielov, A., "Projections of semianalytic sets," *Funct. Anal. Appl.* **2**, No. 4 (1968), pp. 282-291.

[11] Hardt, R., "Stratifications of real analytic maps and images," *Inventiones Math.* **28** (1975), pp. 193-208.

[12] Hermann, R., "On the accessibility problem in control theory," in *International Symposium on Nonlinear Differential Equations and Nonlinear Mechanics*, J.P. LaSalle and S. Lefschetz Eds., Academic Press, New York (1963).

[13] Hironaka, H., *Subanalytic Sets*, Lect. Notes Istituto Matematico "Leonida Tonelli," Pisa, Italy (1973).

[14] Krener, A. J., "A generalization of Chow's Theorem and the bang-bang principle," *S.I.A.M. J. Control* **12** (1974), pp. 43-52.

[15] Kupka, I., "The ubiquity of Fuller's phenomenon," to appear in *Finite Dimensional Controllability and Optimal Control*, H.J. Sussmann ed., M. Dekker, Inc.

[16] Lee, E. B., and L. Markus, *Foundations of Optimal Control Theory*, J. Wiley, New York, 1967.

[17] Lojasiewicz, S., *Ensembles semianalytiques*, Notes Inst. Hautes Études, Bures-sur-Yvette (1965).

[18] Lojasiewicz Jr., S. and H. J. Sussmann, "Some examples of reachable sets and optimal cost functions that fail to be subanalytic," *S.I.A.M. J. Control and Optimization* **23**, No. 4 (1985), pp. 584-598.

[19] Marchal, C., "Chattering arcs and chattering controls," *J. Optim Theory Appl.* **11** (1973), pp. 441-468.

[20] Nagano, T., "Linear differential systems with singularities and an application to transitive Lie algebras," *J. Math. Soc. Japan* 18: 398-404 (1966).

[21] Pontryagin, L.S., V.G. Boltyansky, R.V. Gamkrelidze and E.F. Mischenko, *The Mathematical Theory of Optimal Processes*, J. Wiley (1962).

[22] Schättler, H., "On the local structure of time-optimal bang-bang trajectories in $\mathbb{R}^3$," *S.I.A.M. J. Control Opt.* **26** (1988), pp. 186-204.

[23] Schättler, H., "The local structure of time-optimal trajectories in $\mathbb{R}^3$ under generic conditions," *S.I.A.M. J. Control Opt.* **26** (1988), pp. 899-918.

[24] Schättler, H., and H.J. Sussmann, "On the regularity of optimal controls," *J. Appl. Math. Physics (ZAMP)* **38** (1987), pp. 292-301.

[25] Sussmann, H. J., and V. Jurdjevic, "Controllability of nonlinear systems," *J. Diff. Equations* **12** (1972), pp. 95-116.

[26] Sussmann, H. J., "Orbits of families of vector fields and integrability of distributions," *Trans. Amer. Math Soc.* **180** (1973), pp. 171-188.

[27] Sussmann, H. J., "A generalization of the closed subgroup theorem to quotients of arbitrary manifolds," *J. Diff. Geometry* **10** (1975), pp. 151-166.

[28] Sussmann, H. J., "Existence and Uniqueness of minimal realizations of nonlinear systems," *Math. Systems Theory* **10** (1977), pp. 263-284.

[29] Sussmann, H. J., "Subanalytic sets and feedback control," *J. Diff. Equations* **31**, No.1 (1979), pp. 31-52.

[30] Sussmann, H. J., "Analytic stratifications and control theory," in *Proc. 1978. Int. Congress of Mathematicians*, Helsinki (1980), pp. 865-871.

[31] Sussmann, H. J., "A bang-bang theorem with bounds on the number of switchings," *S.I.A.M. J. Control and Optimization* **17**, No. 5 (1979), pp. 629-651.

[32] Sussmann, H. J., "Single-input observability of continuous-time systems," *Math. Systems Theory* **12** (1979), pp. 371-393.

[33] Sussmann, H. J., "Les sémigroupes sousanalytiques et la régularité des commandes en boucle fermée," *Astérisque (Soc. Math. de France)* **75-76** (1980), pp. 219-226.

[34] Sussmann, H. J., "Bounds on the number of switchings for trajectories of piecewise analytic vector fields," *J. Diff. Equations* **43** (1982), pp. 399-418.

[35] Sussmann, H. J., "Lie brackets and local controllability: a sufficient condition for scalar input systems," *S.I.A.M. J. Control and Optimization* **21**, No. 5 (1983), pp. 686-713.

[36] Sussmann, H. J., "Time-optimal control in the plane," in *Feedback Control of Linear and Nonlinear Systems*, D. Hinrichsen and A. Isidori Eds., Springer-Verlag Lect. Notes in Control and Information Sciences, No. 39, New York (1982), pp. 244-260.

[37] Sussmann, H. J., "Subanalytic sets and optimal control in the plane," in *Proc. 21st I.E.E.E. Conference on Decision and Control*, Orlando, Fla. (Dec 1982), pp. 295-299.

[38] Sussmann, H. J., "Lie brackets, real analyticity and geometric control theory," in *Differential Geometric Control Theory*, R.W. Brockett, R.S. Millman and H.J. Sussmann Eds., Birkhäuser Boston Inc. (1983), pp. 1-115.

[39] Sussmann, H. J., "Lie Brackets and real analyticity in control theory," in *Mathematical Control Theory*, C. Olech Ed., Banach Center Publications, Volume 14, PWN-Polish Scientific Publishers,Warsaw, Poland, 1985, pp. 515-542.

[40] Sussmann, H. J., "Resolution of singularities and linear time-optimal control," in *Proceedings of the 23rd I.E.E.E. Conference on Decision and Control*, Las Vegas, Nevada (Dec. 1984), pp. 1043-1046.

[41] Sussmann, H. J., "A general theorem on local controllability," *S.I.A.M. J. Control and Optimization* **25**, No. 1 (1987), pp. 158-194.

[42] Sussmann, H. J., "The structure of time-optimal trajectories for single-input systems in the plane : the C^∞ nonsingular case," *S.I.A.M. J. Control and Optimization* **25**, No. 2 (1987), pp. 433-465.

[43] Sussmann, H. J., "The structure of time-optimal trajectories for single-input systems in the plane: the general real-analytic case," *S.I.A.M. J. Control and Optimization* **25**, No. 4, 1987,pp. 868-904.

[44] Sussmann, H. J., "Regular synthesis for time-optimal control of single-input real-analytic systems in the plane," *S.I.A.M. J. Control and Optimization* **25**, No. 5, 1987, pp. 1145-1162.

[45] Sussmann, H. J., "Envelopes, conjugate points and optimal bang-bang extremals," in *Algebraic and Geometric Methods in Nonlinear Control Theory*, M. Fliess and M. Hazewinkel Eds., D. Reidel Publishing Co., Dordrecht, The Netherlands, 1986, pp.325-346.

[46] Sussmann, H. J., "A weak regularity theorem for real analytic optimal control problems," *Revista Matemática Iberoamericana* **2**, No. 3, 1986, pp. 307-317.

[47] Sussmann, H. J., "Recent developments in the regularity theory of optimal trajectories," in *Linear and Nonlinear Mathematical Control Theory*, Rendiconti del Seminario Matematico, Università e Politecnico di Torino, Fascicolo Speciale 1987, pp. 149-182.

[48] Sussmann, H. J., "Trajectory regularity and real analyticity: some recent results," in *Proceedings of the 25th I.E.E.E. Conference on Decision and Control*, Athens, Greece, Dec. 1986, pp. 592-595.

[49] Sussmann, H. J., "Real analytic desingularization and subanalytic sets: an elementary approach," to appear in *Transactions of the A.M.S.*

[50] Sussmann, H. J., "Hamiltonian lifts and optimal trajectories," Rutgers Center for Systems and Control Technical Report 88-05, *Proceedings of the 27th I.E.E.E. Conference on Decision and Control*, Austin, Texas, Dec. 1988, pp. 1182-1186.

[51] Sussmann, H. J., "A counterexample on optimal control regularity for real analytic systems," Rutgers Center for Systems and Control Technical Report 88-07, September 1988.

[52] Sussmann, H. J., "Envelopes, higher-order optimality conditions, and Lie brackets," Rutgers Center for Systems and Control Technical Report Sycon 89-18, September 1989. To appear in the Proceedings of the 1989 IEEE CDC.

[53] Sussmann, H. J., "Synthesis, presynthesis, sufficient conditions for optimality and subanalytic sets," Rutgers Center for Systems and Control Technical Report 89-20, September 1989. To appear in *Finite Dimensional Controllability and Optimal Control*, H.J. Sussmann ed., M. Dekker, Inc.

Department of Mathematics, Rutgers University, New Brunswick, NJ 08903, U.S.A.

E-mail: sussmann@math.rutgers.edu .

SOME PROBLEMS ARISING IN CONNECTION WITH THE THEORY OF OPTIMIZATION

V. M. Tikhomirov

The theory of extremal problems consists of the following chapters:

1. Mathematical apparatus.
2. Conditions for an extremum.
3. Existence theorems.
4. Numerical methods.

Basically in this paper we will touch only the first three of them. Our purpose is to retrace the fundamental ideas on which the most important results of the general theory are based. These ideas are formulated in the form of short principles, which are then illustrated by some theorems and applications to specific classes of extremal problems.

Principle One: *The mathematical apparatus of the theory of extremal problems is the differential calculus in infinite-dimensional spaces and convex analysis.*

In recent years these two branches of functional analysis have been further developed in the theory of multivalued mappings and in nonsmooth analysis, but also in these new fields the fundamental connecting element is the interaction of smoothness and convexity. Here we would like to present an example of a theorem which is traditional in its form, but whose statement contains a not altogether standard combination of smooth and convex structures.

Theorem 1 (a smooth-convex covering theorem).

Let X and Y be Banach spaces, U — a topological space, W — a neighborhood of a point $\widehat{x}$ in X, and let F: $W \times U \to Y$ have the following properties:

a) $F(\widehat{x}, \widehat{u}) = 0$,

b) *the mapping $F(\widehat{x}, \cdot)$ is continuous at $\widehat{u}$,*

c) *the mapping $F(\cdot, \widehat{u})$ is strongly differentiable at $\widehat{x}$,*

d) *the set $F(\widehat{x}, U) := \{y \mid y \in F(\widehat{x}, u),\ u \in U\}$ is convex in Y,*

e) *the codimension of the subspace $L := F_x(\widehat{x}, \widehat{u})X$ in Y is finite and the zero of the space Y/L belongs to the interior of the set $\pi F(x\, U)$ (where π: $Y \to Y/L$ is the canonical projection).*

Then there exists a neighborhood $\mathcal{W} \subset W$, *mapping* $\xi\colon \mathcal{W} \times Y \times U \to X$, $\vartheta\colon \mathcal{W} \times Y \times U \to U$ *and a number* $K > 0$ *such that* $F\big(\xi(x,y,u),\ \vartheta(x,y,u)\big) = 0$, $\|\xi(x,y,u) - x\| \le K\,\|F(x,u) - y\|$.

If U does not appear and $F'(\widehat{x})$ is a homeomorphism of X onto Y, then we obtain the classical inverse function theorem. Theorem 1 is an example of a "covering theorem", well suited to applications in mathematical programming, calculus of variations and optimal control theory.

The proof of Theorem 1 uses a modified Newton method, and takes advantage of the convexity of the set $F(\widehat{x},\ U)$.

Principle Two: *Necessary conditions for an extremum in the theory of extremal problems correspond to one general principle, namely, the Lagrange principle of elimination of constraints.*

We have often put forward this principle ([1], [2]).

Here we present two results which confirm the general idea. The first is a further development of the smooth-convex principle, discussed in the books [1] and [2]. But the following two points should be emphasized. First, the formulation of Theorem 2 and its proof have been substantially simplified and clarified in comparison with analogous theorems in [1, 2]. Secondly, Theorem 2 can be directly applied to general optimal control problems (and not via artificial tricks of the type of change of time, as was done, say, in [1]). Theorem 2 can also be applied to the majority of the so-called problems with distributed parameters. Thus, a variety of theorems concerning necessary conditions for optimality are encompassed in one, fairly easily provable result: from the simplest facts proved by the founders of mathematical analysis, up to the new ideas connected with the maximum principle for problems with distributed parameter.

Theorem 2 (Lagrange principle for smooth approximately convex problems).

Let X *and* Y *be Banach spaces,* U *a topological space,* $\widehat{W}$ *a neighborhood of a point* $\widehat{x}$ *in* X, f_0 — *a real-valued function defined on* W *and Fréchet differentiable at* $\widehat{x}$. *Let* $F\colon W \times U \to Y$, $F(\widehat{x},\ \widehat{u}) = 0$, *be a mapping with the following properties:*

a) *continuity in* u ($F(\widehat{x},\ \cdot)$ *is continuous at* $\widehat{u}$),

b) *smoothness in* x (*there exists a continuous linear operator* $\Lambda\colon X \to Y$ *such that* $\forall \varepsilon > 0$, $\exists \delta > 0$ *and a neighborhood* $\mathcal{U}$ *of* $\widehat{u}$ *such that* $\|x - \widehat{x}\| < \delta$, $\|x' - \widehat{x}'\| < \delta$ *implies*

$$\big\|F(x,u) - F(x',u) - \Lambda(x'-x)\big\| \le \varepsilon\,\|x - x'\| \quad \forall u \in \mathcal{U} \qquad (\Lambda = F_x(\widehat{x},\ \widehat{u}))),$$

c) *approximate convexity with respect to* u ($\forall \overline{u} := (u_1, \dots, u_s)$, $\overline{\alpha} := (\alpha_0, \dots, \alpha_s)$, $\alpha_j \geq 0$, $\sum_{j=0}^{s} \alpha_j = 1$, $x \in W$, $\beta > 0$ $\exists u_{\overline{\alpha}\beta} = u_{\overline{\alpha}\beta}(x, \overline{u})$ *such that* $u_{\overline{\alpha}\beta} \to \widehat{u}$ *as* $\overline{\alpha} \to (1, 0 \dots, 0)$, $F(x, u_{\overline{\alpha}\beta}) \to \alpha_0 F(x, \widehat{u}) + \sum_{j=1}^{s} \alpha_j F(x, u_j)$),

d) *finite codimensionality* ($\operatorname{codim} F_x(\widehat{x}, \widehat{u}) < \infty$).

Then the Lagrange principle holds at the point $(\widehat{x}, \widehat{u})$ *of strong minimum for the problem*

$$f_0(x) \to \inf, \qquad F(x, u) = 0, \qquad u \in U. \tag{ζ}$$

Remark. In [1] and [2] a version of Theorem 2 was proved (actually, a particular case), where U is an arbitrary set and, accordingly, condition (a) is omitted, the smoothness condition is replaced by the strong differentiability at $\widehat{x}$ of the mapping $F(\cdot, u)$, $\forall u \in U$, and approximate convexity is replaced by the usual convexity.

Here, we say that in (ζ) a strong minimum is attained at $(\widehat{x}, \widehat{u})$ if there exists $\delta > 0$ such that for any feasible pair (x, u) (i.e. $F(x, u) = 0$, $u \in U$), $\|x - \widehat{x}\| < \delta$ implies $f_0(x) \geq f_0(\widehat{x})$.

It remains to explain what the "Lagrange principle" means here.

Lagrange's idea is to "eliminate constraints" using the Lagrange function.

In the process of developing linear programming it has become clear that it is sometimes convenient to divide the constraints into two groups, where one of them is used in forming the Lagrange function and the other is not. Also here, in Problem (ζ), there are two types of constraints (one of equality type: $F(x, u) = 0$, and the second, $u \in U$, of "inclusion type"), and only the first is used in forming the Lagrange function. Namely, the Lagrange function of Problem (ζ) has the form

$$\mathcal{L}\big((x, u), y^*, \lambda_0\big) = \lambda_0 f_0(x) + \langle y^*, F(x, u)\rangle. \tag{1}$$

Now, following Lagrange's idea we are led to investigate two problems (considering the Lagrange multipliers, i.e. (y^*, λ_0), to be fixed):

$$\mathcal{L}\big((x, \widehat{u}), y^*, \lambda_0\big) \to \inf, \tag{ζ_x}$$

$$\mathcal{L}\big((\widehat{x}, u), y^*, \lambda_0\big) \to \inf, \qquad u \in U. \tag{ζ_u}$$

Problems of similar type were studied in detail in [1] and [2]. In (ζ_x) we follow exactly the "general principle" of Lagrange, according to which we have to construct a Lagrange function (Lagrange himself investigated only problems with equality type constraints) and "later search for a maximum or a minimum of the constructed sum, as if the variables were independent". In (ζ_u) we slightly extend Lagrange's idea and minimize the Lagrange function *with respect to those variables whose constraints are not included in this function.* Problems (ζ_x) and (ζ_u) are of different nature. Problem (ζ_x) belongs to the class of *smooth* (and elementary, i.e. unconstrained) problems. Problem (ζ_u) is a typical *convex programming* problem

$(\langle y^*, y\rangle \to \inf,\ y \in \mathcal{A} := F(\widehat{x}, U))$, provided that the set $F(\widehat{x}, U)$ is assumed to be convex (which is in fact true in a particular case under our assuptions; in general, the *closure* of $F(\widehat{x}, U)$ is convex).

A necessary condition for an extremum in Problem (ζ_x) goes back to Fermat — it is the *stationarity condition* $\mathcal{L}_x = 0$. For Problem (ζ_u) a necessary condition for a minimum is $-y^* \in N\mathcal{A}(\widehat{x})$ ($N\mathcal{A}(\widehat{x})$ denotes the normal cone to the set $\mathcal{A}$ at $\widehat{y}$, i.e. $\langle y^*, y - \widehat{y}\rangle \geq 0$, $\forall y \in \mathcal{A}$, $\widehat{y} = F(\widehat{x}, \widehat{u}) = 0$), which is equivalent to the *minimum principle*:

$$\min_{u\in U} \mathcal{L}\big((\widehat{x}, u), y^*, \lambda_0\big) = \mathcal{L}\big((\widehat{x}, \widehat{u}), y^*, \lambda_0\big) = 0.$$

Summing up, the expression "the Lagrange principle holds" in Theorem 2 means that the stationary condition and the minimum principle (with $\lambda_0 \geq 0$) are satisfied:

$$\mathcal{L}_x\big((\widehat{x}, \widehat{u}), y^*, \lambda_0\big) = 0, \tag{2_x}$$

$$\min_{u\in U} \mathcal{L}\big((\widehat{x}, u), y^*, \lambda_0\big) = \mathcal{L}\big((\widehat{x}, \widehat{u}), y^*, \lambda_0\big) = 0. \tag{2_u}$$

Now, let us illustrate the presented result (hence also our Principle Two) by examples of different classes of extremal problems.

1) Mathematical programming.

A fairly broad class of problems, traditionally connected with this branch of optimization, can be incorporated in the following scheme.

Let X and Y be Banach spaces, f_i, $0 \leq i \leq m$, functions defined on X, and let F: $X \to Y$. Assume that f_i and F are continuously differentiable on some neighbourhood of a point $\widehat{x}$ at which a local minimum of the following problem is attained:

$$\begin{aligned} &f_0(x) \to \inf, \\ &f_i(x) \leq 0, \quad 1 \leq i \leq m', \quad f_i(x) = 0, \quad m'+1 \leq i \leq m, \quad F(x) = 0. \end{aligned} \tag{ζ_1}$$

This is the class of the so-called *smooth mathematical programming problems*. Problem (ζ_1) easily reduces to (ζ):

$$(\zeta_1) \iff f_0(x) \to \inf, \qquad \mathcal{F}(x, u) = 0, \qquad u \in U,$$

where

$$\mathcal{F}(x, u) = \big(f_1(x) + u_1, \ldots, f_{m'}(x) + u_{m'}(x),\ f_{m'+1}(x), \ldots,\ f_m(x),\ F(x)\big),\ U = R_+^{m'}.$$

In order to be able to apply Theorem 2, we have to add assumptions guaranteeing the finite codimensionality condition d) of Theorem 2. To this end it is enough to require that the mapping F be *regular* at $\widehat{x}$, i.e.

$$F'(\widehat{x})X = Y. \tag{3}$$

Applying Theorem 2 yields automatically the following conditions for (ζ_1):

— stationarity: $\mathcal{L}_x = 0 \iff \sum_{i=0}^{m} \lambda_i f_i'(\widehat{x}) + F'^*(\widehat{x})y^* = 0,$

— nonnegativity: $\lambda_i \geq 0,\ 0 \leq i \leq m',$

— complementary slackness: $\lambda_i f_i(\widehat{x}) = 0,\ 1 \leq i \leq m'.$

This unifies the classical results developed by Fermat, Newton, Leibniz, Euler and Lagrange, and numerous fundamental ones obtained more recently (Lyusternik, John and others). All conditions for a weak extremum in classical calculus of variations are particular cases, including such traditional classes of problems as the simplest one, the Bolza problem, the isoperimetric problem, problems with higher derivatives, with moving-boundary conditions, with phase-space constraints, Lagrange problems with holomorphic and nonholomorphic constraints, etc. etc.

Let us formulate one of the most important reasons of such universality.

Principle Three: *The mapping given by a nonlinear integral equation of the second kind with smooth integrand satisfies the condition of finite codimension.*

Let us check this principle against an example of a differential relation in *Pontryagin's form*:

$$\dot{x} = \varphi(t,\, x\, u) \iff x(t) = x(t_0) + \int_{t_0}^{t} \varphi(\tau,\, x(\tau),\, u(\tau))\, d\tau. \tag{4}$$

Here we consider the mapping

$$F\big(x(\cdot),\, u(\cdot)\big)(t) = x(t) - x(t_0) - \int_{t_0}^{t} \varphi(\tau,\, x(\tau),\, u(\tau))\, d\tau$$

with regard to its "smooth" part, i.e. to $x(\cdot)$, with $u(\cdot)$ treated as fixed and with the function φ smooth in x. Then

$$F'_{x(\cdot)}\left(\widehat{x}(\cdot),\, \widehat{u}(\cdot)\right)\left[x(\cdot)\right](t) = x(t) - x(t_0) - \int_{t_0}^{t} \varphi_x\big(\tau,\, \widehat{x}(\tau),\, \widehat{u}(\tau)\big)x(\tau)\, d\tau.$$

Note that here a stronger condition is satisfied than that of finite codimension, namely the regularity condition (3).

For a relation of the type

$$x(t) - \int_T K(t,\,\tau)\,\varphi(\tau,\,x(\tau),\,u(\tau))\,d\mu(\tau), \tag{5}$$

under general assumptions on the measure space (T, Σ, μ) and the kernel $K(\cdot\,,\,\cdot)$, after differentiation with respect to $x(\cdot)$ we obtain a Fredholm integral equation of the second kind which satisfies the condition of finite codimension.

The next class of problems which reduces to (ζ) is the following class of *convex programming* problems:

$$\begin{aligned} &g_0(u) \to \inf, \\ &g_i(u) \le 0, \quad 1 \le i \le m', \quad g_i(u) = 0, \quad m'+1 \le i \le m, \quad u \in A \subset U, \end{aligned} \tag{ζ_x}$$

where U is a linear space, A is a convex set and the functions g_i are convex for $0 \le i \le m'$ and affine for $i \ge m' + 1$.

Here we have to put

$$\begin{aligned} f_0(u) &= \xi, \\ \mathcal{F}(u,\,\alpha) &= \big(g_0(u) + \alpha_0 - \xi,\, g_1 + \alpha_1, \ldots, g_{m'}(u) + \alpha_{m'},\, g_{m'+1}(u), \ldots, g_m(u)\big). \end{aligned}$$

Application of Theorem 2 leads to the well-known Kuhn–Tucker theorem. As a particular case we obtain a criterion for the existence of a solution to a linear programming problem.

2) Optimal control problems.

As we remember, in Theorem 2 there are two essential conditions: finite codimensionality and approximate convexity. In the third principle we tried to clarify the reasons why the condition of finite codimension is so often satisfied. Here we start with the clarification of the reasons for which the convexity requirement is satisfied under quite general conditions.

Principle Four: *Mappings associated with integration have either the property of convexity or of approximate convexity.*

Strong support to this principle is given by the following famous

Lyapunov's Theorem.

Let $(T,\,\Sigma)$ be a measurable space and $\mu = (\mu_1, \ldots, \mu_n)$ a continuous finite vector measure defined on it. Then the range of μ $(= \{\, y = \int_A d\mu,\ A \in \Sigma \,\})$ is convex and compact.

This result allows Theorem 2 to be applied to the class of the so-called Lyapunov problems, which — in some respects — are the simplest optimal control problems.

Let us consider the problem

$$\mathcal{F}_0\big(u(\cdot),\, v\big) = \int_{t_0}^{t_1} \varphi_0\big(t,\, u(t)\big)\, d\mu + g_0(v) \to \inf,$$

$$\mathcal{F}_i\big(u(\cdot),\, v\big) = \int_{t_0}^{t_1} \varphi_1\big(t,\, u(t)\big)\, d\mu + g_i(v) \begin{cases} \le 0, & 1 \le i \le m', \\ = 0, & 1 + m' \le i \le m, \end{cases} \qquad (\zeta_3)$$

$$v \in A, \qquad u(t) \in U \quad \text{for a.a. } t \in [t_0,\, t_1],$$

where V is a linear space, the g_i are convex for $0 \le i \le m'$, and affine, finite-valued for $i \ge m' + 1$, $A \subset V$ is a convex set and U a topological space. Problem (ζ_3) can be reduced to (ζ) owing to the following result:

Lemma on convexity of the range (see [3], p. 355).

Let $\mathcal{U}$ be the set of measurable mappings $u(\cdot)$: $[t_0,\, t_1] \to U$ such that $t \to f_i\big(t,\, u(t)\big) \in L_1\big([t_0,\, t_1]\big)$. Then

$$\operatorname{Im} \mathcal{F} = \big\{\, \xi = (\xi_0, \dots, \xi_m) \mid \xi_i = \mathcal{F}_i\big(u(\cdot)\big),\, u(\cdot) \in \mathcal{U} \,\big\}$$

is a convex set in $\mathbb{R}^{m+1}$.

Mappings similar to (4) have no such property. However, the following result supporting our principle is valid.

Let $(T,\, d)$ be a compact metric space, B the σ-algebra of Borel sets on T and μ a continuous measure on the measurable space $(T,\, B)$. Then it is easy to see that for any $\beta > 0$ and $\overline{\alpha} \in \Sigma^s := \big\{\, (\alpha_0, \dots, \alpha_s),\ \alpha_i \ge 0,\ \sum \alpha_i = 1 \,\big\}$ the set T can be split into a finite number of subsets: $T = \bigcup_{i=1}^{r} T_i(\beta)$, $T_i(\beta) \cap T_j(\beta) = \emptyset$, $i \ne j$, with each $T_i(\beta)$ split in turn into subsets $T_{i\alpha_j}(\beta)$, such that the diameter of each $T_i(\beta)$ is at most β and $\mu\big(T_{i\alpha_j})(\beta)\big) = \alpha_j \mu\big(T_i(\beta)\big)$.

For example, if $T = [0, 1]$ and $s = 1$, then for $T_i(\beta)$ we can take $T_i(\beta) = [\, i\beta,\, (i+1)\beta\, [$ and put $T_{i\alpha_0}(\beta) = [i\beta,\, (i + \alpha_0)\beta]$, $T_{i\alpha_1}(\beta) = T_i(\beta) \setminus T_{i\alpha_0}(\beta)$.

The set of $\mathbb{R}^r$-valued measurable functions defined on T is denoted by $\mathcal{S}(T;\, \mathbb{R}^r)$. For a set of $s + 1$ functions $\{u_j(\cdot)\}_{j=0}^{s}$ belonging to $\mathcal{S}(T;\, \mathbb{R}^r)$ and for $\overline{\alpha} \in \Sigma^s$ we define

$$u_{\overline{\alpha}\beta}(t) = u_j(t), \qquad t \in T_{i\alpha_j}(\beta).$$

The function $u_{\overline{\alpha}\beta}(\cdot)$ is called the $(\overline{\alpha},\, \beta)$-*mix* of the controls $\{u_j(\cdot)\}$.

Theorem 3 (on mix).

Let K: $T \times T \to \mathcal{L}(\mathbb{R}^n,\, \mathbb{R}^n)$ be a continuous kernel, g: $T \times \mathbb{R}^r \to \mathbb{R}^n$ — a continuous mapping, and

$$\mathcal{F}\ u(\cdot) \ = \int_T K(t,\, \tau)\, g(\tau,\, u(\tau)\ \ d\mu.$$

Then the operator $\mathcal{F}$ satisfies the condition of approximate convexity (as an operator from $L_\infty(T, \mathbf{R}^r)$ into $C(T; \mathbf{R}^m)$):

$$\mathcal{F}(u_{\overline{\alpha}\beta}(\cdot)) \underset{C(T,\mathbf{R}^n)}{\longrightarrow} \sum_{j=0}^{s} \alpha_j \mathcal{F}(u_j(\cdot)) \qquad \text{as} \quad \beta \to 0,$$

$$u_{\overline{\alpha}\beta}(\cdot) \underset{L_1(T,\mathbf{R}^r)}{\longrightarrow} u_0(\cdot) \qquad \text{as} \quad \overline{\alpha} \to (1, 0, \ldots, 0).$$

On the basis of this result optimal control problems of the following form can be reduced to problem (ζ):

$$B_0\big(x(\cdot), u(\cdot)\big) \to \inf, \quad B_i\big(x(\cdot), u(\cdot)\big) \begin{cases} \le 0, & 1 \le i \le m' \\ = 0, & m'+1 \le i \le m, \end{cases} \qquad \dot{x} = \varphi(t, x, u),$$

where $B_i\big(x(\cdot), u(\cdot)\big) = \int_{t_0}^{t_1} f_i(t, x, u)\, dt + \psi_i\big(x(t_0), x(t_1)\big)$.
This is so since, according to our comments on Principle Three, the condition of finite codimension is satisfied for the problems of the above type, and the condition of approximate convexity holds by Theorem 3.

Then the stationary condition directly implies the adjoiont equation, while the minimum principle reveals itself at once to be the maximum principle.

To be more precise, relation (ζ_u) leads to the following equation:

$$\min_{u(\cdot)\in\mathcal{U}} \mathcal{L}\Big(\big(\widehat{x}(\cdot), u(\cdot)\big), p(\cdot), \lambda\Big) = \mathcal{L}\Big(\big(\widehat{x}(\cdot), \widehat{u}(\cdot)\big), p(\cdot), \lambda\Big) \iff$$

$$\iff \min_{u(\cdot)\in\mathcal{U}} \int_{t_0}^{t_1} L(t, \widehat{x}(t), \dot{\widehat{x}}(t), u(t))\, dt = \int_{t_0}^{t_1} \widehat{L}(t)\, dt, \tag{5}$$

where

$$\mathcal{L} = \sum_{i=0}^{m} \lambda_i B_i + \int_{t_0}^{t_1} \langle p(t), \dot{x}(t) - \varphi\big(t, x(t), u(t)\big)\rangle\, dt,$$

$$\mathcal{U} = \Big\{ u(\cdot) \in L_\infty\big([t_0, t_1], \mathbf{R}^r\big) \mid u(t) \in U \Big\}.$$

Here we apply the measurable selection theorem, which allows us to change the order of minimization and integration. The relation

$$\min_{u\in U} L\big(t, \widehat{x}(t), \dot{\widehat{x}}(t), u\big) = \widehat{L}(t) \qquad \text{almost everywhere}$$

is equivalent to the maximum principle.

Certainly, in the same way we can obtain conditions for a strong extremum in problems of classical calculus of variations. Such conditions have been derived in numerous investigations starting with the work of Weierstrass (Bolza, Graves, Mac Shane and others).

In many optimal control problems, besides control constraints of inclusion type, there are also mixed constraints. This makes it natural to consider the following class of problems.

Let X, Y, Z, V be normed spaces, $\mathcal{U}$ a set, $\mathcal{V}$ an open set in X, $\mathcal{W}$ an open set in Z, f_0: $\mathcal{V} \to \mathbf{R}$, F: $\mathcal{V} \times \mathcal{U} \to \mathbf{R}$, G: $\mathcal{V} \times \mathcal{W} \times \mathcal{U} \to Z$. Consider

$$f_0(x) \to \inf, \qquad F(x, u) = 0, \qquad G(x, v, u) = 0, \qquad u \in \mathcal{U}, \qquad (\zeta')$$

$$\mathcal{L} = \lambda_0 f_0 + \langle y^*, F\rangle + \langle z^*, G\rangle.$$

The following result is formulated without detailed comments.

Theorem 4 (Lagrange principle for Problem (ζ')).

Suppose that the mapping $\mathcal{F}((x,v), u) = (F(x,u), G(x,v,u))$ has the following properties: continuity in u, smoothness in (x, v), approximate convexity in u (i.e., $G(\widehat{x}, u_{\overline{\alpha}\beta}, \widehat{v}) \to 0$ and $u_{\overline{\alpha}\beta} \to \widehat{u}$ as $\overline{\alpha} \to (1, 0, \dots, 0)$, and $F(x, u_{\overline{\alpha}\beta}) \to \sum_{j=0}^{s} \alpha_j F(x, u_j)$ $(u_0 := \widehat{u})$ as $\beta \to 0$) and regularity in (x, v). Then, for Problem (ζ'), the Lagrange principle holds in the following form:

stationarity with respect to x: $\mathcal{L}_x\big((\widehat{x}, \widehat{v}, \widehat{u}), y^*, z^*, 1\big) = 0,$

stationarity with respect to v: $\mathcal{L}_v\big((\widehat{x}, \widehat{v}, \widehat{u}), y^*, z^*, 1\big) = 0,$

minimum principle with respect to u: $\displaystyle\min_{\{u \mid u\in\mathcal{U}, G(\hat{x},u,\hat{v})=0\}} \mathcal{L}\big((\widehat{x}, \widehat{v}, u), y^*, z^*, 1\big) = \widehat{\mathcal{L}}.$

Let us pass to the third chapter. Here we will consider only the problems of classical calculus of variations and of optimal control. A feature of these problems is their inherent convexity, which, however, does not necessarily appear in the initial formulation of the problem. This fact is expressed in the following general statement.

Principle Five.

Theoretically, the integrands of functionals of classical calculus of variations can be considered convex (quasiconvex in multidimensional cases) with respect to the derivatives. Any optimal control problem reduces to the problem of minimization of an integral functional with the integrand being convex with respect to the derivatives, subject to constraints in the form of a convex-valued differential inclusion.

The following result supports this assertion.

Theorem 5.

Let $\mathcal{D}$ be a Lipschitz domain in $\mathbf{R}^n$ (i.e. with Lipschitz continuous boundary) and let $f: \mathcal{D} \times \mathbf{R}^m \times B_r N^{mn} \to \mathbf{R}$ (where $B_r M^{mn} = \{ u \in M^{mn} = \mathcal{L}(\mathbf{R}^n, \mathbf{R}^m) \mid \|u\| \leq r \}$), be a Carathéodory function (i.e. $f(t, \cdot, \cdot)$ is continuous and $f(\cdot, x, u)$ is measurable). Suppose

$$t \to \sup \{ \, |f(t, x, u)| \, \mid \, |x| \leq r, \; \|u\| \leq r \, \} \in L_1(\mathcal{D}),$$

$t \to f(t, x(t), \dot{x}(t)) \in L_1(\mathcal{D})$ $\forall x(\cdot) \in W^1_\infty(\mathcal{D}, \mathbf{R}^n)$. Moreover, let $U(t, x)$ be a closed-valued mapping onto M^{mn}, continuous in the Hausdorff metric. Then there exists a Carathéodory function φ: $\mathcal{D} \times \mathbf{R}^m \times B_r M^{mn}$, quasiconvex in $\dot{x}$, such that the problem

$$f(x(\cdot)) = \int_{\mathcal{D}} \varphi(t, x(t), \dot{x}(t)) \, dt \to \inf, \quad \dot{x}(t) \in \overline{\mathrm{qco}}\, U(t, x(t)), \quad x\big|_{\partial\mathcal{D}} = \xi,$$

is a lower semicontinuous extension of the problem

$$J(x(\cdot)) = \int_{\mathcal{D}} f(t, \, x(t), \, \dot{x}(t)) \, dt \to \inf, \quad \dot{x}(t) \in U(t, \, x(t)), \quad x\big|_{\partial\mathcal{D}} = \xi.$$

In the case where either n or m is one, $\overline{\mathrm{qco}}$ coincides with the closed convex hull and quasiconvexity reduces to convexity.

Note that the standard constraints in optimal control problems have the form $\dot{x} = \varphi(t, x, u)$, $u(t) \in U$, which is a particular case of constraints of the type $\dot{x} \in U(t, x)$. Hence any optimal control problem in Pontryagin's form reduces to a problem with a convex-valued differential inclusion.

In principle this allows a complete investigation of the problems of classical calculus of variations and those of optimal control by means of introducing suplementary constraints.

Stating his twentieth problem D. Hilbert wrote: "An important problem ... is the question concerning the existence of solutions of partial differential equations when the values on the boundary of the region are prescribed. ... It is my conviction that it will be possible to prove these existence theorems by means of a general principle whose nature is indicated by Dirichlet's principle. This general principle will then perhaps enable us to approach the question: Has not every regular variation problem a solution, provided certain assumptions regarding the given boundary conditions are satisfied..., and provided also if need be that the notion of a solution shall be suitably extended?"

Introducing the suplementary constraint allows us to give an affirmative answer to Hilbert's question in maximal generality.

Principle Six:

Any regular problem of calculus of variations with compact constraints on the derivative has a solution.

As an example, consider the problem which, in principle, was meant by Hilbert:

$$\int_{\mathcal{D}} L(t, x, \dot{x})\, dt \to \inf, \qquad x\big|_{\partial\mathcal{D}} = \xi, \tag{ζ_w}$$

where $\mathcal{D}$ is a domain in $\mathbf{R}^n$ (we always assume it to have a Lipschitz continuous boundary), and L is a regular integrand (i.e. the function $L(t, x, \cdot)$ is convex). Certainly, problems of this type do not always have solutions. However, we will add the requirement $\|\dot{x}\| \le r$, where r is some, in general large, number. Moreover, we assume that there exists at least one function $x(\cdot)$ satisfying all constraints. We have

Theorem 6.

Under the above assumptions Problem (ζ_w) has a solution.

Independent considerations show that (in general) in this case the differential equation mentioned by Hilbert will always be satisfied.

Problem (ζ_w) with the constraint $\|\dot{x}\| \le r$ can be solved using the direct method. Letting r tend to infinity we either find that the infimum in (ζ_w) is $-\infty$, or obtain the convergence to the desired solution.

Thus, the idea of optimal control offers new possibilities both in investigating and in effective solution of problems in calculus of variations.

Finally, we would like to discuss still another problem connected with the first chapter. Namely, we will discuss one of the forms of the well-known principle related to duality.

Principle Seven: *Convex objects allow double description.*

It seemed interesting to me to investigate the application of this idea to problems where the convex and differential–topological dualities are combined.

Let M be a compact n-dimensional oriented smooth Riemannian manifold. By $\mathcal{L}_p^k(M)$ we denote the space of k-forms whose coefficients in each chart belong to $\mathcal{L}_p$. Let $\mathcal{W}_p^{1,k}(M)$ denote the space of k-forms $\omega \in \mathcal{L}_p^k(M)$ for which $d\omega \in \mathcal{L}_p^{k+1}(M)$, and $\mathcal{W}_p^{1,k}(M)$ — the space of k-forms whose coefficients in each chart belong to the Sobolev space W_p^1. The metric tensor allows one to construct the duality operator $*$ connecting k-forms and $(n-k)$-forms. The following holds:

Theorem 7.

Let F be a convex function defined on $\mathcal{L}_p^k(M)$ and continuous at zero. Then the problems:

$$F(\omega) \to \inf, \qquad \omega \in \widetilde{\mathcal{W}}^{l,k}(M), \qquad d\omega = 0 \tag{ζ_{pk}}$$

$$-F^*(*\omega) \to \sup, \qquad \omega \in d\mathcal{W}_{p'}^{1,n-k-1}(M), \qquad \frac{1}{p} + \frac{1}{p'} = 1, \tag{ζ_{pk}^*}$$

are dual to each other.

In conclusion it is tempting to formulate still another principle, which although it has numerous partial confirmations, has not yet been stated in the form of a unified result. One may believe that reasonable (i.e., closed to being necessary) sufficient conditions for an extremum provide a tool for constructing a field of extremals and give a key to the Hamilton–Jacobi theory.

This idea has been realized in the form of a rigorous theorem for constraints of equality type and for classical calculus of variations. For problems with inequalities and for optimal control problems some points still remain unclear.

The theorems discussed in this paper have been the field of investigations of the author and his students in the last three years. They continue and, in some respects, conclude the research developed in [1], [2]. Theorems 3 and 4 were obtained jointly by the author and E. A. Sambisavam (India), Theorem 5 was proved by O. A. Kruzhalov and Theorem 7 by I. I. Dzhioeva.

References

[1] A. D. Ioffe, V. M. Tikhomirov, Theory of extremal problems, North-Holland, Amsterdam, 1979.

[2] V. M. Tikhomirov, The general principles in the theory of extremal problems, Wiley, New York, 1986.

[3] V. M. Alexeev, V. M. Tikhomirov, S. V. Fomin, Optimal control theory, Nauka, Moscow, 1979 (in Russian).

V. M. Tikhomirov
Moscow State University
Department of Mathematics and Mechanics
Leninskie Gory, 117–234 Moscow, USSR

LIBRARY
THE UNIVERSITY OF TEXAS
PAN AMERICAN AT BROWNSVILLE
Brownsville, TX 78520-4991